Biodiversity Conservation for Sustainable Management

The Editor

Renowned environmentalist and Vice-Chancellor of Periyar University, Salem **Prof. Dr. K. Muthuchelian** has been working with dedication since 1982 for the development of India by harnessing both traditional and frontier technologies. He earned his M.Sc., (Botany) from Madurai Kamaraj University with distinction and Ph.D. from the School of Biological Sciences, Madurai Kamaraj University, Madurai, India and then he did his Post-Doctoral research at the University of Ancona, Italy. Dr.K.Muthuchelian was a faculty member of the School of Energy, Environment and Natural Resources, Madurai Kamaraj University, Madurai. He was conferred the prestigious *"Doctor of Science"* (D.Sc.) in recognition of his research accomplishment on biomass technology. He became the Director of the *'Centre for Biodiversity and Forest Studies'* in Madurai Kamaraj University in 2001. His laboratory has got International Recognition (Nitrogen Fixing Tree Associations, Hawaii, USA) for 'Erythrina Research'.

He served as a member in various prestigious professional bodies such as The New York Academy of Sciences - USA, Rural Development Forestry Network – UK, American Association for the Advancement of Science - USA and International Union for Conservation of Nature and Natural Resources (IUCN), Switzerland (an affiliated body of UNO) for Southeast Asia. He has been an expert member of *'Man and Biosphere (MAB) programme'* of Ministry of Environment and Forests, Government of India, New Delhi, India. He has been nominated as *"Fellow of International Energy Foundation, Saskatchewan, Canada"* in recognition of his outstanding International contributions to the 'Biomass production and energy transfer technology'. He has been elected as *"Fellow of National Academy of Biological Sciences, India"* for his excellent contribution to the 'Environmental Sciences'. He also served as a member and Chairman of the Peer Team of NAAC (National Assessment and Accreditation Council, Bangalore) in many premier academic institutions in India.

Dr.Muthuchelian's scientific contribution has been recognized through numerous awards such as *'Tamil Nadu Best Scientist in Environmental Sciences (1999)'* by TNSCST, Government of Tamil Nadu, Chennai, India, *'Best Scientist in Environmental Sciences (2001-2002)'* by Nehru Yuvakendra, Government of India and Tamil Nadu Sports Authority, *'Best Scientist in Environmental Management (2005) (Karma Veerar Kamarajar Award)'* by Department of Environment, Government of Tamil Nadu, Chennai, India. He has been awarded the *'Best Teacher Gold Medal (First Prize)'* by the Madurai Kamaraj University, Madurai in 2008. He was awarded the prestigious *'76th Indian Science Congress Endowment – Eminent Scientist Award in Natural Sciences'* by Madurai Kamaraj University in 2010. He has been honored twice the *"Merit of Excellence Award"* in recognition of his outstanding contributions in the field of medicinal plants conservation and in plant science at international level. For his remarkable contribution to the society and dedicated involvement in the enlistment of the downtrodden, he has been awarded twice *'Best Vice-Chancellor Award'* by the Indian Red Cross Society, Tamil Nadu branch, Chennai.

Dr.K.Muthuchelian is the First Indian Scientist honored with the prestigious award of the University of Ancona, Italy for his outstanding contribution in 'Bioenergy production'. He was an expert invitee of Nitrogen Fixing Tree Association, Hawaii, USA. He has attended several International Conferences and Symposia in USA, Germany, Hungary, Bangladesh and Italy. He has published more than 150 research papers in refereed National and International Journals and presented many papers in National and International Symposia / Conferences. He has organized 32 National Scientific Meetings, Workshops, Seminars and Conferences. Recently, he has published a book in Tamil entitled *"Uyir Virimam"* (Biodiversity: Current Status and Management) widely appreciated in both the academic and other professionals. He has produced 26 Ph.D's and currently guiding 12 Ph.D. scholars.

While ensuring accelerated environmental development through applications of science and technology, Dr.K.Muthuchelian has been relentlessly focusing attention on ecological conservation and on sustainable development. The integrated approach has remained one of the distinguishing aspects of Dr. K. Muthuchelian's original and path breaking contributions towards the revolutionary growth for ecological security and environmental sustainability in India and indeed the world.

Biodiversity Conservation for Sustainable Management

Editor

Dr. K. Muthuchelian
Ph.D., D.Sc., FNABS., FIEF (Canada)
Vice Chancellor
Periyar University
Periyar Palkalai Nagar
Salem – 636 011

2013
Daya Publishing House®
A Division of
Astral International Pvt. Ltd.
New Delhi – 110 002

Published by : **Daya Publishing House®**
A Division of
Astral International Pvt. Ltd. –
ISO 9001:2008 Certified Company –
4760-61/23, Ansari Road, Darya Ganj
New Delhi-110 002
Ph. 011-43549197, 23278134
E-mail: info@astralint.com
Website: www.astralint.com

Laser Typesetting : **Classic Computer Services**
Delhi - 110 035

Printed at : **Salasar Imaging Systems**
Delhi - 110 035

PRINTED IN INDIA

Preface

India being one of the 17 megadiversity nation in the world with unique biogeographical location, diversified climatic conditions has been known for its rich biodiversity. India possesses a diversity of ecological habitats ranging from tropical, sub-tropical, alpine to desert. Biogeographically, India represents the two of the major realms (the Paleo-arctic and Indo-Malayan), five biomes [tropical humid forests, tropical dry/deciduous forests (including monsoon forests), warm/semi-deserts; coniferous forests and alpine meadows], ten biogeographic zones (Trans-himalayan, Himalayan, Indian desert, Semi-arid, Western Ghats, Deccan Peninsula, Gangetic Plain, Northeast India, Islands and Coasts) and twenty-seven biogeographic provinces. The Indian landmass is bounded by the Himalayas in the north, Bay of Bengal in the east and the Arabian sea in the west and the Indian Ocean in the south with a total geographical area of about 3,029 million hectares. In India, the Western Ghat forests and the Eastern Himalayan forests have been identified as the megadiversity hot spots. In fact, within the 2.4 per cent of the world's total land area, India is known to have 11 per cent and 7.5 per cent of the flora and fauna, respectively.

The year of 2010 is considered as the "*International Year of Biodiversity*" which includes a series of programme to protect the biodiversity. India is hosting the Eleventh Conference of Parties (CoP) to the Convention on Biodiversity (CBD) in Hyderabad on 1-9 October, 2012. India with a strong commitment is contributing towards achievement of three objectives of the CBD, the 2010 target and the strategic plan. The CBD will provide India an opportunity to consolidate, scale-up and demonstrate our initiatives and strengths on biodiversity. The Ministry of Environment and Forests (MoEF) framed and formulated the strategy for conservation and sustainable utilization of biodiversity in India. With a strong institutional, legal and policy framework, India has the potential and capability to emerge as the world leader in conservation and sustainable use of biodiversity.

The broad vision for biodiversity in Agenda 21 in its conservation and sustainable use accompany by equitable benefit sharing mechanism. This includes a focus on enhancing national biodiversity protection measures involving the development of national strategies; mainstreaming of biodiversity concerns; ensuring the fair and equitable sharing of the benefits acquiring from biodiversity; country-wide studies on biodiversity; fostering traditional methods and indigenous knowledge; encouraging biotechnological innovations along with the suitable sharing of their benefits and promoting regional and international cooperation.

The countries adopted an ambitious Strategic Plan at the Convention of Biological Diversity (CBD) 2011-2020 in October, 2010 at Nagoya, Japan with a 20 time bound target to halt the extinction rate by 2020. These target popularly known as 'Aichi Targets' clearly defines that by 2020, the extinction of threatened species should be prevented and their conservation status of those most decline has been improved and sustained. According to UN FAO prediction, 24 per cent of world's mammalian population will be extinct by 2020. Nearly 12 per cent of Indian bird population also faced the extinct condition. The Bird Life International and IUCN have listed 12 species of birds as 'critically endangered' category. The conservation strategies and the utilization of biodiversity for sustainable development of nation includes special protection to biodiversity by establishment of national parks, wildlife sanctuaries, biosphere reserves, ecologically fragile and sensitive areas. Others strategies include afforestation of degraded areas and wastelands and establishment of gene banks as part of *ex-situ* conservations. The four forest types (tropical, sub-tropical, temperate and alpine) were further divided into sixteen major types and 232 sub-types for the purpose of conservation. The major *in-situ* conservation of India lies in its impressive Protected Areas (PAs) network. A network of 668 PAs has been established, extending over 1, 61, 221.57 sq. kms (4.90 per cent of total geographic area), comprising 102 National Parks, 515 Wildlife Sanctuaries, 47 Conservation Reserves and 4 Community Reserves. This network resulted in significant restoration of large population of mammals such as tiger, lion, rhinoceros, crocodiles, and elephants. In addition, for special flagship programmes of specific management of tiger and elephant habitats, overall 40 Tiger Reserves and 28 Elephant Reserves have been established. The Ministry of Environment and Forests (MoEF) established the National Afforestation and Eco-development Board (NAEB) in August 1992 to promote afforestation and eco-development strategies for join forest management and microplanning. The various central Acts related to biodiversity includes Forest Act, 1927, Wildlife (Protection) Act, 1972, Forest (Conservation) Act, 1980 and the Environment (Protection) Act, 1986 and the policies and strategies directly relevant to biodiversity includes National Forest Policy (1988), National Conservation Strategy and Policy Statement for Environment and Sustainable Development, National Agricultural Policy, National Land Use Policy, National Fisheries Policy, National Policy and Action Strategy on Biodiversity, National Wildlife Action Plan and Environmental Action Plan.

I hope this book will bring out the different biodiversity conservation strategies in various places in and around India and this will pave the way for sustainable environment.

Dr. K. Muthuchelian

Contents

2013, Biodiversity Conservation for Sustainable Management   *Pages 1–11*
Editor: **Dr. K. Muthuchelian,** *Vice Chancellor, Periyar University, Salem*
Published by: **Daya Publishing House, NEW DELHI**

Chapter 1

Biodiversity and its Conservation

R. Ramanujam and R. Seenivasan

Department of History,
Sri S. Ramasamy Naidu Memorial College,
Sattur – 626 203, Tamil Nadu

Introduction

In a world where we are losing species year on year, many of which we have perhaps not even recognised, it is essential to understand the existing biodiversity of the earth in order to conserve it. To document what exists at the moment scientists have developed methods for studying species in the field, for identifying them, and for counting them. Only once a wild population is identified and numbers estimated is there any possibility of demonstrating whether that species is increasing, stable or decreasing. Only by understanding the environment in which a declining species lives is it possible to decide on appropriate action to reverse the decline. Whilst it is usually scientists who are learning about biodiversity and the factors affecting it, it is rarely they who make the decisions which determine whether it will increase or decrease, so it is important that they present their data in a form which demonstrates to policy makers just how and why biodiversity should be preserved. We have tried to organise this guide to give you a taste of the myriad online resources about the whole process from understanding biodiversity to implementing conservation, both at the local and at the global levels.

The types of resource that immediately spring to mind would perhaps include the websites of university departments doing research into biodiversity and government bodies trying to implement conservation measures, but there are many others. Museums, botanic gardens, zoos, software manufacturers, amateur enthusiasts, and local societies all produce relevant websites, and Web 2.0 technology is now providing the means for much more networking between organisations as well as allowing amateur enthusiasts to contribute directly to databases and discussions.

A fascinating example of this mix of professional and amateur expertise is the Encyclopedia of Life (EOL) http://www.eol.org/, launched in May 2007, which will eventually provide for each known species a webpage in two parts: a controlled scientific view by an expert along with a blog-like section to which anyone can contribute.

To give this guide some structure we have presented examples of resources under a series of headings but of course many, if not most, of them could equally well appear elsewhere. The local wildlife trusts, for example, often use their websites to raise awareness of biodiversity and to educate and involve the public, but most also receive species records and undertake practical conservation by managing reserves.

What is Biodiversity?

Biodiversity is a modern term which simply means "the variety of life on earth". This variety can be measured on several different levels. Genetic–variation between individuals of the same species. This includes genetic variation between individuals in a single population, as well as variations between different populations of the same species. Genetic differences can now be measured using increasingly sophisticated techniques. These differences are the raw material of evolution.

Species–species diversity is the variety of species in a given region or area. This can either be determined by counting the number of different species present, or by determining taxonomic diversity. Taxonomic diversity is more precise and considers the relationship of species to each other. It can be measured by counting the number of different taxa (the main categories of classification) present. For example, a pond containing three species of snails and two fish, is more diverse than a pond containing five species of snails, even though they both contain the same number of species. High species biodiversity is not always necessarily a good thing. For example, a habitat may have high species biodiversity because many common and widespread species are invading it at the expense of species restricted to that habitat.

Ecosystem–Communities of plants and animals, together with the physical characteristics of their environment (*e.g.* geology, soil and climate) interlink together as an ecological system, or 'ecosystem'. Ecosystem diversity is more difficult to measure because there are rarely clear boundaries between different ecosystems and they grade into one another. However, if consistent criteria are chosen to define the limits of an ecosystem, then their number and distribution can also be measured.

Losses of Biodiversity

Extinction is a fact of life. Species have been evolving and dying out ever since the origin of life. One only has to look at the fossil record to appreciate this. (It has been estimated that surviving species constitute about 1 per cent of the species that have ever lived). However, species are now becoming extinct at an alarming rate, almost entirely as a direct result of human activities. Previous mass extinctions evident in the geological record are thought to have been brought about mainly by massive climatic or environmental shifts. Mass extinctions as a direct consequence of the activities of a single species are unprecedented in geological history.

The loss of species in tropical ecosystems such as the rain forests, is extremely well-publicised and of great concern. However, equally worrying is the loss of habitat and species closer to home in Britain. This is arguably on a comparable scale, given the much smaller area involved.

Predictions and estimates of future species losses abound. One such estimate calculates that a quarter of all species on earth are likely to be extinct, or on the way to extinction within 30 years. Another predicts that within 100 years, three quarters of all species will either be extinct, or in populations so small that they can be described as "the living dead".

It must be emphasised that these are only predictions. Most predictions are based on computer models and as such, need to be taken with a very generous pinch of salt. For a start, we really have no idea how many species there are on which to base our initial premise. There are also so many variables involved that it is almost impossible to predict what will happen with any degree of accuracy. Some species actually benefit from human activities, while many others are adversely affected. Nevertheless, it is indisputable that if the human population continues to soar, then the ever increasing competition with wildlife for space and resources will ensure that habitats and their constituent species will lose out.

It is difficult to appreciate the scale of human population increases over the last two centuries. Despite the horrendous combined mortality rates of two World Wars, Hitler, Stalin, major flu pandemics and Aids, there has been no dampening effect on rising population levels. In 1950, the world population was 2.4 billion. Just over 50 years later, the world population has almost tripled, reaching 6.5 billion.

In the UK alone, the population increases by the equivalent of a new city every year. Corresponding demands for a higher standard of living for all, further exacerbates the problem. It has been estimated that if everyone in the world lived at the UK standard of living (and why should people elsewhere be denied this right) then we would either need another three worlds to supply the necessary resources or alternatively, would need to reduce the world population to 2 billion. The only possible conclusion is that unless human populations are substantially reduced, it is inevitable that biodiversity will suffer further major losses.

Some species are more vulnerable to extinction than others. These include:

Species at the Top of Food Chains, such as Large Carnivores
Large carnivores usually require fairly extensive territories in order to provide them with sufficient prey. As human populations increasingly encroach on wild areas and as habitats shrink in extent, the number of carnivores which can be accommodated in the area also decreases. These animals may also pose a threat to people, as populations expand into wilder areas inhabited by large carnivores. Protective measures, including elimination of offending animals in the area, further reduces numbers.

Endemic Local Species (Species found only in one geographical area) with a Very Limited Distribution
These are very vulnerable to local habitat disturbance or human development.

Species with Chronically Small Populations

If populations become too small, then simply finding a mate, or interbreeding, can become serious problems.

Migratory Species

Species which need suitable habitats to feed and rest in widely spaced locations (which are often traditional and 'wired' into behaviour patterns) are very vulnerable to loss of these 'way stations'.

Species with Exceptionally Complex Life Cycles

If completion of a particular lifecycle requires several different elements to be in place at very specific times, then the species is vulnerable if there is disruption of any single element in the cycle.

Specialist Species

Specialist species with very narrow requirements such as a single specific food source, *e.g.* a particular plant species.

Why Conserve Biodiversity?

Ecological Reason

Individual species and ecosystems have evolved over millions of years into a complex interdependence. This can be viewed as being akin to a vast jigsaw puzzle of inter-locking pieces. If you remove enough of the key pieces on which the framework is based then the whole picture may be in danger of collapsing. We have no idea how many key 'pieces' we can afford to lose before this might happen, nor even in many cases, which are the key pieces. The ecological arguments for conserving biodiversity are therefore based on the premise that we need to preserve biodiversity in ord Two linked issues which are currently of great ecological concern include world-wide deforestation and global climate change.

Forests not only harbour untold numbers of different species, but also play a critical role in regulating climate. The destruction of forest, particularly by burning, results in great increases in the amount of carbon in the atmosphere. This happens for two reasons. Firstly, there is a great reduction in the amount of carbon dioxide taken in by plants for photosynthesis and secondly, burning releases huge quantities of carbon dioxide into the atmosphere. (The 1997 fires in Indonesia's rain forests are said to have added as much carbon to the atmosphere as all the coal, oil and gasoline burned that year in western Europe). This is significant because carbon dioxide is one of the main greenhouse gases implicated in the current global warming trend.

Average global temperatures have been showing a steadily increasing trend. Snow and ice cover have decreased, deep ocean temperatures have increased and global sea levels have risen by 100–200 mm over the last century. If current trends continue, scientists predict that the earth could be on average 1°C warmer by 2025 and 3°C warmer by 2100. These changes, while small, could have drastic effects. As an example, average temperatures in the last Ice Age were only 5°C colder than current temperatures.

Rising sea levels which could drown many of our major cities, extreme weather conditions resulting in drought, flooding and hurricanes, together with changes in the distribution of disease-bearing organisms are all predicted effects of climate change.

Forests also affect rainfall patterns through transpiration losses and protect the watershed of vast areas. Deforestation therefore results in local changes in the amount and distribution of rainfall. It often also results in erosion and loss of soil and often to flooding. Devastating flooding in many regions of China over the past few years has been largely attributed to deforestation.

These are only some of the ecological effects of deforestation. The effects described translate directly into economic effects on human populations.

Environmental disasters such as floods, forest fires and hurricanes indirectly or directly caused by human activities, all have dire economic consequences for the regions afflicted. Clean-up bills can run into the billions, not to mention the toll of human misery involved. Susceptible regions are often also in the less-developed and poorer nations to begin with. Erosion and desertification, often as a result of deforestation, reduce the ability of people to grow crops and to feed themselves. This leads to economic dependence on other nations.

Non-sustainable extraction of resources (*e.g.* hardwood timber) will eventually lead to the collapse of the industry involved, with all the attendant economic losses. It should be noted that even if 'sustainable' methods are used, for example when harvested forest areas are replanted, these areas are in no way an ecological substitute for the established habitats which they have replaced.

Large-scale habitat and biodiversity losses mean that species with potentially great economic importance may become extinct before they are even discovered. The vast, largely untapped resource of medicines and useful chemicals contained in wild species may disappear forever. The wealth of species contained in tropical rain forests may harbour untold numbers of chemically or medically useful species. Many marine species defend themselves chemically and this also represents a rich potential source of new economically important medicines. Additionally, the wild relatives of our cultivated crop plants provide an invaluable reservoir of genetic material to aid in the production of new varieties of crops. If all these are lost, then our crop plants also become more vulnerable to extinction.

There is an ecological caveat here of course. Whenever a wild species is proved to be economically or socially useful, this automatically translates into further loss of natural habitat. This arises either through large-scale cultivation of the species concerned or its industrial production/harvesting. Both require space, inevitably provided at the expense of natural habitats.

Perhaps the rain forests and the seas should be allowed to keep their secrets.

Ethical Reasons

Do we have the right to decide which species should survive and which should die out?

Do we have the right to cause a mass extinction?

Most people would instinctively answer 'No!'. However, we have to realise that most biodiversity losses are now arising as a result of natural competition between humans and all other species for limited space and resources.

If we want the luxury of ethics, we need to reduce our populations.

Aesthetic Reasons

Most people would agree that areas of vegetation, with all their attendant life forms, are inherently more attractive than burnt, scarred landscapes, or acres of concrete and buildings. Who wouldn't prefer to see butterflies dancing above coloured flowers, rather than an industrial complex belching smoke?

Human well-being is inextricably linked to the natural world. In the western world, huge numbers of people confined to large urban areas derive great pleasure from visiting the countryside. The ability to do so is regarded not so much as a need, but as a right. National governments must therefore juggle the conflicting requirements for more housing, industry and higher standards of living with demands for countryside for recreational purposes.

How do we Conserve Biodiversity?

There are two main ways to conserve biodiversity. These are termed ex-situ (i.e. out of the natural habitat) and in-situ (within the natural habitat)

Ex-situ Conservation–Out of the Natural Habitat

☆ **Zoos**–These may involve captive breeding programmes

☆ **Aquaria**–Research, public information and education

☆ **Plant Collections**–Breeding programmes and seed storage

Zoos

In the past, zoos were mainly display facilities for the purpose of public enjoyment and education. As large numbers of the species traditionally on display have become rarer in the wild, many zoos have taken on the additional role of building up numbers through captive breeding programmes.

Although comparatively far more invertebrates than vertebrates face extinction, most captive breeding programmes in zoos focus on vertebrates. Threats to vertebrate extinction tend to be well publicised (*e.g.* Dormouse, Panda). People find it easier to relate to and have sympathy with animals which are more similar to ourselves, particularly if they are cute and cuddly (at least in appearance, if not in fact!). Not many visitors to zoos are likely to get excited over the prospect of the zoo 'saving' a tiny beetle, which they can barely see, let alone spiders or other invertebrates which often invite horror rather than wonder. Vertebrates therefore serve as a focus for public interest. This can help to generate financial support for conservation and extend public education to other issues. This is a very important consideration, as conservation costs money and needs to be funded from somewhere.

The focus on vertebrates is not solely pragmatic. Many of the most threatened vertebrates are large top carnivores, which the world stands to lose in disproportionate numbers. Such species require extensive ranges to provide sufficient prey to sustain them. In many cases, whole habitats for these predators have all but disappeared. Some biased expenditure on their survival may therefore be justified.

Several species are now solely represented by animals in captivity. Captive breeding programmes are in place for numerous species. At least 18 species have been reintroduced into the wild following such programs. In many cases the species was actually extinct in the wild at the time of reintroduction (Arabian Oryx, Pere David Deer, American Bison). In some cases, all remaining individuals of a species, whose numbers are too low for survival in the wild, have been captured and the species has then been reintroduced after captive breeding (California Condor).

The role of zoos in conservation is limited both by space and by expense. At population sizes of roughly 100-150 individuals per species, it has been estimated that world zoos could sustain roughly 900 species. Populations of this size are just large enough to avoid inbreeding effects. However, zoos are now shifting their emphasis from long-term holding of species, to returning animals to the wild after only a few generations. This frees up space for the conservation of other species.

Genetic management of captive populations via stud records is essential to ensure genetic diversity is preserved as far as possible. There are now a variety of international computerised stud record systems which catalogue genealogical data on individual animals in zoos around the world. Mating can therefore be arranged by computer, to ensure that genetic diversity is preserved and in-breeding minimised (always assuming the animals involved are prepared to co-operate).

Research has led to great advances in technologies for captive breeding. This includes techniques such as artificial insemination, embryo transfer and long-term cryogenic (frozen) storage of embryos. These techniques are all valuable because they allow new genetic lines to be introduced without having to transport the adults to new locations. Therefore the animals are not even required to co-operate any longer. However, further research is vital. The success of zoos in maintaining populations of endangered species is limited. Only 26 of 274 species of rare mammals in captivity are maintaining self-sustaining populations.

Reintroduction of Species to the Wild Poses Several Different Problems.

Diseases

The introduction of new diseases to the habitat, which can decimate existing wild populations. Alternatively, the loss of resistance to local diseases in captive-bred populations.

Behaviour

Behaviour of captive-bred species is also a problem. Some behaviour is genetically determined and innate, but much has to be learned from other adults of the species, or by experience. Captive-bred populations lack the *in-situ* learning of their wild relatives and are therefore at a huge disadvantage in the wild. In one case of reintroduction, a number of monkeys starved because they had no concept of having to search for food to eat–it had always been supplied to them in captivity. In the next attempt, the captive monkeys were taught that they had to look for food, by hiding it in their cages, rather than just supplying it.

Genetic Races

Reintroduced populations may be of an entirely different genetic make-up to original populations. This may mean that there are significant differences in reproduction habits and timing, as well as differences in general ecology. Reintroduction of individuals of a species into an area where the species has previously become extinct, is in many cases just like introducing a foreigner. The Large Copper Butterfly is a good example of this. Although extinct in Britain, it persists in continental Europe. There have been over a dozen attempts to re-establish it in Britain over the last century, but none have been successful. This is probably due to the differing ecology of the introduced races. Replacement of extinct populations by reintroduction from other areas may not therefore be an option.

Habitat

The habitat must be there for reintroduction to take place. In many cases, so much habitat has been destroyed, that areas must first be restored to allow captive populations to be reintroduced. Suitable existing habitats will also (unless the species is extinct in the wild) usually already contain wild members of the species. In this case, it is likely that within the habitat, there are already as many individuals as the habitat can support. The introduction of new individuals will only lead to stress and tension as individuals fight for limited territory and resources such as food. In this case, nothing positive has been accomplished by reintroduction, it has merely increased the stress on the species. It may even in some cases result in a decrease in numbers. In contrast, the provision of additional restored habitat nearby can allow wild populations to expand into it without the need for reintroduction.

Aquaria

The role of aquaria has largely been as display and educational facilities. However, they are assuming new importance in captive breeding programmes. Growing threats to freshwater species in particular, are leading to the development of *ex-situ* breeding programmes. The World Conservation Union (IUCN) is currently developing captive breeding programmes for endangered fish. Initially this will cover those from Lake Victoria in Africa, the desert fishes of N. America and Appalachian stream fishes. Natural habitats will be restored as part of the programme.

Marine, as well as freshwater species are also the subject of captive breeding programmes. For example, The National Marine Aquarium, in South West England, is playing an important role in the conservation of sea horse species through their captive breeding programme.

Plant Collections

Populations of plant species are much easier than animals to maintain artificially. They need less care and their requirements for particular habitat conditions can be provided more readily. It is also much easier to breed and propagate plant species in captivity.

There are roughly 1,500 botanic gardens world-wide, holding 35,000 plant species (more than 15 per cent of the world's flora). The Royal Botanic Gardens of England (Kew Gardens) contains an estimated 25,000 species. IUCN classifies 2,700 of these as rare, threatened or endangered. Many botanic gardens house collections of particular taxa which are of major conservation value. There is however, a general geographic imbalance. Only 230 of the world's 1,500 gardens are in the tropics. Considering the greater species richness of the tropics, this is an imbalance that needs to be addressed.

A more serious problem with *ex-situ* collections involves gaps in coverage of important species, particularly those of significant value in tropical countries. One of the most serious gaps is in the area of crops of regional importance, which are not widely traded on world markets. These often have recalcitrant seeds (unsuited to long-term storage) and are poorly represented in botanic collections. Wild crop relatives are also under-represented. These are a potential source of genes conferring resistance to diseases, pests and parasites and as such are a vital gene bank for commercial crops.

Plant genetic diversity can also be preserved *ex-situ* through the use of seed banks. Seeds are small but tough and have evolved to survive all manner of adverse conditions and a host of attackers. Seeds can be divided into two main types, orthodox and recalcitrant. Orthodox seeds can be dried and stored at temperatures of -20°C. Almost all species in a temperate flora can be stored in this way. Surprisingly, many tropical seeds are also orthodox. Recalcitrant seeds, in contrast, die when dried and frozen in this manner. Acorns of oaks are recalcitrant and it is believed that so are the seeds of most tropical rain forest trees.

The result of storing seeds under frozen conditions is to slow down the rate at which they lose their ability to germinate. Seeds of crop plants such as maize and barley could probably survive thousands of years in such conditions, but for most plants, centuries is probably the norm. This makes seed banking an attractive conservation option, particularly when all others have failed. It offers an insurance technique for other methods of conservation.

All of the *ex-situ* conservation methods discussed have their role to play in modern conservation. Generally, they are more expensive to maintain and should be regarded as complementary to *in-situ* conservation methods. For example they may be the only option where *in-situ* conservation is no longer possible.

Conservation at the National Level

National governments are vital to the preservation of biodiversity through the passing of laws requiring protection of species and habitats. If national laws do not protect species, then there is little hope of preserving them. However, it is not enough just to have laws, there must also be the will and the resources to enforce them. Even in economically developed nations, the necessary resources to properly enforce laws are not always made available. In under-developed nations, even the most basic resources for enforcement may be lacking. In addition, national laws may not in the end translate into local action, in which case they do not accomplish much. In democratic nations, national laws are also driven to a large extent by public opinion.

They may in some cases be drafted more as a response to emotion than by actual scientific need.

Several international conventions exist for the preservation of biodiversity. These include such conventions as the Ramsar Convention (1976) which provides for the conservation of internationally important wetlands and the Bern Convention (1979) which requires the protection of endangered and vulnerable species of flora and fauna in Europe and their habitats. There are many others. Signatory nations to these conventions must ratify national laws to ensure compliance with the conventions.

In Britain, the main piece of legislation covering conservation is the Wildlife and Countryside Act 1981 and 1985, which implements preceding EU conventions. It protects both species and sites of UK importance. Enforcement of conservation directives is the responsibility of the Environment Agency, a government organisation. English Nature, a government funded watchdog, is also responsible for the promotion of the conservation of England's wildlife.

In addition to the enforcement of laws, the Environment Agency is also responsible for data collection and monitoring. Environmental monitoring and biodiversity surveys are important because they provide information on the condition of ecosystems and the changes that are taking place within them. They therefore provide the scientific information on which to base environmental policy decisions. Similarly, assessments of the environmental impact of large development projects are vital before relevant authorities can either grant permission to proceed, or require that changes be made to development designs.

Conclusion

We are interfering with biodiversity on a great many levels, from the molecular (genetic modification), all the way through habitats and possibly global climate change as well. However, the many predictions made about species and habitats losses need to be carefully examined in each case and not just taken at face value. Many are based on computer simulations and emotions can get in the way of clear practical thinking. Anyone concerned about conservation needs to question whether the innumerable strategies and policies in place are actually being delivered. International conventions and national laws are in the end only ideas on pieces of paper. These must be translated into concrete action in local situations for anything to be truly accomplished. On the plus side, it is possible to restore some habitats which have been lost or degraded. This is not to imply that it is permissible to destroy habitats in the first place. This causes the local extinction of all the species in the habitat and it can take hundreds of years for complex ecosystems to become re-established. The species which have recolonised the restored habitat will also not necessarily be of the same genetic make-up as the original inhabitants. However, restoration does mean that action can be taken to repair damage. The natural world given half a chance is amazingly resilient. All it needs is space and time.

References

Chopra, K., 2000. *Economic Aspects of Biodiversity Conservation*. Tata Mc-Graw Hill Publishing, New Delhi, p. 571–579.

Kumar, A.S., Walker and Molur, S., 2004. *Setting Biodiversity Conservation Priorities in India.* Tata Mc-Graw Hill Publishing, New Delhi, p. 341–425.

Negi, S.S., 2005. *Biodiversity and its Conservation in India: Forests and Forestry*– A Handbook of the Himalayas and Himalayan Wildlife.

Panigrahi, Ashok Kumar, 2009. *Biodiversity and its Conservation.* New Delhi, p. 215–219.

2013, Biodiversity Conservation for Sustainable Management   Pages *12–17*

Editor: **Dr. K. Muthuchelian,** *Vice Chancellor, Periyar University, Salem*

Published by: **Daya Publishing House, NEW DELHI**

Chapter 2

Biodiversity Loss: Causes and Consequences

R. Ganesan, S. Venkatesan and K. Muthuchelian

School of Energy, Environment and Natural Resourses,
Madurai Kamaraj University, Madurai

Biodiversity is a species richness (plants, animals, microorganisms) an interacting system in a given habitat or on the entire Earth. Biodiversity is often used as a measure of the health of biological systems. The biodiversity found on Earth today consists of many millions of distinct biological species. The Rapid environmental modifications typically cause extinctions. Of all species that have existed on Earth, 99.9 percent are now extinct. Since life began on Earth, five major mass extinctions have led to large and sudden drops in the biodiversity of species. The Phanerozoic eon (the last 540 million years) marked a rapid growth in biodiversity in the Cambrian explosion–a period during which nearly every phylum of multicellular organisms first appeared. The next 400 million years was distinguished by periodic, massive losses of biodiversity classified as mass extinction events. The most recent, the Cretaceous–Tertiary extinction event, occurred 65 million years ago, and has attracted more attention than all others because it killed the nonavian dinosaurs. The year 2010 has been declared as the International Year of Biodiversity.

Today there is concern that the period since the emergence of humans is part of a mass reduction in biodiversity, the Holocene extinction, caused primarily by the impact humans are having on the environment, particularly the destruction of plant and animal habitats. In addition, human practices have caused a loss of genetic biodiversity. The relevance of biodiversity to human health is becoming a major international issue, as scientific evidence is gathered on the global health implications of biodiversity loss. Biodiversity is not distributed evenly on Earth. It is consistently richer in the tropics.

As one approaches polar regions one generally finds fewer species. Flora and fauna diversity depends on climate, altitude, soils and the presence of other species. In the year 2006 large numbers of the Earth's species were formally classified as rare or endangered or threatened species; moreover, many scientists have estimated that there are millions more species actually endangered which have not yet been formally recognized. About 40 per cent of the 40,177 species assessed using the IUCN Red List criteria, are now listed as threatened species with extinction–a total of 16,119 species. biodiversity declines from the equator to the poles in terrestrial ecoregions, whether this is so in aquatic ecosystems is still a hypothesis to be tested, especially in marine ecosystems.

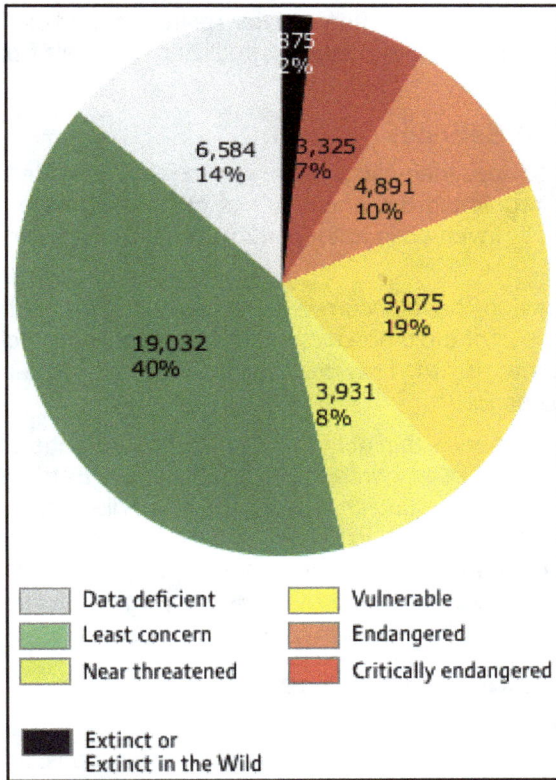

Figure 2.1: Proportion of all assessed species in different threat categories of extinction risk on the IUCN Red List, based on data from 47,677 species.

Source: IUCN, pie chart compiled by Secretariat of the Convention on Biological Diversity (2010) Global Biodiversity Outlook 3, May 2010.

Human Benefits

Biodiversity provides food for humans. Although about 80 per cent of our food supply comes from just 20 kinds of plants, humans use at least 40,000 species of plants and animals a day. Many people around the world depend on these species for their food, shelter, and clothing.

Human Health

One of the key health issues associated with biodiversity is that of drug discovery and the availability of medicinal resources. A significant proportion of drugs are derived, directly or indirectly, from biological sources; Chivian and Bernstein report that at least 50 per cent of the pharmaceutical compounds on the market in the US are derived from natural compounds found in plants, animals, and microorganisms, while about 80 per cent of the world population depends on medicines from nature (used in either modern or traditional medical practice) for primary healthcare.

Business and Industry

A wide range of industrial materials are derived directly from biological resources. These include building materials, fibers, dyes, resirubber and oil. There is enormous potential for further research into sustainably utilizing materials from a wider diversity of organisms.

Other Ecological Services

Biodiversity provides many ecosystem services that are often not readily visible. It plays a part in regulating the chemistry of our atmosphere and water supply. Biodiversity is directly involved in water purification, recycling nutrients and providing fertile soils.

Massive extinctions have occurred *five times* during the earth's history, the last one was the extinction of the dinosaurs, 65 million years ago. Scientists are calling what is occurring now, the *sixth mass extinction*. The loss of species is about losing *the very web of life* on Earth.

Although they are uncertain of the numbers, most scientists believe the *rate of loss* is greater now than at any time in the history of the Earth. Within the next 30 years as many as *half of the species on the earth* could die in one of the fastest mass extinctions in the planet's 4.5 billion years history. Dr Leakey, author of "The Sixth Extinction," believes that 50 per cent of the earth's species will vanish within 100 years and that such a dramatic and overwhelming mass extinction threatens the entire, complex fabric of life, including *Homo sapiens*.

Causes of the Extinction of Species

Scientists have identified the key causes of the crisis. In particular, the loss of species is caused by as the growing size of human populations, and the rate at which humans consume resources and cause changing climate. Extinction is a natural process that has occurred for millions of years so why does it deserve so much attention now? The problem is that the rate of extinction has increased dramatically in recent years due to our impact as humans. The rate of change is perhaps as damaging as the effects of the changes. *Humans create all of these causes.*

Climate Change and the Loss of Species

At the end of the Permian period, 251 million years ago, global warming caused the worst mass extinction in the history of the planet. That time a 6°C increase in the global temperature was enough to kill up to 95 per cent of the species that were alive

Table 2.1: According to IUCN Red List.

Group	Critically Endangered							
	1996/98	*2000*	*2002*	*2003*	*2004*	*2006*	*2007*	*2008*
Mammals	169	180	181	184	162	162	163	188
Birds	168	182	182	182	179	181	189	190
Reptiles	41	56	55	57	64	73	79	86
Amphibians	18	25	30	30	413	442	441	475
Fishes	157	156	157	162	171	253	254	289
Insects	44	45	46	46	47	68	69	70
Molluscs	257	222	222	250	265	265	268	268
Plants	909	1014	1046	1276	1490	1541	1569	1575

on Earth. This extinction is called the "Great Dying." Gigantic volcanic eruptions caused this warming by triggering a "runaway greenhouse effect" that nearly put an end to life on Earth. It took *100 million years* for species diversity to return to former levels. In this case, the carbon dioxide buildup which created this greenhouse effect came from massive volcanic eruptions. Today the build up of carbon dioxide is coming from our life style and industrial activity.

Rising ocean temperatures cause corals to become stressed, and they expel the zooxanthellae and turn white or "bleach". If zooxanthellae do not return to the coral's tissue, the coral will die. As little as a 1° Celsius (1.8°F) increase in temperature above the summer maximum can cause corals to bleach. Tropical sea temperatures have increased by 1° Celsius over the past 100 years and are predicted to continue rising.

Habitat Loss as a Cause of the Loss of Species

Other than global warming, the greatest threat to biodiversity is habitat loss and fragmentation by deforestation and urbanization. *Urbanization* has dramatically increased the rate of habitat loss and change. Sprawling development is consuming land at a rate of five or more times the rate of population growth, destroying wildlife habitat and degrading water quality. Dredging, draining, bulldozing, and paving the land for housing developments, malls, business parks, and new roads, all destroy habitat.

Deforestation is also one of the leading causes of habitat loss. For centuries, humans have altered landscapes, through deforestation, fire and over-use. Already, around half of the world's original forests have disappeared. As tropical forests contain at least half the Earth's species, the clearance of some 17 million hectares each year is causing a dramatic loss of biodiversity. Habitat loss is identified as a main threat to 85 per cent of all species described in the IUCN's *Red Lists.*

Table 2.2: Habitat and human disturbance by continent.

Total Area	Per cent Human	Per cent Partial	Per cent Undisturbed (km²)	Dominated Disturbed
Europe	5 759321	64.9	19.6	15.6
Asia	53 311 557	29.5	27.0	43.5
Africa	33 985316	15.4	35.8	48.9
North America	26179907	24.9	18.8	56.3
South America	20120346	15.1	22.5	62.5
Australia	9 487 262	12.0	25.8	62.3
Antarctica	13 208 983	0	0	100.0

Pollution Leads to a Loss of Species

Pollution is found everywhere in the world—chemicals have been found in animals even in the Arctic and Antarctic. Chemicals can cause mutations and fertility problems, already seen in the reproductive organs of fish, alligators, and polar bears. A recent EPA report noted that nearly 40 per cent of the nation's rivers, lakes, and estuaries are too polluted for safe fishing and swimming. Fifty percent of freshwater species populations, from fish and frogs to river dolphins, are declining from pollution by pesticides, fertilizers and other agricultural chemicals.

Consequences of Biodiversity Loss

The loss of biodiversity has often been seen as an aesthetic or bioethical issue. The lack of a broader understanding of the consequences of the declining diversity of our living resources has been an important gap in our scientific understanding of the world. A large-scale study called the BIODEPTH project has shown that reduced plant diversity impairs important aspects of ecosystem functioning. The demonstrated that reduced biodiversity of grassland plants also lowers the productivity of the land. These findings have important implications for agriculture, grassland management, water quality and sustainable land use. Wild animals and plants make up an essential part of nature so product nature.

Conclusion

Therefore, future changes in biodiversity are likely to result in major alterations in ecosystem services, with potentially considerable social and economic implications for human societies. The problem is not just the loss of species. There is also the loss of the genetic diversity *within* species, as well as the loss of diversity of different types of *ecosystems,*which can contribute to or hasten whole species extinction. Preserving the wider gene pool diversity in subdivisions of species, such as subspecies and populations, offers the raw material for the evolution of new species in the future. If we loss species diversity, source of new drugs are food crops might be wiped out before they are discovered. Biodiversity also should be maintained for it's beauty and essential part of nature, so conserve our ecosystem and adds loveliness of our world.

References

Hill, Marquita K., 2004. *Understanding Environmental Pollution*. Cambridge University Press.

Rana, S.V.S., 2005. *Essentials of Ecology and Environmental Science,* 2nd Edn.

The World Book Encyclopedia (International). Scott Felzer Company.

2013, Biodiversity Conservation for Sustainable Management   *Pages 18–27*
Editor: **Dr. K. Muthuchelian,** *Vice Chancellor, Periyar University, Salem*
Published by: **Daya Publishing House, NEW DELHI**

Chapter 3

Indian Biodiversity: Proud to Say but Pathetic to See

V. Kumar[1]* and S. Vinoth Kumar[2]
[1]Department of Pathology,
Directorate of Rice Research, Hyderabad
[2]Department of Environmental Health Engineering,
Sri Ramachandra University, Chennai

Introduction

Biodiversity is a worthless resource and part of nature, which is the key feature of nature for holding its ever amazing beauty. It is very proud to say India is one of 12 megabiodiversity region, has 2 hotspots, 5 world heritage sites, 12 biosphere reserves, etc. But the fact is all are under severe declination in size and population. Especially Western Ghats extending up to 180,000 sq.kms, is facing much critical problems such as habitat loss and fragmentation, water and food scarcity, due to urbanization, poaching, transport and their needs, pollution, etc.

Wealth of Western Ghats

Western Ghats, proudly contributing to Global Biodiversity, is one of the 34 world hotspots. It is the motherland to many species of Plants (Bryophytes, Pteridophytes, Gymnosperms and Angiosperms) and Animals (Most of the Phyla including Arthropods, Chordates). The forest region also shows its diversity as evergreen, semi-evergreen, moist deciduous and dry deciduous forests.

---

* Corresponding Author: E-mail: vkumar.apr@gmail.com

Floral Fantasy of Western Ghats

Western Ghats is the green backbone of Plant diversity in India as well as contributing to World Flora. There are several centers of plant endemism and species richness within the Western Ghats. The Western Ghats harbors approximately 5,000 species of vascular plants belonging to nearly 2,200 genera; about 1,700 species (34 percent) are endemic. There are also 58 endemic plant genera, and, while some are remarkably specific (like *Niligrianthus*, which has 20 species), nearly three-quarters of the endemic genera have only a single species. Approximately 63 percent of India's woody evergreen taxa are endemic to the Western Ghats. The tree genera endemic to the Western Ghats include *Blepharistemma*, *Erinocarpus*, *Meteromyrtus*, *Otenophelium*, *Poeciloneuron*, and *Pseudoglochidion*. Other plant genera endemic to the Western Ghats include *Adenoon*, *Griffithella*, *Willisia*, *Meineckia*, *Baeolepis*, *Nanothamnus*, *Wagatea*, *Campbellia*, and *Calacanthus*.

Animals-A Special Care to Chordates

Chordates, the higher phyla of animals, have more diversified class of organism which highly differs in habit, habitat (aquatic, terrestrial, aquatic and terrestrial), feeding behavior (herbivore, carnivore, omnivore, scavengers).

Pisces

Over 218 species of fishes are found in primary and secondary fresh waters of Western Ghats, in which 116 are endemic including genera *Brachydanio*, *Lepidopygopsis*, *Bhavania*, *Travancoria*, *Horabagrus*, *Horaglanis and Horaichthys*.

Aves

A total of 508 species of birds found in Western Ghats, out of which 324 are resident species. Majority of avian diversity found in Central part of Western Ghats *i.e.* Uttara Kannada District.

Reptiles

Nearly 157 genera has been reported in western ghats representing 36 genera with 2 genera of tortoise/turtles, 14 genera of lizards and 20 genera of snakes. Out of this 50 per cent are endemic. Evergreen forest alone supports 130 species of reptiles.

Amphibians

About 126 Amphibians reported under 24 genera from Western Ghats. Western Ghats account for Highest Amphibian Endemicity in India.

Mammals

Mammals, the higher chordates reported in Western Ghats are 137 species. Out of these, 16 are endemic to Western Ghats and not found any other place on earth. Among 16, 13 species are endangered/threatened, within the 13 species, 3 are Critically Endangered. Wroughton's free-tailed bat (*Otomops wroughtonii*) is one of the critically endangered species.

Chronic Condition of Western Ghats

The International Union for Conservation of Nature (IUCN) notes that many species are threatened with extinction. In addition, 75 per cent of genetic diversity of agricultural crops has been lost, 75 per cent of the world's fisheries are fully or over exploited,

Up to 70 per cent of the world's known species risk extinction if the global temperatures rise by more than 3.5°C, 1/3rd of reef-building corals around the world are threatened with extinction, Every second a parcel of rainforest the size of a football field disappears, Over 350 million people suffer from severe water scarcity.

Among all, 332 globally threatened species occur in the Western Ghats. The globally threatened flora and fauna in the Western Ghats are represented by 229 plant species, 31 mammal species, 15 bird species, 52 amphibian species, four reptile species, and one fish species. Of the total of 332 globally threatened species in the Western Ghats, 55 are Critically Endangered, 148 are endangered, and 129 are Vulnerable (Table 3.1).

Table 3.1: Globally threatened species occurring in Western Ghats.

Taxonomic Group	Critically Endangered	Endangered	Vulnerable	Total
Plants	39	111	79	229
Fish	-	-	1	1
Birds	2	1	12	15
Reptiles	-	1	3	4
Amphibians	11	28	13	52
Mammals	3	7	21	31
Total	55	148	129	332

Why Pathetic and What to Do?

One of the main reasons for today's pathetic biodiversity is improper human intervention and utilization of biodiversity and natural resources. Nature made human as the higher most organisms in the earth, but human having lost his/her highness, injured the nature heavily for his/her need, sophistication, desire and joy.

In-situ Conservation

In-situ conservation, which is the best method of conservation, is not properly implemented in many places for many plants and animals. It is to be strictly followed that the organism is to be maintained under its own native habitat. Provided the physico-chemical properties, water sources, prey-predator ratio and all the elements of that ecosystem should be in a properly balanced state and basic line of food chain of that ecosystem should be rich in population. Population of all preys from preferred to optional prey for the organism which is being conserved should be high, at the same time predator's population should be very less or even absent.

Impact of One spp. Decline on the Other

Population decline in one species (1° spp) will lead to population decline of the other; due to break in food chain and also because it may be a most preferred pray for the other (2° spp), leading to population decline of that 2° organism. In that case the 2° spp. should be conserved by providing optional prey populations (ex. Nilgiri Tahr is the preferred prey for Tiger, both are endangered). To overcome this, both *in-situ* and *ex-situ* conservation and developing the population of common wild and domesticated animals and plants, getting them back gradually to wild forest habitat would be very much helpful in maintaining their population size from decline and for utilizing them as an optional prey population for next class of wild organisms in food chain, which are less in population (fodder grass and edible shrubs, rabbit, goat, cow, etc.).

Reforestation with Edible Fruit Trees

All available wild varieties of both common and uncommon edible fruit varieties like Ficus, Neem, Manila Tamarind, Mango, Jackfruit, Banana, Jamun, Sapodilla, Guava, Custard Apple, Fig, Pomegranate, Grapes, Amla, Jujube, etc., are to be propagated in biosphere reserve areas and also in forests. It will be very much useful for feeding various species of birds, squirrels, monkeys, insects, bats, ants, etc,. Which in turn act as the important seed dispersers of the wild forest tree varieties, so this will be a combined way of conservation of both wild fruit tree varieties as well as related wild animal species.

Figure 3.1: Giant Squirrel on fruit tree.

Figure 3.2: Great Hornbill feeding on.

Figure 3.3: Fruit eating Bat.

Worth of Weed Plant

Most of the weed plants, south Indian weeds in particular are rich in useful phytochemicals and are used as medicinal plants. Most of the herbs mentioned in ayurvedic and herbal medicines are nothing but common weed plants. Many weed plants have nectar rich flowers which acts as an important food source for various butterfly, honey bee, insect and ant species. Ruminants feed on some selected weed plants, which will provide some essential nutrients and may also acts as food as medicine improving the immunity of cattle. Most of these useful weed plants are declining severely, that should be conserved and propagated in herbivores' conservation areas. Some example for worthy weed plant–Leuca indica (Thumba), Senna auriculata (Senna), Euphorbia cyathophora (Poinsettia), etc.

Colourful spp. in Dark Status

Butterflies and honey bees have a great ecological importance that they act as pollinators, there by facilitates and improve propagation of wide range of plant species. There is a great decline in butterflies' and honey bees' population in recent years and the major reason for this is the decline in nectar plants and common weed plants. There are many plant specific butterfly species, which will lay eggs only on a particular host plat and whose leaves only the pupa will eat. More number of nectar rich plant species should be propagated in the borders of zoos, sanctuaries, biosphere reserves and even in forests.

Figure 3.4: Butterfly on Manila tamarind.

Figure 3.5: Honey Bee on Poinsettia.   **Figure 3.6**: Guate Butterfly on Poinsettia.

Co-Conservation

Conservation of more than one herbivore or the other in a same place can be done, provided no two animals should have same food source/feeding behavior and no animal should be a prey or predator for the other. For example, Deer, Elephant, Squirrel and Birds can be conserved in a same area, in which deer will feed on grass and shrubs, elephants feed on trees, squirrels feed on fruits and birds feed on insects and fruits, so there won't be any competition for food. Also their population can be increased.

Way to Care Young Orphan Animals

The young orphan animals, after rescue are often taken to nearest Zoo park, kept in a cage or the other, fed with readymade food, grown and later released into forest. Instead it has to be taken to biosphere reserve area along with a practiced adult animal (as step mother), so that it may get trained how to hunt and catch the prey, safeguard from predators, places of water source, routs for migration, etc.

Endangered Animals–Ever Displaying Should Be Never Displayed In Zoo Parks

All the zoological parks by name sake are modernized; but in actual most of the animals in most of the zoos are not maintained in their habitat, instead they are kept in a small land space, surrounded by polluted water and provided with readymade food.

Figure 3.7: Orangutan kept lonely.

Many endangered animals like Lion tailed Macaque, Great Hornbill, Orangutan, Giant Squirrel, King Cobra, etc., were kept in individual (single) merely for the sake of displaying for recreation of visitors. Instead all these individual animals of same spp from each zoo park can be collected, combined and conserved in their natural habitat, so that they will be happy, feed well and breed well. After some years on getting improvement in their population, a pair can be taken back to zoo parks providing their habitat.

Selective Mating

Most of the conserved animals from herbivore like elephant to carnivore like tiger, leopard were allowed mate with an individual opposite sex, the one which is available and decided by concern person in that sanctuary.

Figure 3.8: White Tiger on polluted water.

The common mating behavior for most of the wild animals is that, many male approaches a female, whereas that female will allow only a selected single male individual which is dominating and healthier among the other in that particular location.

This kind of selective mating behavior is an important strategy in evolution and adaptation because, the selective male and female undergoing matting must be well adopted to their environmental conditions, thereby dominating the other and thus the resulting offspring will be one step more dominating

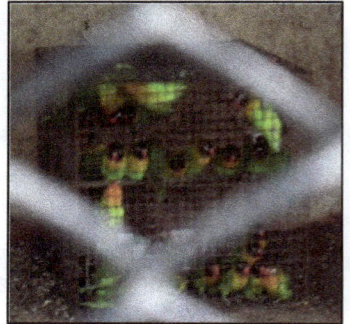

Figure 3.9: Parrots kept in a very small cage.

Photographs at Nehru Zoological Park-Hyderabad

and capable of adopting better to that environmental conditions. Such a chance of selective matting should be given to all the animals having such matting habit.

Animal Death Due To Carelessness of Government Departments Other Than Forest Department

Improperly planned, controlled and low constructed electricity lines and equipments are contributing to death of animals especially Elephants and Birds. Railways department, the biggest government department proudly contributes a bigger number of death of animals especially Elephants and also tigers. Road transports with lesser number of incidences shares its part with other department. Both railways and roadways should limit their speed and also avoid further enlarging or construction in forest areas.

Figure 3.10: Elephant found dead by train collision.

Figure 3.11: Elephant died after touching Transformer kept at just 4 feet height.

Urbanization

According to Census of India 2001, the level of Urbanization in Nilgiris district was 59.51 per cent (ranking 4th in the state following Chennai, Coimbatore and Kanyakumari). Also Now its planned to construct Central Information Technology

Reasearch Institute (CITRI) at Nilgiris District in 40 acres of land. It should be always remembered that overall district comes under Nilgiris Biosphere Reserve.

It has been recorded that 95 per cent of urban stray cattle in India suffering from various ailments due to hazardous material inside their abdomen, out of them 90 per cent are plastic bags (*The Indian Cow Oct-Dec, 2005*). Plastic bags that end up at sea are easily swallowed by marine life that mistakes them for food. An estimated 1,00,000 whales, seals, turtles and other marine life die every year after swallowing plastic bags.

Unnecessary Domestication of Wild Animals

Many wild animals which are not necessary for present human life are still domesticated and being used like elephant, camel, horse etc merely for the sake of using them as an ornamental thing in temples, marriages and other festivals and celebrations. Unnecessary domestication of these wild animals, by the way changing their habitat, lifestyle and in particular food style has lead to metabolic problems, diseases and health disorders (TB and Diabetics reported in domesticated elephants). This should be strictly banned and they all should be re-habituated to their own wild habitat.

Human Intervention on Natural Resources

In Kodaikanal, near Dum Dum rock and all over the borders' interior areas of the forest, it is reported that people often come and use to have alcohol, some other eatables and leave the used alcohol bottles, plastic cups, poly bags etc., into the forest as such. Also many couples are getting into the forest, hidden for having sex and the used condoms were spotted in many places along with that many non-degradable things lying within (reported in a tamil TV channel on 1.8.10). All these happens due to lack of check posts and carelessness of both forest and police departments.

To avoid all these, tourists should be allowed only up to a particular limit to go by their own vehicle. Later they should be allowed to go only by a common government vehicle, so that no one can enter into the forest wherever they wish, by the way vehicle based pollution can also be reduced in the forest area.

Conclusion

Reforestation with all rare wild trees, edible leaf and fruit bearing wild tree varieties in particular, introducing water sources by digging ponds and small lakes in required places of the reserved and forest area, re-habituation of wild animals in various habitats accordingly, allowing them to eat and mate by their own selection, banning of eco-forest tourism, maintenance and updating of complete data on all the aspects for all the conserved plants and animals in a biosphere reserve area or sanctuary and proper monitoring; all these should be implemented and followed strictly by the higher most organism, human not for loosing his highness. It's Our Great Duty to Realize, Protect and Conserve Our Evolutionary Seniors because Earth is for each and every Organism.

References

Champion, H.G. and Seth, S.K., 1968. *A Revised Survey of Forest Types of India.* Forest Research Institute, New Delhi.

Ecosystem Profile of Western Ghats Region in Western Ghats and Sri Lanka Biodiversity Hotspot. Final Version 2007. Critical Ecosystem Partnership Fund.

Myers, N., Mittermeier, R.A., Mittermeier, C., da Fonseca, G.A.B. and Kent, J., 2000. Biodiversity hotspots for conservation priorities. *Nature* , 403(24): 853–857.

Nameer, P.O., Sanjay Molur and Sally Walker, 2001. Mammals of Western Ghats: A simplistic Overview. *Zoos' Print Journal,* 16(11): 629–639.

Nair, S.C., 1991. The Southern Western Ghats: A biodiversity conservation plan. In: *Studies in Ecology and Sustainable Development,* V. 4. Indian National Trust for Art and Cultural Heritage INTACH, New Delhi.

Nair, N.C. and Daniel, P., 1986. The floristic diversity of the Western Ghats and its conservation: A review. In: *Proc. Indian Acad Sci. (Animal Sci./Plant Sci). Suppl:* 127–163.

Sukumar, R., 1989. *The Asian Elephant: Ecology and Management.* Cambridge University Press, Cambridge, pp. 251.

2013, Biodiversity Conservation for Sustainable Management   Pages 28–33
Editor: Dr. K. Muthuchelian, Vice Chancellor, Periyar University, Salem
Published by: Daya Publishing House, NEW DELHI

Chapter 4

Approaches and Strategy for Biodiversity Conservation

J. Pushpa
Department of Agricultural Extension and Rural Sociology,
Madurai–625 104, T.N.

Introduction

The 1992 United Nations Conference on Environment and Development, held in Rio de Janeiro, brought the topic of biodiversity conservation into the living rooms of the world and helped place this critical issue on the agendas of world leaders. While the ranks of those concerned with biodiversity seem to have diversified and increased, a basic understanding of what it is, what it means to mankind, and how it can be protected is still lacking.

In an effort to solve these problems, the World Conservation Union has attempted to clarify the definition and show the value of "biodiversity." Going beyond "genetic makeup," the IUCN interprets biodiversity to encompass all species of plants, animals, and microorganisms and the ecosystems (including ecosystem processes) to which they belong. Usually considered at three different levels–genetic diversity, species diversity, and ecosystem diversity–it is the complicated mosaic of living organisms that interact with abiotic substances and gradients to sustain life at all hierarchical levels (McNeely, 1990). Furthermore, each of these levels extends enormous, often immeasurable, economic and social benefits to mankind. Although it is recognized that a very high percentage of the total biodiversity exists in a small number of tropical countries, significant diversity also occurs in temperate zones and in aquatic ecosystems as well.

Biodiversity conservation is accomplished in a number of ways. *Ex-situ* methods focus on species conservation in botanic gardens, zoos, gene banks, and captive breeding programs. *In-situ* methods use conservation areas as "warehouses" of biological information. Many scientists and conservationists feel that until methods are available to discern easily which of the millions of species and varieties will have economic value, *in-situ* conservation through the protection of natural areas should be the primary means for the maintenance of these resources. However, a rigid preservation approach is virtually impossible to implement and even less likely to be maintained over time. Considering trends in population growth and the urgency of economic development–especially in the developing countries–a more appropriate response would be to pursue proactive alternatives to high-impact development activities, and to implement carefully formulated strategies for *in-situ* methods that would include protected areas in the development mix. Unfortunately, the formulation of that development mix is not easy, because moral, as well as technical and economic, choices are involved. According to Wilson (1984):

To choose what is best for the near future is easy. To choose what is best for the distant future is easy. To choose what is best for both the near and distant futures is a hard task, often internally contradictory, and requiring ethical codes yet to be formulated.

Although integrated regional development planning makes no claim to moral superiority, it does provide a framework for making such very difficult choices. That biodiversity conservation must be a part of development planning efforts is clear.

Key Factors of Successful Responses to Biodiversity Loss

☆ *Mobilize knowledge.* Ensure that the available knowledge is presented in ways that can be used by decision-makers.

☆ *Recognize complexity.* Responses must serve multiple objectives and sectors; they must be integrated.

☆ *Acknowledge uncertainty.* In choosing responses, understand the limits to current knowledge, and expect the unexpected.

☆ *Enable natural feedbacks.* Avoid creating artificial feedbacks that are detrimental to system resilience.

☆ *Use an inclusive process.* Make information available and understandable to a wide range of affected stakeholders.

☆ *Enhance adaptive capacity.* Resilience is increased if institutional frameworks are put in place that allow and promote the capacity to learn from past responses and adapt accordingly.

☆ *Establish supporting instrumental freedoms.* Responses do not work in a vacuum, and it is therefore critical to build necessary supporting instrumental freedoms–enabling conditions like transparency, markets, education–needed in order for the responses to work efficiently and equitably.

☆ *Establish legal frameworks.* A legally binding agreement is generally likely to have a much stronger effect than a soft law agreement.

☆ *Have clear definitions.* Agreements with clear definitions and unambiguous language will be easier to implement.

☆ *Establish principles.* Clear principles can help guide the parties to reach future agreement and guide the implementation of an agreement.

☆ *Elaborate obligations and appropriate rights.* An agreement with a clear elaboration of obligations and rights is more likely to be implemented.

☆ *Provide financial resources.* Availability of financial resources increases the opportunities for implementation

What Governance Approaches can Promote Biodiversity Conservation

☆ To promote biodiversity conservation, *strong institutions are needed at all levels.* The principle that biodiversity should be managed at the lowest appropriate level has led to decentralization in many parts of the world. However, all levels of government need to be involved, with laws and policies developed by central governments in order to support the authority at the lower levels of government enabling them to provide incentives for sustainable resource management. Neither complete centralization nor complete decentralization of authority always results in better management.

☆ In some countries local norms and traditions regarding property rights and ecosystems are much stronger than the law on paper. In that case, local knowledge, integrated with other scientific knowledge, becomes critical in managing local ecosystems.

☆ It is well documented that many of the structural adjustment programs of the mid–to late 1980s aiming for economic stability, sectoral growth and poverty reduction caused deterioration in ecosystem services and a deepening of poverty in many developing countries. More efforts are needed in integrating biodiversity conservation and sustainable use activities within such large decision-making frameworks.

☆ International cooperation requires increased commitments to conserve biodiversity and promote sustainable use of biological resources. Indeed, to be most effective, multilateral re financial resources are not sufficient, market mechanisms may increase the potential for implementation.

☆ Establish implementing and monitoring agencies. The establishment of subsidiary bodies with authority and resources to undertake specific activities to enhance the implementation of the agreements is vital to ensure continuity, preparation, and follow-up to complex issues.

☆ Establish good links with scientific bodies. As ecological issues become more complex, it becomes increasingly important to establish good institutional links between the legal process and the scientific community.

☆ Integrate traditional and scientific knowledge. Identify opportunities for incorporating traditional and local knowledge in designing responses.

Ecological Approaches

☆ Take an agricultural landscape perspective

☆ Empower multi-stakeholder processes for landscape management

☆ Build farming communities' landscape management capacity

☆ Adapt the role of conservationists in agricultural landscapes

☆ Coordinate environmental and agricultural policies for more effective landscape planning

In-situ Conservation of Biodiversity and Protected Areas

Although viable populations of some organisms can be maintained *ex-situ* either under cultivation or in captivity, these methods are far less effective than *in-situ* methods, and, generally, they are extremely costly. Likewise, although *ex-situ* methods are important under a number of conditions, *in-situ* methods are generally recognized as being more secure and financially efficient. The challenge in using *in-situ* methods is to expand our vision of protected areas to include multiple use and extractive reserves and to develop new models for conservation including, for example, such innovative proposals as using damaged ecosystems to preserve rare, endangered, and threatened species (Cairns, 1986) and to expand the range of options available for economic development.

As of 1993 nearly 7, 000 parks and protected areas covering in excess of 650 million acres had been established worldwide (WRI, 1992). When combined with smaller areas such as state parks and private reserves, a large portion of the planet's land surface is receiving some degree of protection. All eight Natural Realms and 14 Biomes, as categorized by Udvardy (1975), are represented. Nevertheless, the participants in the IVth World Congress on National Parks and Protected Areas and the 1992 Earth Summit concluded that although progress had been made in conserving samples of these biogeographic provinces, coverage was still insufficient. Indeed, there is scientific consensus that the total expanse of protected areas needs to be increased by a factor of three in order to maintain the earth's biotic resources (McNeely *et al.,* 1990). Properly conserving these underrepresented provinces will require the establishment of additional areas that are properly funded and managed to ensure that the broadest possible range of biotic resources are protected and available to support future economic development (UNEP, 1992).

Advantages, Risks, and Opportunities

In-situ maintenance of biodiversity through the establishment of conservation and multiple-use areas offers distinct advantages over off-site methods in terms of coverage, viability of the resource, and the economic sustainability of the methods:

1. *Coverage.* A worldwide system of protected and multiple-use areas would allow a significant number of indigenous species and systems to be protected, thus taking care of the unknowns until such time as methods are found for their investigation and utilization (Burley, 1986).

2. *Viability.* Natural selection and community evolution continue and new communities, systems, and genetic material are produced (World Conservation Monitoring Center, 1992; Soulé, 1986).

3. *Economic sustainability.* A country that maintains specific examples of biodiversity stores up future economic benefits. When the need develops and this diversity is thoroughly examined, commercially valuable genetic and biochemical material may be found (Eisner, 1990, 1992, and Reid, 1993).

It is not sufficient to establish a conservation area and then assume its biodiversity is automatically protected and without risk. Many risks, both natural and man-created, remain. An extreme example was the near-obliteration of the entire remaining habitat of the golden lion tamarin (*Leontopithecus r. rosalia*) in 1992 by fire (Castro, 1995). Shaffer (1981) cites four broad categories of natural risk:

1. *Demographic uncertainty* resulting from random events in the survival and reproduction of individuals.

2. *Environmental uncertainty* due to random, or at least unpredictable, changes in weather, food supply, and the populations of competitors, predators, parasites, etc.

3. *Natural catastrophes* such as floods, fires, or droughts, which may occur at random intervals.

4. *Genetic uncertainty or* random changes in genetic make-up due to genetic drift or inbreeding that alter the survival and reproductive probabilities of individuals.

The greatest uncertainties, however, are often anthropogenic. The elimination of habitat to make way for human settlement and associated development activities is the most important factor contributing to the diminishing mosaic of biodiversity. These uncertainties can only be met with a full array of conservation programs, including those that use *ex-situ* methods.

Despite the long list of uncertainties and risk, there is hope for progress. In the last decade not only have pressures from the scientific community and the efforts of non-governmental organizations led to stronger language in international agreements, but segments of the development community have accepted the idea that a large degree of compatibility exists between the need to develop and the need to maintain biodiversity. Further acceptance depends, however, on a number of attitudinal adjustments on the part of many who call for *in-situ* conservation, as well as on a clearer understanding of the rationale behind it by those whose activities conflict with it. The success of conservation also requires a modification of how we cost economic goods and services in the short, medium, and long term (McNeely, 1988).

Globally, the possibilities for undertaking *in-situ* programs such as national parks, biological reserves, and other conservation areas appear to be somewhat favorable. However, the status of these protected areas is often not healthy and unforeseen problems repeatedly arise. The establishment of the Gurupi Biological Reserve in the eastern Brazilian Amazon, for example, significantly increased the level of threat by causing a rush of illegal extraction of forest resources. This site is probably the most endangered conservation unit in the Amazon basin (Rylands, 1991; Oren, 1988). Worldwide, the list of endangered protected areas is growing in number, and additional human-dominated activities such as water development, mining, road construction and resulting development, livestock grazing, poaching, logging, and other removal of vegetation continue to threaten their integrity (IUCN/UNEP/WWF, 1991).

References

Agarwal, S.K., 2000. *Environment Scenario for 21st Century.* APH Publication Corporation, New Delhi.

Khan, M.A., 2000. *Environment, Biodiversity and Conservation.* APH Publication Corporation, New Delhi.

Maheswari, J.K., 1977. Conservation of rare plants: Indian scene vis-a-vis world scene. *Bull. Bot. Survey,* India, 19(1–4): 167–173.

Nair, N.C., 1984. Endemism and pattern of distribution of endemic genera in india. *Indian J. Econ. Tax. Bot,* 1: 99–110.

Rao, C.K. Angiosperm genera endemic to the Indian Floristic Region and its neighbouring areas. *Ind. For.*

2013, Biodiversity Conservation for Sustainable Management   Pages *34–36*
Editor: Dr. K. Muthuchelian, *Vice Chancellor, Periyar University, Salem*
Published by: Daya Publishing House, NEW DELHI

Chapter 5

Biodiversity Measurement, Approaches and Challenges

P. Selvam, D. Mitursan, and R. Kannan

Department of Microbiology,
PKN Art's and Science College, Thirumangalam

Introduction

"Study nature, love nature, stay close to nature, it will never fail you"

— Frank Lloyd Wright

While bio diversity is a hot word the conceptual scientific issues undergoing scientific issues undergoing its qualification have received scant attention. what is bio diversity? Biodiversity is the variety of species, their genetic makeup and the natural communities in which they occur and we see about the certain challenge's to protect of biodiversity we want a new approaches for optimising the multiple use's of biodiversity considering possible trace off's and conflict's. In the measurement of biodiversity is an essential and new technology should be developed. The developers need to work well with planners to integrate bio diversity into their development proposals. According to their proposals understand the motivation and incentives that under lie biodiversity loss or protection. and 30 per cent plants were prepared from plants so destroying the forest will give the way for destroying the world.

What is Biodiversity?

Biodiversity is the variety of species their genetic makeup communities which they occur. It includes all of the native plants and animals, the air we breath is a product of photosynthesis by green plants more than 90 per cent of the calories consumed by people worldwide are produced from 80 per cent plant species, almost 30 per cent medicines are developed from plant's and animal's

Challenges in Biodiversity

Biodiversity changing coordinated observation system and standardized methods to monitor biodiversity. Integrated analysis and models ecological evolutionary process to predict future biodiversity changes. Social consequences of bio diversity changes integrated ecological economic moderns to access the multiple effects of bio diversity changes on human societies. The scientific challenges are enormous we need a research efforts of the size of the space exploration programs for the exploration of earth's biodiversity. The causes of it's loss and the best needs to conserve and use it.

Measurement

Diversity may be measure of different scales. These are three indices used by ecologist. There are three types of diversity.

1. Alpha diversity
2. Beta diversity
3. Gamma diversity

Alpha Diversity

Alpha diversity refers to diversity within a particular area community are ecosystem and is measured by countries the number of taxa within the ecosystem.

Beta Diversity

Is a species diversity between ecosystems, this involves comparing the number of taxa that are unique to each of the ecosystems.

Gamma Diversity

Is the measure of the overall diversity for different ecosystems within a region.

Other Measures of Diversity

Species diversity, ecological diversity, morphological diversity, genetic diversity. a few studies have attempted to quantitatively clarify the relationship between types of diversity.

Table 5.1: Example of measurement in diversity.

	World	*India*
Fishes	31000 (1)	2439 (1)
Amphibians	6184 (3)	277 (4)
Reptiles	8734 (5)	408 (3)
Birds	9782 (6)	1179 (7)
Mammals	5416 (9)	410 (3)

Conclusion

Despite the number of animal species that have become extinct we want to their life because 30 per cent medicines were prepared from plants and animals. Integrated

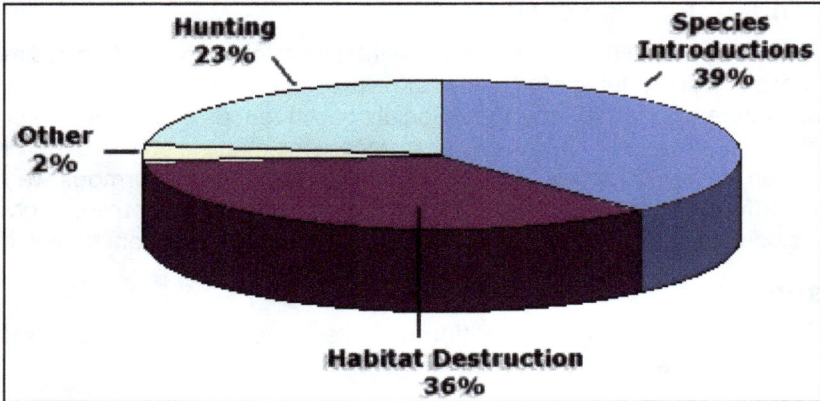

Figure 5.1: Known causes of animal extinctions since 1600.

ecological economic to access the multiple efforts of bio diversity changes on human societies. destroying the forest will give the way for destroying the world.

References

Anasas, P.T. and Zimmerman, J.B., 2003. Design through the 12 principles of green engineering. Env. Sci. Tech., 37(5).

Reid, W.V. and Whitemore, T. Sayer, 1992. *Tropical Deforestation and Species Extinction.* Chapman and Hall, London.

2013, Biodiversity Conservation for Sustainable Management   *Pages 37–49*
Editor: Dr. K. Muthuchelian, *Vice Chancellor, Periyar University, Salem*
Published by: Daya Publishing House, NEW DELHI

Chapter 6

Biosphere Reserves: Concept, Design and Functions in Protection and Maintenance of Biodiversity in India

*A.G. Pandurangan**

Tropical Botanical Garden and Research Institute,
Palode, Thiruvananthapuram–695 562, Kerala

Introduction

The idea of 'Biosphere Reserves' was initiated by UNESCO in 1971 under its Man and Biosphere (MAB) Programme. The MAB, launched by UNESCO, is a broad based ecological programme aimed to develop within the natural and social sciences a basis for the rational use and conservation of the resources of the biosphere and for the improvement of the relationship between man and the environment; to predict the consequences of today's actions on tomorrows world and thereby to increase man's ability to manage efficiently the natural resources of the biosphere. The approach emphasizes the importance of the structure and functioning of ecological systems and their mode of reaction when exposed to human intervention including impact of man on the environment and vice-versa. The MAB is primarily a programme of research and training and seeks scientific information to find solution of concrete practical problems of management and conservation. MAB's field projects and Biosphere reserves constitute the main goal of the whole programme.

---

* Corresponding Author: E-mail: agpandurangan@mail.com

The Indian National Man and Biosphere (MAB) Committee identifies and recommends potential sites for designation as Biosphere Reserves, following the UNESCO's guidelines and criteria. So far 16 Biosphere reserves have been established in India and some additional sites are under consideration (www.envfor.nic.in).

Definition

Biosphere Reserve (BR) is an international designation by UNESCO for representative parts of natural and cultural landscapes extending over large area of terrestrial or coastal/marine ecosystems or a combination thereof. BRs are designated to deal with one of the most important questions of reconciling the conservation of biodiversity, the quest for economic and social development and maintenance of associated cultural values. Therefore, BRs are thus special environments for both people and the nature and are living examples of how human beings and nature can co-exist while respecting each others' needs. These areas are internationally recognized within the framework of UNESCO's Man and Biosphere (MAB) programme, after receiving consent of the participating country. The world's major ecosystem types and landscapes are represented in this network.

Characteristics of Biosphere Reserve

The characteristic features of Biosphere Reserves are:

1. Each Biosphere Reserves are protected areas of land and/or coastal environments wherein people are an integral component of the system. Together, they constitute a world wide network linked by International understanding for exchange of scientific information.

2. The network of BRs include significant examples of biomes throughout the world.

3. Each BR includes one or more categories.

4. BRs are representative examples of natural biomes.

5. BRs conserve unique communities of biodiversity or areas with unusual natural features of exceptional interest. These representative areas may also contain unique features of landscapes, ecosystems and genetic variations. For example one population of a globally rare species; their representativeness and uniqueness may both be characteristics of an area.

6. BRs have examples of harmonious landscapes resulting from traditional patterns of land-use.

7. BRs have examples of modified or degraded ecosystems capable of being restored to more natural conditions.

8. BRs generally have a non-manipulative core area, in combination with areas in which baseline measurements, experimental and manipulative research, education and training is carried out. Wherever these areas are not contiguous, they can be associated in a cluster.

Functions of Biosphere Reserves

Conservation

☆ To ensure the conservation of landscapes, ecosystems, species and genetic variations;

☆ To encourage the traditional resource use systems;

☆ To understand the patterns and processes of functioning of ecosystems;

☆ To monitor the natural and human-caused changes on spatial and temporal scales.

Development

☆ To promote, at the local level, economic development which is culturally, socially and ecologically compatible and sustainable;

☆ To develop the strategies leading to improvement and management of natural resources.

Logistics Support

☆ To provide support for research, monitoring, education and information exchange related to local, national and global issues of conservation and development;

☆ Sharing of knowledge generated by research through site specific training and education; and

☆ Development of community spirit in the management of natural resources.

Beneficiaries

Direct Beneficiaries of the Biosphere Reserves are the local people and the ecological resources and *indirect beneficiaries* are scientists, government decision makers and the world community at large. In nutshell, the biosphere reserve offers to conserve one genetic resources as renewable resources by keeping the genetic stock as 'capital reserve' on which the world community will certainly be depended for future requirements.

International Status of Biosphere Reserves

Like Minded Mega-diverse Countries of the World

Of the 17 like minded mega-diverse countries namely Brazil, Bolivia, China, Columbia, Democratic Republic of Congo, Ecuador, Indonesia, Kenya, Mexico, Madagascar, Malaysia, Peru, Philippines, South Africa, India, Venezuela, Zaire, which collectively constitute 60-70 per cent of the world's biodiversity, Australia alone is considered developed. The pressure on wild lands and forests for economic development is too high in these countries taking priority over environmental or ecological issues. Myers (1990) identified 18 Hotspots and later expanded to 34 that feature exceptional concentrations of species with high levels of endemism and that face exceptional threat of destruction. Of these, 10 Hotspots are in the Asia-Pacific region, (13 per cent of the land area of earth) out of which, India harbours two hotspots viz., 'Eastern Himalayas' and 'Western Ghats'. Demands for economic growth are high in these regions and the rapidly increasing population–60 million annually, *i.e.* 50 per cent of the world's population exerts too much pressure on the biological resources. Unless immediate decisive steps are taken to counter the effects of deforestation, fragmentation and degradation of the remaining wilderness areas, pragmatic assumptions foretell that much of the biodiversity of Asia will be lost within a few decades.

International Efforts

The International Co-ordinating Council (ICC) of UNESCO in its first meeting in Paris held during 9-19 November, 1971, introduced the designation 'Biosphere Reserve' for natural areas. Future functions of BRs were given concrete shape in MAB Project area of "Conservation of natural areas and of the genetic material they contain" (UNESCO, 1972). The concept of Biosphere Reserves was refined by a Task Force of UNESCO's MAB Programme in 1974, and BR network was formally launched in 1976. During September 26–October 2, 1983 first international Biosphere Reserve congress was convened in Minsk (Belarus) which gave rise to an 'Action Plan for BRs'. At the twelfth session of ICC held in Paris from 25 to 29 January 1993, following five priority areas were identified to enable BRs to implement results of United Nations Conference on Environment and Development, UNCED held in Rio de Janeiro in June 1992.

☆ Conservation of biological diversity and ecological processes

☆ Development of sustainable use strategies

☆ Promotion of information dissemination and environmental education.

☆ Establishment of a training structure

☆ Contribution to the establishment and implementation of global environmental monitoring system.

Network of Biosphere Reserves

In order to facilitate cooperation, BRs are admitted into International network by International Coordinating Council (ICC) of the Man and Biosphere (MAB) Programme of UNESCO on the request of the participating country subject to their fulfillment of prescribed criteria. The BRs remain under the sole sovereignty of the concerned country/state where it is situated, and participation in World Network is voluntary. Delisting from international Network is done as an exception on ground of violation of obligation for conservation and sustainable development of Biosphere Reserves after consulting the concerned Government. As on 12th September, 2007 there were 507 Biosphere Reserves on World Network in 102 countries recognized by UNESCO which include Nilgiri, Sunderbans, Gulf of Mannar and Nanda Devi BRs from India.

Biosphere Reserves: Indian Approach

Bio-geographical Regions in India

The geographical location of India between 8° 4' N and 37° 6' N provides a wide latitudinal spread and permits a wide range of variations in temperature. The topographical diversity marked by mountainous regions covering an area close to 100 million hectares, arid and semi-arid zones spreading over 30 million hectare and the long coast line over 7000 kms, coupled with varied precipitation constitute a rich landscape diversity. Accordingly India consists of :

1. Two 'Realms–the Himalayan region represented by Palearctic Realm and the rest of the sub-continent represented by Malayan Realm

2. Five Biomes–(i) Tropical Humid Forests (ii) Tropical Dry or Deciduous Forests (including Monsoon Forests) (iii) Warm deserts and semi-deserts (iv) Coniferous forests and (v) Alpine meadows.

3. Ten Bio-geographic Zones–1. Trans Himalayan, 2. Himalayan, 3. Indian Desert, 4. Semi-Arid, 5. Western Ghats, 6. Deccan Peninsula, 7. Gangetic Plain, 8. North-East India, 9. Islands, 10. Coasts, and

4. Twenty five Bio-geographic provinces.

It is the ecological diversity that makes India as one of the mega-diversity regions on the globe. Efforts are on to designate at least one Biosphere Reserve in each of the Bio-geographic Provinces.

National Biosphere Reserve Programme

India has created a network of *protected areas* in the form of 96 National Parks, 510 Wildlife Sanctuaries and 28 Tiger Reserves and 25 Elephant Reserves. The area covered under protected area network accounts for around 5 per cent of the total geographical area of the country. The rich biodiversity in India has given shape to variety of cultural and ethnic diversity which includes over 550 tribal communities of 227 ethnic groups spread over 5,000 forest villages.

The national Biosphere Reserve Programme was initiated in 1986 and its aims and objectives are detailed.

Aims of the Scheme

☆ To serve as wider base for conservation of entire range of living resources and their ecological foundations in addition to already established protected area network system.

☆ To bring out representative ecosystems under conservation and sustainable use on a long term basis.

☆ To ensure participation of local inhabitants for effective management and devise means of improving livelihood of the local inhabitants through sustainable use.

☆ To integrate scientific research with traditional knowledge of conservation, education and training as a part of the overall management of BR.

The Core Advisory Group of Experts, constituted by Indian National MAB Committee identified and prepared a preliminary inventory of 14 potential sites for recognition as BRs in 1979. Subsequently additional BR sites were proposed by the National Committee/State Governments, Experts etc.

Objectives

It may be noted that BRs are not a substitute or alternative, but a re-enforcement to the existing protected areas. The objectives of the Biosphere Reserve programme, as envisaged by the Core Group of Experts, are as follows:

☆ To conserve the diversity and integrity of plants and animals within natural ecosystems;

☆ To safeguard genetic diversity of species on which their continuing evolution depends;

☆ To provide areas for multi-faceted research and monitoring;

☆ To provide facilities for education and training; and

☆ To ensure sustainable use of natural resources through most appropriate technology for improvement of economic well-being of the local people.

These objectives should be oriented in such a way that the BRs are the Units wherein the Biological, socio-economic and cultural dimension of conservation are integrated together into a realistic conservation strategy.

Criteria

The criteria for selection of sites for BRs as laid down by the Core Group of Experts in 1979 are listed below:

Primary Criteria

☆ A site that must contain an effectively protected and minimally disturbed core area of value of nature conservation and should include additional land and water suitable for research and demonstration of sustainable methods of research and management.

☆ The core area should be typical of a biogeographical unit and large enough to sustain viable populations representing all tropic levels in the ecosystem.

Secondary Criteria

☆ Areas having rare and endangered species

☆ Areas having diversity of soil and micro-climatic conditions and indigenous varieties of biota.

☆ Areas potential for preservation of traditional tribal or rural modes of living for harmonious use of environment.

Structure and Design of Biosphere Reserves

In order to undertake complementary activities of biodiversity conservation and development of sustainable management aspects, Biosphere Reserves are demarcated into three inter-related zones. These are (i) natural or core zone (ii) manipulation or buffer zone and (iii) transition zone outside the buffer zone.

The Core Zone

The core zone is kept absolutely undisturbed. It must contain suitable habitat for numerous plant and animal species, including higher order predators and may contain centres of endemism. Core areas often conserve the wild relatives of economic species and also represent important genetic reservoirs. The core zones also contain places of exceptional scientific interest. A core zone secures legal protection and management and research activities that do not affect natural processes and wildlife are allowed. Strict nature reserves and wilderness portions of the area are designated as core areas of BR. The core zone is to be kept free from all human pressures external to the system.

The Buffer Zone

In the Buffer Zone, which adjoins or surrounds core zone, uses and activities are managed in ways that protect the core zone. These uses and activities include restoration, demonstration sites for enhancing value addition to the resources, limited recreation, tourism, fishing and grazing, which are permitted to reduce its effect on core zone. Research and educational activities are encouraged. Human activities, if natural within BR, are likely to be permitted to continue if these do not adversely affect the ecological diversity and their evolutionary process.

The Transition Zone

The Transition Zone is the outermost part of a Biosphere Reserve. This is usually not delimited one and is a zone of cooperation where conservation, knowledge and management skills are applied and uses are managed in harmony with the purpose of the Biosphere Reserve. This includes settlements, crop lands, managed forests and area for intensive recreation, and other economic uses characteristic of the region.

In Buffer Zone and the Transition Zones, manipulative macro-management practices are used. Experimental research areas are used for understanding the patterns and processes in the ecosystem. Modified or degraded landscapes are included as rehabilitation areas to restore the ecology in a way that it returns to sustainable productivity.

Legal Framework

Rules and regulations provide a broad planning approach to conservation and wise use of resources of BRs. These aims to ensure that:

☆ National land and water use planning measures take full account if the functions nad values of BRs, and

☆ Conservation of their biodiversity is guaranteed for sustainable use of benefits of BRs.

The Action Plan of BRs must therefore, be developed and implemented in conformity with other relevant national policies affecting BRs, relevant sections of Plan documents, National Conservation Strategy and Policy Statement on environment and Development (1992), the National Action Plan on Biodiversity (1997), the National Forest policy (1988), the National Water Policy (1987), Coastal Regulation Zones (CRZ), Environmental Protection Act (1986), Wildlife (Protection) Act, 1972 and its amendment (1991) and other relevant acts together with the relevant planning documents. After review of existing laws, institutions and practices, National planning related to BR is required to be reviewed. Appropriately amended economic valuation of BRs should be applied and the role of the stakeholders in the process should be ensured (so that the plan incorporates both top-down and bottom-up approaches). The corporate sector should be included and Environmental Impact Assessment (EIA) and restoration of degraded ecosystems within BRs must be integrated in to the planning process.

At present BRs are established within the framework of existing laws including Wildlife (Protection) Act, 1972. Rules and Regulations specifically for BRs can be examined by the respective State Governments, of existing laws are inadequate to deal with the requirements if the Biosphere reserves.

How Biosphere Reserves are Different from Protected Areas?

BR is not intended to replace existing protected areas but it widens the scope of conventional approach of protection and further strengthens the Protected Area Network. Existing legally protected areas (National Parks, Wildlife Sanctuary, Tiger Reserve and reserve/protected forests) may become part of the BR without any change in their legal status. On the other hand, inclusion of such areas in a BR will enhance their national value. It, however, does not mean that Biosphere Reserves are to be established only around the National Parks and Wildlife Sanctuaries. However, the Biosphere Reserves differ from protected areas due to their emphasis on:

☆ Conservation of overall biodiversity and landscape, rather than some specific flagship species, to allow natural and evolutionary processes to continue without any hindrance.

☆ Different components of BRs like landscapes, habitats, and species and land races.

☆ Developmental activities, and resolution/mitigation of conflicts between development and conservation.

☆ Increase in broad-basing of stakeholders, especially local people's participation and their Training, compared to the features of scheme on Wildlife Sanctuaries and National Parks.

☆ Sustainable environment friendly development, and sustained coordination amongst different development organizations and agencies.

☆ Research and Monitoring to understand the structure and functioning of ecological system and their mode of reaction when exposed to human intervention.

How Biosphere Reserves are Designated?

At the initiative of the central/state governments, detailed study is carried out and a project report is prepared by the concerned state following the UNESCO's criteria adopted for designation of BRs. The land and forest being the state concerns, the respective state governments have to agree to designate the identified area as Biosphere Reserve. The Central Govt. provides financial assistance for management and research activities in these BRs. The Management of Biosphere Reserves is the responsibility of the concerned State/UT with necessary technical input and training facilities provided by the Central Government.

The Government: Role and Responsibilities

The Central Government

At the national level, the Central Govt. assumes responsibility of overall coordination at international and national level. The Central Government is responsible for the following:

- ☆ Financial assistance for implementation of the approved items of the programme.
- ☆ Technical expertise and know-how including training of personnel.
- ☆ Detailed guidelines covering all aspects of management for implementation by the State/UTs machinery.
- ☆ Periodic evaluation and monitory of implementation of conservation and research programmes.

The State Governments/UT Administration and other Stakeholders

As per the constitutional framework, the States' are the proprietors and custodians of 'Land' and 'Forests'. Accordingly, the local management of the BRs is the responsibility of the concerned State Government/UT Administration.

Other Stakeholders

The management activities are to be implemented involving effectively the local communities, local govt. agencies, Scientists, economic interest groups, cultural groups and other stakeholders.

Policy/Planning and Management Mechanism

The State Government ensure that each BR will have effective and long term management policy or plan and an appropriate 'Authority' or 'mechanism' to implement it. The management of a BR should include:

- ☆ A mechanism to protect the core zone.
- ☆ Appropriate facilities to undertake research and monitoring. The management authority must ensure encouragement to research and monitoring by Research Institutions.
- ☆ Adequate provision for people's participation by enlisting their cooperation. (Local and regional understanding in planning and managing the area for conservation and sustainable development is important for human benefit).

Conservation, Development and Logistics Support

Development of Biosphere Reserves will have 3 components *i.e.* Management Action Plans for Conservation and Development, Research and Monitoring and Education and Training for work support which are generally eligible for Central Government assistance.

Management Action Plans (MAPs) for Conservation and Development

Management which includes the management of the buffer zone, and in a way that ensures local community participation in conservation and utilization of the resources in a sustainable way as well as evolve ways and means by which economic well-being of local people is secured. It also involves development of management measures that protect the core by relieving pressures on its natural resources. Since the thrust of the programme is on creation of Supplementary and Alternate livelihoods to reduce biotic and anthropogenic pressure, synergy should be developed among the employment-generating programmes of other Departments, and involvement of various line Departments such as Agriculture, Rural Development, Tribal Affairs, Irrigation, Rural and Khadi Village Industry, Soil and Water Conservation, Women and Child Development, Horticulture, Animal Husbandry, Fisheries and Tourism should be ensured. Assistance for value addition and marketing local produce should also be provided.

At present the MAP for each BR is prepared by the concerned State Government. In general, the responsibility to manage Biosphere Reserves is given to the Forest Departments. This has resulted in inadequate participation of other relevant departments. It is therefore desirable that a Biosphere Reserve Management Authority is to be established as an autonomous body for effective coordination, management and development of BRs on a scientific basis involving various stakeholders which is expected to include officers and staff from Forest Departments and other line Departments as mentioned above. The staff handling this subject in/respective Departments can be pooled to constitute the proposed body. This is necessary to facilitate more effective participation of various stakeholders in the programme. Depending on local socio-economic features, involvement of Eco–Development Committees (EDCs), Panchayats, Forest Protection Committees (FPCs), Self Help Groups(SHGs), Biodiversity Management Committees (BMCs), Joint Forest Management Committees (JFMCs) could be ensured in various management interventions which may not only facilitate people's participation, but also lead to greater transparency.

Although items of intervention shall differ in each Biosphere Reserve, generally the following components are eligible for financial assistance:

Value Addition Activities

Formulation of comprehensive resource inventory and augmentation of required expertise and prioritization of activities with reference to additional income-generating activities will be given priority. These include popularization of energy alternatives, range land and grassland management, habitat improvement, animal husbandry, aquaculture, apiculture and encouragement for continuance of traditional crops including wild relatives of cultivated species for agro-biodiversity conservation, adoption of technologies that make resource utilization sustainable, and cottage industries based on local raw material with eco-friendly processing and production process.

Setting Up of Pilot Plots

Among other preferred activities are ecologically appropriate forestry, production of biomass, cultivation of medicinal plants, traditional agriculture and horticulture, facilities for ex-situ conservation measures and development of practices for sustainable use of threatened economically important species.

Rehabilitation of Landscapes of Threatened Species and Ecosystems

Pockets within BR harboring threatened species should be demarcated for special attention.

Socio-economic Upliftment of Local Communities

Creation of facilities for improved health care such as immunization, supply of drinking water, establishment of schools and development of small-scale household industries for manufacturing crafts based on local resources. The agencies such as Khadi Gram Udyog, SIDBI, KVIC, NABARD etc. may be involved to promote activities such as pisci-culture, apiary, mushroom cultivation, duckery, poultry, medicinal plants cultivation among other cottage industries.

Facilitating and Associating Conservation of Critical Habitats in Buffer Zones

There are many Critical habitats in buffer zones which are sometimes privately owned but important for long term survival of the eco-system. These habitats should be given special appropriate attention and even can be marked as "Entities of incomparable values".

Maintenance and Protection of Corridor Areas

To augment continuity of ecological processes and regulate movement of wild animal population from one habitat to the other in search of water, food and shelter, corridor areas in buffer zones should be critically monitored. Appropriate viable livelihoods provided to residents in the vicinity of the corridor areas.

Development of Communication System and Networking

Development of viable linkages between various Biosphere Reserves, stakeholders and government and non-government agencies operating in the region to facilitate protection measures and exchange of information for monitoring of movement displacement of animals, poachers local climatic changes etc.

Development of Eco-tourism

The thrust of the management is to augment appreciation of people for nature, generate income through eco-tourism, provide means for the people who live and work within and around BR, to attain a balanced relationship with the natural world and to show a more sustainable future while contributing towards the needs of the society. Local community participation in planning and management of BR must be ensured. Development of management practices that ensure the maintenance of high species diversity, establishment of research, education and training units should be given priority so as to create research facilities for undertaking research by concerned experts/organizations.

Research and Monitoring

Interface with research institutions involved in research activities in the BR area and designated lead institution should be ensured to incorporate relevant research-based recommendations in Management Action Plans. Research and monitoring in existing Biosphere Reserves and Potential sites not only crucial but constitutes the very basis of designing development strategies and solutions for Management of relevant problems. Infact research is very crucial to understand the impact of the management practices on ecosystem health. The universities, colleges, research institutions, non-governmental organizations, etc. are encouraged to formulate and implement research projects in BRs. Such proposals are considered by the Central Government for funding. Various relevant organizations are encouraged to develop innovative, inter-disciplinary research proposals for BRs, including modeling system for integrating social, economic and ecological data. The Central Government has designated lead/co-coordinating research institutions for each existing BR. These institutions are entrusted with a responsibility to collate and disseminate research based information and identify gap areas in research and will serve as focal point for formulation of research projects. These centers will also advise BR Managers on research inputs to be incorporated in management plans. The Project Investigators shall interact with Lead Centers and Project Managers while formulating research proposals.

Thrust Areas for Research and Monitoring

The following thrust areas are recognized for research and monitoring in BRs:

☆ The design of BR requires integrated knowledge on eco-geographical aspects, socio-economic aspects of local communities, magnitude of biodiversity, political and economic factors and categories of people who use the Reserve.

☆ Determination of monitoring regimes which include the identification of indicators, the frequency at which monitoring should be done is an important component of the management of BR.

☆ The role of species in the maintenance of ecosystem health and their response to natural and man-made disturbance regime are critical inputs for management of BRs.

☆ Ecological rehabilitation of degraded habitats is of prime importance in the maintenance of biodiversity as well as in the sustainable use of landscapes and species for economic benefit of the local communities. Research in the area of ecological restoration should be given priority. This may also include propagation technique for rare endemic species.

☆ Valuing of biodiversity may provide the basis for the economic management of the BRs. Consequently, natural resource accounting form an important component of research and development.

☆ Identification of appropriate technologies compatible with the goals of conservation and evaluation of environmental and socio-economic efficiency of the identified technologies.

☆ Applied researches for increasing the efficiency of food crops, animal husbandry and other domestic sectors that bring down the local pressure on forests.

☆ Identification of factors that lead to environmental degradation and unsustainable use of biological resources.

☆ Development of alternative means of livelihood for local populations when existing activities are limited or prohibited within the Biosphere Reserve.

☆ Identification of institutional mechanisms that ensure equitable sharing of benefits from resources available in buffer zone.

Education and Training

Education and training among the local communities, public and visitors is an essential component of the management of Biosphere Reserves. Audio-visuals depicting the role of Biosphere Reserves in protecting life-supporting systems and the need to caring for earth through sustainable use of resources should be given priority. Attempt should be made to present to our people the true value of our plant and animal diversity and make them accept its relevance in their own life. Training to the local youth in skills that enable them to undertake participation roles in the management of BRs is also crucial for the long-term maintenance of BRs. Designing of training package is also a priority area and imparting training is an aspect of management. Development and demonstration of integrated resource management with people's participation in buffer zone villages is also a part of the training. In addition, training and education of the personnel responsible for management of BRs may also be assured in order to assimilate modem concept and understanding about conservation and sustainable use of Biological resources.

Conclusion

Biosphere reserves are the principal means for achieving the objectives of the MAB programme and the visible instrument through which UNESCO as whole could demonstrate its commitment to sustainability through policy relevant site–based research, capacity enhancement and demonstration. In order to respond to new and emerging environmental and economic challenges at all space, the management and co-ordination among biosphere reserves in India is vital since these challenges are focusing on genetic resources based on which development depends. By implementing BR programmes we may be in a better bargaining position as per the provisions of GATT and IPR regimes which extensively dealt genetic resources as tools for future development.

2013, Biodiversity Conservation for Sustainable Management   Pages 50–61
Editor: Dr. K. Muthuchelian, Vice Chancellor, Periyar University, Salem
Published by: Daya Publishing House, NEW DELHI

Chapter 7

Role of Corporates in Conservation of Biodiversity

B. Anbazhagan, K.P. Ganesan and S.S. Vignesh
Department of Business Administration,
Sourashtra College, Madurai

Introduction

Businesses affect biodiversity in many ways, the most obvious being activities such as mining, timber extraction or oil exploration. If big business could be persuaded to set aside even a small percentage of the money in its corporate coffers for investment in biodiversity-friendly activities, the benefits to conservation would be staggering. Changes in production methods and the development of new, sustainable products can lead to substantially higher profitability.

The more enlightened companies are looking beyond environmental compliance and actively cultivating partnerships with non-governmental organizations (NGOs) and other stakeholders. By working in partnership with NGOs, businesses can ensure that the social and environmental dimensions are managed effectively. At first sight, business and biodiversity may appear to be unsuitable bedfellows. In reality, they each have much to gain from working in partnership. (Mark Rose)

Biodiversity and Economic Development

Wikipedia observes that the current text book definition of 'biodiversity' is "variation of life at all levels of biological organisation", that is, genes, species and habitations existing on land, water and air. According to the HIPPO hypothesis of Edward O. Wilson (1988), the threat factors for biodiversity are: Habitat destruction (H), Invasive species (I), Pollution (P), Human over population (P) and Over harvesting (O). Economic development has led to industrialisation, greater use of fossil fuel, population explosion and a high degree of regional inequality. A deeper analysis will show that it is not economic development as such but market led economic development and market failures which are at the root of the environmental crisis of the present world. Interventions by government and national and international level have thus become important. (Amitabha Sinha)

Biodiversity Climbing Corporate Agenda

No sector or business in the economy will escape unaffected by changes to the availability of environmental resources for business and consumers, according to analysis by PricewaterhouseCoopers as part of a landmark study by the UN Environment Programme (UNEP) examining the economics of biodiversity and ecosystem loss for business. The Economics of Ecosystems and Biodiversity (TEEB) estimates the global economic impact annually of biodiversity loss at between $2-4.5 trillion water used in food and drink production, timber for packaging, furniture and paper, productive land for fruit and vegetables, and fibres for clothes, are amongst just some of the biodiversity and ecosystem 'services' whose economic value and protection is examined in the study. Unpriced, and largely unaccounted for in business life, the flow and use of natural resources is embedded in the global economy every day we need to start thinking about ecosystems as an extension of our asset base, part of the plant and machinery, and accounting for the value they deliver. (www.teebweb.org)

Biodiversity and Corporate Realities

A review by PricewaterhouseCoopers (PwC) of the annual reports of the 100 largest companies in the world by revenue in 2008 only found 18 companies mentioned biodiversity. On top of that PwC estimated the cost of soil erosion in Europe at €53 per hectare each year. Annual economic losses caused by introduced agricultural pests in the US, UK, Australia, South Africa, India and Brazil exceed $100bn, PwC estimated. But depressingly, only two of the world's largest 100 companies identified biodiversity loss as a strategic issue. (http://us.asiancorrespondent.com/green-business-blog/biodiversity-and-corporate-realities)

Biodiversity and Business—Experts' Views (Mike Scott)

Climate change is no longer a concern just for environmentalists, but for business and governments. Biodiversity is following a similar trajectory. Biodiversity loss is speeding up. Ahmed Djoghlaf, executive secretary of the Convention on Biological Diversity, says extinction rates may be up to 1,000 times higher than the historical background rate. This matters because biodiversity provides crucial services. "The ecological infrastructure of the planet is generating services to humanity worth by some estimates over $70,000bn [£46,000bn, €55,000bn] a year, perhaps substantially more," says Achim Steiner, UNEP executive director. PwC, the professional services group, says business is already being affected by risks related to declining biodiversity and loss of ecosystem services.

While extractives industries, such as forestry, farming and fishing, are affected most broadly, all sectors of the economy are affected in some way—and impacts are being felt now. Soil erosion in Europe is estimated to cost €53 per hectare every year, while the annual economic losses caused by introduced agricultural pests in the US,

UK, Australia, South Africa, India and Brazil exceed $100bn. The loss of ecosystems and biodiversity can lead to reduced productivity, scarcity or increased cost of resources. Businesses need to start thinking about ecosystems as an extension of their asset base, part of their plant and machinery, says Jon Williams, a partner at PwC. And ecosystem protection, or the impact of ecosystem loss, needs to be factored into investment appraisal and capital allocation decision-making.

"Mainstream investors can no longer ignore trends such as loss of biodiversity and the overuse of ecosystem services such as clean air, fresh water and fertile soil," says James Gifford, executive director of the UN's Principles for Responsible Investment. Resource constraints and pollution are increasingly becoming important factors in corporate profitability, and investors need to consider the implications of how companies manage these issues, and encourage them to improve.

However, biodiversity and ecosystems are even more complex to deal with than climate change, where there is at least one global unit of measurement–the tonne of CO2 emitted–around which action can coalesce. There is no equivalent for biodiversity.

At a recent Zematt Summit, Colin Melvin, chief executive of Hermes Employee Ownership Services (a socially active UK pension group), argued that the best way of persuading companies to behave more responsibly was for shareholders to push them. Already the Dutch institution, Rabobank, has specific requirements for financing projects and needs impact reports on biodiversity for palm oil and soya and works with borrowers to improve their environmental performance.

The first market opportunity that those behind TEEB see is the development of a market for reducing emissions from deforestation and degradation and related land-based carbon offset initiatives (REDD+). The argument is this will also support biodiversity through the conservation of natural forests. It is something that reflects the global carbon market which was worth over US$140 billion in 2009.

Integrating Biodiversity into Business Strategies–Need of the hour

In 2008, the European Commission undertook a project to assess the costs of inaction if the 2010 target of halting the erosion of biodiversity is not met (*which we know it will not be.*). According to it the degradation of ecological services may represent as much as 7 per cent of world GDP in 2050 or 13,938 billion Euros a year. Accordingly, reconciling economic activity with biodiversity calls for a twofold initiative: encouraging businesses to take action and developing new tools for them to do so. "*Integrating biodiversity into business strategies*" is to be designed to meet this dual need. (Joel Houdet)

Present time calls on them to use their corporate social responsibility and corporate governance foundations to assess and interact with biodiversity issues.

Building the Biodiversity Accountability Framework

An *innovative approach* suggests that biodiversity should become an integral part of business strategy. The challenge, largely sketched out but not yet fully realised at this stage, is to build a *Biodiversity Accountability Framework*, which would be the *biodiversity equivalent of the "Bilan Carbone"* (methodology for greenhouse gas accounting). Financial accounting is not designed to assess and monitor relations between business and biodiversity: this requires the kind of innovation outlined here, to be developed more fully.

Looking at the costs and benefits associated with the reintegration of the economy into biodiversity then becomes a normal way of doing business. This situation also calls for the introduction of a *new accounting system*, complementing the existing framework, which takes account of the relations between business and living systems towards a *taxation system based on all consumption of nature* (Joel Houdet).

Making Simple Steps as Suggested

Remedies Possible to Corporates

1. Creation of Awareness.
   - Conservation knowledge building and knowledge transfer (Study of present status, Survey, documentation, Identification and collection Networking with organization).
2. Conservation and management. Awareness generation, training and exposures, demonstrations.
3. Making Use of different extension/dissemination tools (Exhibitors, cultural events, video documentation, mela's).
4. Promoting organic practices on a scale that will be able to address issues of food/outputs on a 'national' level. Promoting package of practices that will be able to allow 'diversity' to take into account other needs of livelihood.
5. To have Broad views which include
   - Conserve our rich heritage of biodiversity and associated knowledge.
   - Launch biodiversity conservation as a people's movement.
   - Engage folk ecologists in scientific enterprise.
   - Organise a scientific enterprise right colours to the grass roots.
6. Creating Willingness of professionals (Human Resource) to work in communities (Tribal areas).
7. Undertaking Major activities in project
   - Identification of biodiversity resources of potential value.
   - Institutional conservation measures.
   - Organizing restoration activities.
   - Focused short term research.
   - Organizing an information system.
   - Organizing value addition activities.
   Extending Financial support for conservation measures.
8. Biodiversity being seen, again as a money spinner, not as ecosystem service provider to livelihoods.

Based on: (http://www.greenconserve.com/reports2/session_one.html)

The Environmental Capital Report-Balanced Scorecard

In recent decades, consumption of natural resources and the effects of the capital model have contributed to major natural catastrophes. In this context, the awareness and environmental commitment of society has increased, demanding that companies provide information regarding their environmental practices. The problem arises in establishing what information should be reported in order to provide knowledge of issues that are an important part of the value of an organization. In this regard, management can contribute to the emergence of intangible assets to help the company disassociate itself from non-environmentally proactive companies. In short, this information should give a true picture of the company. Enrique Claver *et al.* proposed a theoretical conceptual frame to facilitate the analysis of corporate environmental information reported, using the knowledge map, the environmental capital report and the balanced scorecard (BSC).

Environmental capital report proposed through which information could be revealed about intangibles related to the environment variable, particularly those that create social, environmental and economic value, knowledge of which is especially relevant

EMPLOYEES

- Environmental knowledge to comply with environmental policy and procedures.
- Skills to perform their jobs with a minimum impact on the environment.
- Reasoning and enterprising skills to prevent environmental problems.
- Experience to reduce or prevent significant environmental impacts at their work activities.
- Willingness to learn environmental practices.
- Readiness to share "knowledge" to guarantee an appropriate environmental management.

PRODUCTION

- Adaptation of products and processes to environmental regulations.
- Adaptation to specifics, e.g. through the explicit consideration of alternative materials or specific design for recyclability.
- Efficiency based on cost saving through the reduction of residues, deficiencies and waste, the simplification of the series of materials used, recycling and reuse.
- Permanent quality and improvement, using clean materials and technology, and incorporating a designed, planned and organised Environmental Management System.

CULTURE

- Development of a new group behaviour pattern to reflect the organisation's environmental commitment.
- A policy of incentives and encouragement for knowledge transfer with the aim of keeping staff informed about achievements and new requirements, for example, through a suggestion box or periodical open meetings.
- Successful relationships with stakeholders, based on the exchange of information related to environmental knowledge assets.
- Reputation and image of an ecological firm, not as regards 'green' consumers, but also as regards investors and insurance companies, and even as regards current and potential employees.

R&D

- Innovation capacity helping to develop an internal flow of commercially attractive solutions to problems or environmental ideas related to ecological products and processes.
- Capacity to research environmentally innovative ideas such as new, cleaner technologies.
- Development of 'ecological' products.
- Intellectual ownership with the aim of protecting the firm's environmental leadership to maintain market competitiveness.

MARKETING

- Knowledge of the reasons leading the consumer to a behaviour that reflects environmental concern.
- Ability to influence the behaviour of 'green' consumers'.
- Easy access to distribution channels through the review of potential environmental policies for retailers.

Figure 7.1: BSC.

for shareholders and other third parties concerned. The report would consist of three parts:

1. The vision of the firm, in which the firm would first describe the strategic environmental policy aims it has attained, the value these aims have added for stakeholders, and the objectives set for the immediate future in accordance with the concept of sustainable development. Secondly, it would also include the critical intangibles that make it possible to transform strategic environmental aims into value-creating environmental capital production processes, as well as those the firm counts on for the future.

2. A summary of intangible resources and activities, identifying the intangible environmental resources owned by the firm, the intangible activities developed and the processes that have been completed, in addition to the intangible resources the firm needs to have in order to reach future targets.

3. A system of indicators, allowing the person reading the environmental capital report to check the extent to which the firm is fulfilling its environmental objectives.

Sustainability Report by Corporates

The GRI guidelines (Global Reporting Initiative, 2002) recommend five sections to be included in a sustainability report:

1. Vision and strategy–a statement from the CEO and discussion of the reporting organization's sustainability strategy;

2. Profile–an overview of the reporter's organization, operations, stakeholders, and the scope of the report;

3. Governance structure and management systems–a description of the reporter's organizational structure, policies, management systems, and stakeholder engagement efforts;

4. The GRI Content Index–a cross-referenced table that identifies the location of specific information giving users a clear understanding of the degree to which the reporting organization has covered the content in the GRI guidelines; and

5. Performance indicators–measures of the performance of the reporting organization divided into economic, environmental, and social performance indicators. (Enrique Claver-Corte's *et al.*).

With regard to internal processes, the corporates must use four key sustainable material platforms:

1. Organic–used to describe an agricultural method in which crops are grown without the use of synthetic chemical pesticides (*e.g.* organic cotton);

2. Chemically optimized–materials containing a significantly lower amount of chemicals deemed to be of concern based on Nike's toxic chemical assessment (*e.g.* environmentally preferred rubber);

QUANTITATIVE PERSPECTIVE
- Return on investment.
- FC/VC Rate.
- Added value/employee.
- Income growth.

INTERNAL PROCESSES
- Anticipation of the 'green' consumer's needs.
- Anticipation of competitors' activities.
- Capacity to innovate through the introduction of new technologies (clean productive processes, less polluting machinery, ecological products) and new strategies (environmental audit, certification).
- Evaluation of the environmental management system implemented in the firm through the ISO 14001 Norm, the EMAS Regulation or another procedure.
- Evaluation of the information management systems used, through the Internet or, in a restricted way, through extranets and intranets.
- Reputation and recognition of an ecological brand.

VISION/MISSION AND STRATEGY
- Why should we incorporate the environmental variable into the firm?
- Strategies for gaining environmental knowledge.
- Commitment of all employees to the environment.

RELATIONAL PERSPECTIVE
- Satisfaction level for customers increasingly concerned about the environment.
- Intensification of relationships with suppliers leading to the improvement and constant innovation of 'green' products.
- Interfirm strategic alliances to set in motion environmental systems or programs.
- Encouragement of firm/stakeholder dialogue as a valuation system to accumulate and distribute information.

TRAINING AND LEARNING
- Environmental training and search for continuous improvement.
- Employees' environmental responsibility and increased competence in this matter.
- Align personal objectives with the environmental strategy, and this, in turn, with corporate strategy.

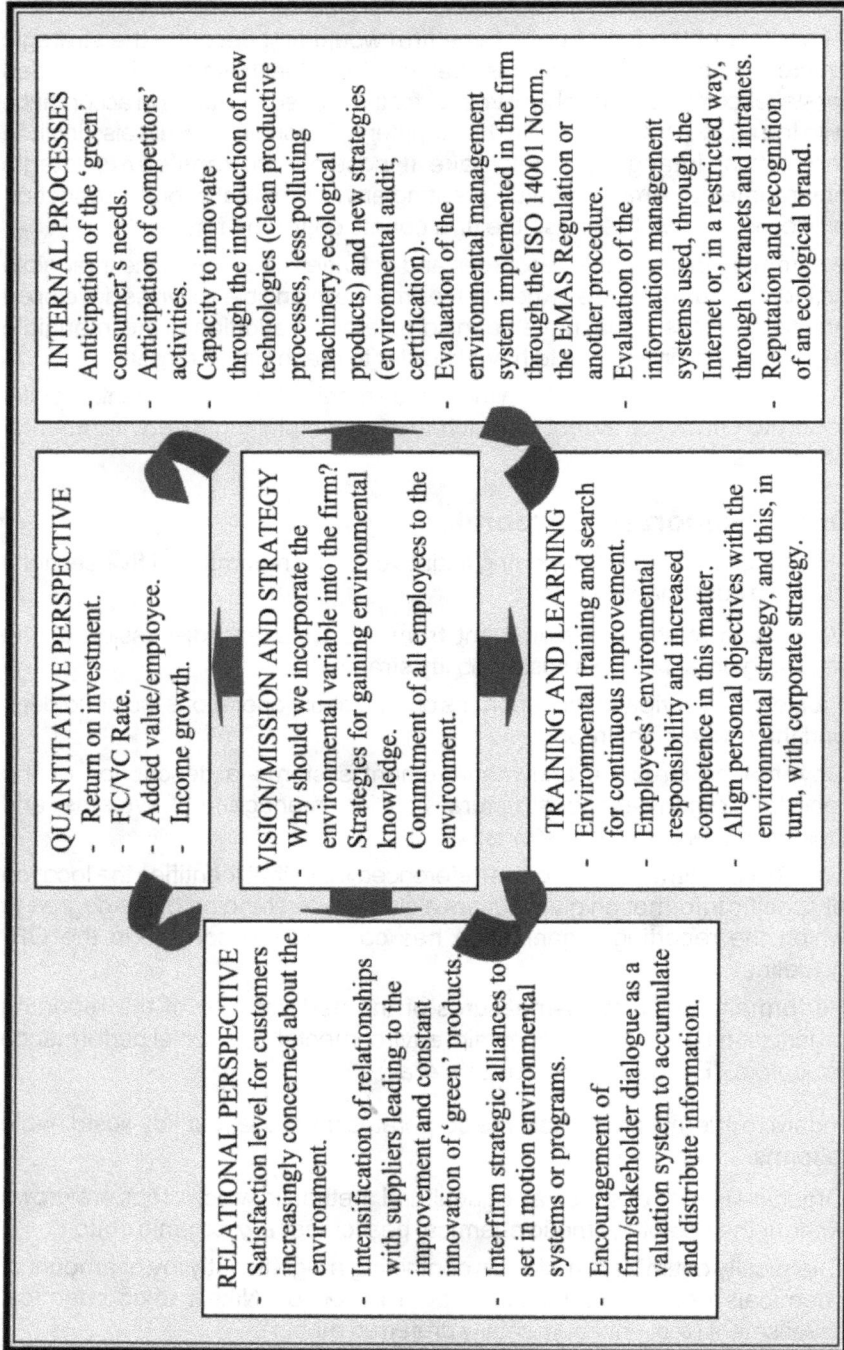

Source: The authors, from Kaplan and Norton (1996)

Figure 7.2: The balanced scorecard from an environmental capital perspective.

3. Regenerated–reprocessed materials or products that can be converted into new products (*e.g.* closed loop materials); and

4. Renewable–(cultivated) plant-based raw material resources that can be used to manufacture natural or textile fibres and polymers (*e.g.* polylactic acid-PLC).

Corporate Responsibilities

When more and more biodiversity is lost, the rate of ecological change will accelerate with unpredictable outcomes (Ehrlich and Wilson, 1991; Naeem *et al.,* 1994; Tilman and Downing, 1994; Wilson, 1992). Our present world is ruled by neoclassical market economy, which exploits nature, cultures and individuals everywhere to maximize short-term financial profits and economic growth. Thus, corporations play the leading role in the rapid decline of biodiversity, cultural diversity, and individual diversity. On the other hand, because of their immense power and influence, corporations could play a leading role in the revival of biodiversity, cultural diversity, and individual diversity. If corporations decided to focus on long-term survival goals, they would find the enrichment of diversity as one of the best ways to ensure continuation of their business in the future because, as diversity enhances the wellbeing of humans and nature, it secures abundant varied, fresh and creative human and natural resources for years to come. (Tarja Ketola). Corporate responsibility (CR) could be used as a tool to manage diversity.

The Role of Decision-Making for Industries in the Climate Change Era

Analysing the role of decision-making economics for industry in the climate change era (Paul E. Hardisty WorleyParsons), Pricing carbon in business decisions including carbon management in effective decision making requires that carbon emissions be given a price. That price can be embedded in financial and economic analysis of projects, and used to understand present and future implications of various capital investment decisions.

Managing GHG (Green House Gases) emissions and understanding the economics of achieving sustainability objectives will become increasingly important for business. Many companies are already establishing their own internal emissions reduction targets, and planning for a carbon-constrained and carbon-impacted future. Significant emissions reductions can be achieved at negative or low cost, in many cases actually reducing overall costs to operators, and improving profitability. How industry responds to these challenges will be an important factor in their future success. In both mitigation and adaptation, there is evidence that risks of inaction far outweigh the costs of well-considered, economically viable action using all of the tools, expertise and market mechanisms currently available to industry. Companies that wait to take action run increasing risks of higher costs, disrupted operations, and stakeholder scrutiny. Climate change carries with it a clear procrastination penalty for industry and the planet.

Risk	Likelihood	Consequence	Mitigation measures
Carbon taxation	High over next 10 to 20 years	Increased production costs, reduced profitability and competitiveness	*Reduce carbon intensity per unit of production*
	Already in place in some jurisdictions		Energy and process efficiency measures
			Fuel switching, supply chain management
			Alterative energy sources and suppliers
			Carbon capture and storage
			Carbon offsetting
			Move production to lower carbon tax jurisdiction
Carbon emissions limits or caps	High over next 10 to 20 years	Limitation on expansion of existing facilities, limitations on new facilities, reduced growth	*Reduce overall carbon emissions*
	Already in place in some jurisdictions		Retrofit existing facilities to lower carbon footing (as above)
			Design and build new facilities for optimal energy efficiency
			Purchase permits or allowances from other firms
			Move production to higher emissions jurisdiction
Shareholder and investor scrutiny	High and rising over next few years	Greater difficulty in securing financing, higher borrowing costs, reduced profits	*Develop strategic plan to reduce shareholder risk from exposure of operations to carbon constraints*
			Implement carbon intensity reduction and overall emissions reduction plans as above
			Participate in carbon disclosure programmes, and other business sustainability indices
Public relations and corporate reputation damage	High and rising over next few years	Declining reputation in the market and with customer, declining sales, reduced profits	*Develop and enact corporate sustainability policies to manage risk of negative public sentiment*
			Take actions that will be seen as part of the solution, not a part of the problem
			Communicate with customers and community stakeholders on climate change, elicit feedback, incorporat into overall mitigation strategy

Technology Transfer for Sustainable Development

In fact, Porter (1991, p. 168) argues that "the conflict between environmental protection and economic competitiveness is a false dichotomy. It stems from a narrow view of the sources of prosperity and a static view of competition". Porter is among a growing group of theorists who believe that environmentalism is good business as well as good social economics.

Economic development does not occur in a social vacuum. Although industrialization around the world has significantly raised living standards and material affluence, it has also created new social problems. Environmental deterioration has recently emerged as a serious concern. Mounting evidence indicates that our current patterns of production and consumption are severely testing the resilience of our planet while rapidly exhausting critical natural resources. Acid rain, global warming, ozone depletion, reduced biodiversity and forecasts of assorted ecological disasters have brought growing calls for a new balance between economic benefits and environmental preservation (Maris G. Martinsons, *et al.*).

Sustainable development requires current market needs, wants and interests to be balanced with longterm social welfare. Consumer and corporate environmentalism can operationalize such a philosophy (Commoner, 1990; Post, 1991). They hold the promise of economic prosperity and a cleaner world (Scott and Rothman, 1991; Silverstein, 1988).

Environmental Education

Environmental education techniques, programmes and strategies may be used as "instruments" through which the ultimate goal of the sound management of environmental resources may be achieved.

Goals of Environmental Education

☆ Foster clear awareness of, and concern about, economic, social, political and ecological interdependence in urban and rural areas;

☆ Provide every person with opportunities to acquire values, attitudes, commitment and skills needed to protect and improve the environment;

☆ Create new patterns of behaviour of individuals, groups and society as a whole towards the environment;

☆ Among some of the goals that need to be commonly pursued, the following may be mentioned;

☆ The need for research aimed at promoting environmental education and ways of efficiently disseminating awareness of the need for sound environmental management;

☆ The need for the systematic development of environmental education and environmental management programmes;

☆ The need to train personnel in both environmental education and environmental management, especially personnel who are able to use environmental education as a tool for environmental management. (Walter Leal Filho)

Conclusion

The loss of biodiversity is occurring worldwide at a rapid rate that has the potential to significantly undermine the prospects for sustainable development. Although the main proximate cause of biodiversity loss is land conversion, the fundamental causes are rooted in economic, institutional, and social factors and include market failures and the lack of property rights. (Collins Ayoo). Biological diversity is also important to society because it provides humans with the necessities of life and forms the basis for the economy. Biological diversity contributes to the quality of life because it yields aesthetic pleasure and holds cultural significance for many different people. (Emerton, 2000). Business houses, a part of society must feel their responsibility in protecting and enhancing the biodiversity even when there is no hard and fast enactments enforced upon them. They must realize their position and act with accountability to the society.

References

Ayoo, Collins, Department of Economics, University of Calgary, Calgary, Canada– www.emeraldinsight.com/1477–7835.html. Economic instruments and the conservation of biodiversity.

Claver, Enrique. Corte´s, Maria Dolores Lo´pez–Gamero, Jose´ Francisco Molina–Azorý´n and Patrocinio Del Carmen Zaragoza–Sa´ez Department of Business Management, University of Alicante, Alicante, Spain–Intellectual and environmental capital. www.emeraldinsight.com/1469–1930.html.

Commoner, B., 1990. "Can capitalists be environmentalists?". *Business and Society Review* 75(Fall): 30–35.

Ehrlich, P. and Wilson, E.O., 1991. "Biodiversity Studies: Science and Policy". *Science*, 253: 758–762.

Emerton, L., 2000. *Using Economic Incentives for Biodiversity Conservation: A Mimeograph.* The World Conservation Union, Gland.

Fauna, Mark Rose and Flora International, Great Eastern House, Tenison Road, Cambridge CB1 2TT, UK 2000 FFI, Oryx, 34(2), 83–84 http://journals.cambridge.org

Houdet, Joel. *Executive Summary–Integrating biodiversity into business strategies.*

IJSE, 1996. *International Journal of Social Economics*, 23(9): 69–96. © MCB University Press, 0306–8293.

JIC, 2007. *Journal of Intellectual Capital,* 8(1): 171–182. Emerald Group Publishing Limited 1469–1930–DOI 10.1108/14691930710715123.

Kaplan, R. and Norton, D., 1996. *The Balanced Scorecard: Translating Strategy into Action.* Harvard Business School Press, Boston, MA.

Ketola, Tarja (Department of Industrial Management, University of Vaasa, Vaasa, Finland) Management of Environmental Quality: An International Journal Vol. 20 No. 3, 2009 pp. 239–254 *q* Emerald Group Publishing Limited 1477–7835 DOI 10.1108/14777830910950649 Corporate responsibility for individual, cultural, and biodiversity.

Martinsons, Maris G., 1996.City University of Hong Kong, Hong Kong and Pacific Rim Institute for Studies of Management, Andy K.Y. Leung Hong Kong University of Science and Technology, Hong Kong and Pacific Rim Institute for Studies of Management, and Christine Loh City University of Hong Kong, Hong Kong– *International Journal of Social Economics*, 23(9): 69–96. © MCB University Press, 0306–8293.

Naeem, S.L., Thompson, S., Lawler, J. and Woodfin, R., 1994. "Declining biodiversity can alter the performance of ecosystems". *Nature,* 368: 734–737.

Paul, E., Hardisty Worley Parsons, Perth, Australia and Imperial College London www.emeraldinsight.com/1477–7835.html.

Porter, M.E., 1991. "America's green strategy". *Scientific American,* 264(4): 168.

Post, J.E., 1991. "Managing as if the earth mattered". *Business Horizons,* pp. 32–38.

Scott, M. and Rothman, H., 1991. *Companies with a Conscience: Intimate Portraits of Twelve Firms that Make a Difference.* Carol Publishing, Secaucus, NJ.

Scott, Mike. Biodiversity concerns rise up the corporate agenda–http://www.ft.com/cms/s/0/a4aa56b8–8b85–11df–ab4d–00144feab49a.html.

Silverstein, M., 1988. "The profits of preservation". *Business and Society Review,* 67(Fall): 29–31.

Sinha, Amitabha. http://apps.fao.org/faostat/forestry/

2013, Biodiversity Conservation for Sustainable Management   Pages 62–76
Editor: Dr. K. Muthuchelian, Vice Chancellor, Periyar University, Salem
Published by: Daya Publishing House, NEW DELHI

Chapter 8

Conservation of Medicinal and Aromatic Plant Resources in Chhattisgarh

V.K. Choudhary
Department of Agricultural Economics, College of Agriculture,
Indira Gandhi Agricultural University, Raipur, Chhattisgarh

ABSTRACT

Chhattisgarh forest covers nearly 43 percent of total geographical area and more than 225 economically important species, with greater potential of their adoptability in the state being used for medicinal and aromatic plant resources (MAPRs) In Chhattisgarh, MAPRs are emerging, which will help in changing the rural scenario by uplifting the standard of living. These plants are grown in about 2000 hectares. Forest areas of Chhattisgarh region are the richest source of biodiversity. All kinds of plant, animal and microorganism are present and tribal people are living under these areas. The ecosystem of forest is balanced but since few years it is disrupted by external interference. This causes erosion of biodiversity. Therefore, this is the right time to understand the importance of conservation of bio diversity. The present study has been undertaken in the sub region Bastar Plateau which is rich in its biodiversity. Topography of the sub region is undulating. Normal Annual rainfall varies from 1200 to 1600 mm. Forest of the sub region are sub-tropical type. The main forest plant species of the region are Sal and Teak. Besides these Tendu, Mahua, Tamarind, Kachnar, Harra, Bahera and Arjun are also present which is also used as medicinal and aromatic plants. Every plant is having its own economic or non-economic importance. Now-a-days it is essential to conserve the flora and fauna of the ecosystem looking to the long term benefit. There is a need to conserve the biodiversity to fulfill the local people's need. Therefore, present study was conducted to work out the quantity collected and value of forest produce, to estimate the disposal pattern of forest, to estimate the change in plant canopy since last one decade and to suggest the policy measures. The study revealed that tendu leaves collection is the main source of forest income followed by Kosa, Bamboo, Tamarind and Mahua. Large proportion of Disposal of forest produce goes to Van Dhan societies except in case of tamarind. Analysis of change since last decade reflects that the availability of forest produce decreases with time except in case of tamarind. On the basis of above study we can conclude that the forest produce are main contributing factor for tribal economy. It is seen that the availability of forest produce decrease since last one decade. The perception of farmers has been taken to find out the causes of decreasing the availability of forest produce. The results show that the main factors are external interference, adverse climate and over exploitation of natural resources. Therefore, the plant having economic importance should be conserve for the interest of tribal welfare. The tribes should be make aware for rational use of forest produce. They should promote for plantation of new plants. The following policy interventions have been suggested for promotion of MAPRs in the state I. systematic research (Agro-climatic zone wise) and development attempt have to be carried out to understand availability, Production marketing and consumption pattern, *ex-situ* cultivation of MAPRs., II. MAPRs may be included in poverty alleviation and rural development programme to promote *ex-situ* conservation. Efficient management of common pool Resources based on local condition to enhance benefit from such resources the financial and Extension support is also required to promote the Productivity of MAPRs.

Introduction

Forest is the richest source of biodiversity. In the forest areas various kinds of plant, animal and microorganism exist. All kinds of plant, animal and tribal people are living under those areas. They have different type of life cycle. Different types of flora and fauna are available in the forest areas. Their ecosystem is balanced but due to some external interference and extra pressure on forest produce the ecosystem is disrupted now a days. This is the right time to understand the importance of biodiversity because it is the Nation's property. Chhattisgarh State is rich in the forest resources. Population of the State is tribal dominant. The economy and culture of the State is forest based. The main forest plant species of the State are Sal (*Sorea robusta*) and Teak (*Tectona grandis*) Besides these *Deptrocarpus marsupium, Terminalia tonentosa, Anagysis latifolia, Madhuca indica, Diaspyrius malengine* spp. also present in the region. These plants make vital role in tribal economy. Looking to the dependency of tribals of forest the present study has been conducted on following specific objectives:

☆ To work out the quantity collected and value of forest produce by tribals,

☆ To estimate the disposal pattern of forest produce,

☆ To estimate the change in availability of forest produce since last decade,

☆ To identify the indirect benefits of forest resources, and

☆ To suggest the policy measures for the management of biodiversity conservation.

Methodology

The present study has been undertaken in the Bastar district of Chhattisgarh State. Tribal contributes 58 per cent of the total population in Bastar district (Census 1991). Therefore, this district has been considered for present investigation. Two villages namely Maachkot and Dhanpunji have been taken for present study, which are situated in Kanger valley. The villages covered by deep forest. The primary data has been collected from 33 respondents through pre-tested questionnaires.

Table 8.1: General description of selected respondents.

Sl. No.	Particulars	Number	Percentage
1.	Total Number of respondents	33	100.00
2.	S C Population (per cent)	11	33.33
3.	ST Population (per cent)	22	66.67
4.	Average Age of Head (in Years)	40.36	
5.	Average Years in School	0.91	
6.	Primary Occupation		
	Agriculture	16	48.48
	Forest produce collection	15	45.45
	Agricultural labour	1	3.03
	Shopkeeper	1	3.03
7.	Secondary Occupation		
	Agriculture	14	42.42
	Forest produce collection	13	39.39
	Agricultural labour	5	15.15
	Service	1	3.03

Table 8.2: Area production and productivity of selected respondents.

Crop	Area (Acres)	Production (Qtl)	Productivity (Qtl/acre)
Paddy	59.38	216.50	3.65
Maize	5.00	4.70	0.94
Nizer	12.00	13.00	1.08
Urd	3.00	2.75	0.92
Mustard	4.00	2.60	0.65
Arhar	3.20	2.05	0.64
Toria	5.50	1.90	0.35
Total	92.08	243.50	2.64

Results and Discussion

Basic Features of Bastar District

Total number of selected respondents is 33 out of which 67 per cent belongs to schedule tribe category. Agriculture is the main occupation of the respondents followed by forest produce collection. The forest is main subsidiary occupation with farming.

Cropping System

Paddy is the main crop. Besides this Nizer, Torai, Arhar, Maize, the respondents have also grown Urd and Mustard. It has been noticed that productivity is very low of all the crops. Crops other than paddy having negligible area. Therefore, forest is the main source of income and employment in off-season.

Forest Product Collection

Quantity and value of forest produce collection is presented in Table 8.3. It reveals that tendu leaves (42.60 per cent) is the main source of forest income followed by Kosa (21.60 per cent), Bamboo (12.20 per cent), Tamarind (10.13 per cent), Mahua (6.97 per cent), Sal seed (4.60 per cent) etc. Above forest produce also generate the source of employment *e.g.* Kosa in Silk preparation, Bamboo in article preparation etc.

Table 8.3: Forest product collection in selected area (Per household).

Sl. No.	Particulars	Quantity	Value	Percentage to Total Value
1.	Tendu leaves (100 No.)	763.64	3263.64	42.60
2.	Mahua (Qtl.)	0.96	534.2	6.97
3.	Tamarind (Qtl.)	1.29	776.36	10.13
4.	Sal seed (Qtl.)	0.88	352.12	4.60
5.	Charota seed (Qtl.)	0.12	17.5	0.23
6.	Siali (100 No.)	4.27	128.18	1.67
7.	Kosa (Numbers)	1103.03	1654.55	21.60
8.	Bamboo (Number)	186.36	934.82	12.20
Total	2060.55		7661.37	100.00

Disposal Pattern of Forest Produce

When we observed the disposal pattern of forest produce we found that most of the share of forest produce goes to society except in case of tamarind. Tamarind has been sold to the traders and society both almost equally. Negligible quantum of other forest resources goes to the traders.

Change in Availability of Forest Produce

Farmer's perception has been taken to estimate the change in availability of forest produce since last decade. It reflects that the availability of forest produce decreases with time except in case of tamarind.

Table 8.4: Disposal pattern of forest product in selected area of Bastar district (Per household).

Sl. No.	Particulars	Quantity Collected	Household Consumption	Sold to Society	Sold to Traders
1.	Tendu leaves (100 No.)	763.64	0	763.64	0.00
2.	Mahua (Qtl.)	0.96	0.05	0.19	0.72
3.	Tamarind (Qtl.)	1.29	0.06	0.6	0.63
4.	Sal seed (Qtl.)	0.88	0	0.88	0.00
5.	Charota seed (Qtl.)	0.12	0	0.12	0.00
6.	Siali (100 No.)	4.27	0	4.27	0.00
7.	Kosa (Numbers)	1103.03	0	203.03	900.00
8.	Bamboo (Number)	186.36	17.88	0	168.48
	Total	2060.55	17.99	972.73	1069.83

Percentage

Sl. No.	Particulars	Quantity Collected	Household Consumption	Sold to Society	Sold to Traders
1.	Tendu leaves (100 No.)	100.00	0.00	100.00	0.00
2.	Mahua (Qtl.)	100.00	5.21	19.79	75.00
3.	Tamarind (Qtl.)	100.00	4.65	46.51	48.84
4.	Sal seed (Qtl.)	100.00	0.00	100.00	0.00
5.	Charota seed (Qtl.)	100.00	0.00	100.00	0.00
6.	Siali (100 No.)	100.00	0.00	100.00	0.00
7.	Kosa (Numbers)	100.00	0.00	18.41	81.59
8.	Bamboo (Number)	100.00	9.59	0.00	90.41
	Total	100.00	0.87	47.21	51.92

Indirect Benefits of Forest Produce

☆ Long-term benefit from the species.

☆ Source of Income and Employment for the tribals.

☆ To make available of needs of tribals.

☆ Medicinal value of plants.

☆ To make refresh the environment.

☆ Accommodation for forest animal, different type of bio-organism.

☆ Recreation of tourists.

☆ It gives raw materials for forest based small-scale industries like apiculture, sericulture, manufacturing of wood arts, bamboo articles etc.

☆ It is the main source of Export of Minor Forest Produce.

Table 8.5: Change in availability of forest produce over last decade.

Sl.No.	Particulars	Increase Availability (per cent)	Decrease Availability (per cent)
1.	Tendu leaves	0.00	21.09
2.	Chironji	0.00	30.00
3.	Mahua	0.00	20.32
4.	Bahera	0.00	20.00
5.	Siali leaves	0.00	38.32
6.	Shikakai	0.00	25.00
7.	Harra	0.00	28.57
8.	Tamarind	7.58	0.91
9.	Sal seed	0.00	12.50
10.	Charota seed	0.00	1.43
11.	Kosa	6.13	6.36
12.	Bamboo	0.00	15.45

Note: Supporting Tables.

Conclusion

The perception of farmers has been taken to find out the causes of decreasing the availability of forest produce. The results show that the main factors are external interference, adverse climate and over exploitation of natural resources. Therefore, the plant having economic importance should be conserve for the interest of tribal welfare. The tribes should be make aware for rational use of forest produce. They should promote for plantation of new plants. The following policy interventions have been suggested for promotion of MAPRs in the state a. systematic research (Agro-climatic zone wise) and development attempt have to be carried out to understand availability, Production marketing and consumption pattern, *ex-situ* cultivation of MAPRs., b. MAPRs may be included in poverty alleviation and rural development programme to promote *ex-situ* conservation. Efficient management of common pool Resources based on local condition to enhance benefit from such resources the financial and Extension support is also required to promote the Productivity of MAPRs.

Suggestions for Biodiversity Conservation

On the basis of results study concludes that the forest produce are main contributing factor for tribal economy. It is seen that the availability of forest produce decrease since last one decade. Therefore, the plants having economic importance should be conserving for the interest of tribal welfare. The tribals should be making aware for rational use of forest produce. Local participation is must for the conservation of biodiversity. The tribals should be enhancing to protect the plants. The alternative use of forest plants should be introduced to minimise the over-exploitation of forest resources. Knowledge of indigenous techniques should be disbursed from one place to another. Therefore, the people can make aware about the utility of particular

Table 8.6: Distribution of some ethnomedicinal tree species in the forest of Bastar district of Chhattisgarh

Sl.No.	Range	Compartment No.	Grid/ Plot No.	Area of Compartment in ha.	Total No. of Trees per ha.	Species	No. of Trees	Percentage of Species (per cent)	Other Species (per cent)
1.	Jagdalpur	123	193	333.367	770	Sal	420	54.45	23.49
						Mahua	10	1.29	
						Char	60	7.79	
						Bija	60	7.79	
						Aonla	40	5.19	
		98	164	260.171	570	Amaltas	10	1.75	85.98
						Arjun	50	8.77	
						Am	10	1.75	
						Bel	10	1.75	
		104	189	231.596	750	Sal	380	50.66	34.69
						Bija	50	6.66	
						Char	40	5.33	
						Aonla	20	2.66	
2.	Chitrakoot	P-58	40	270.825	660	Harra	40	6.06	92.43
						Bija	10	1.51	
		82	30	265.766	170	Bhelwa	10	5.88	94.16
		84	45	205.8	540	Sal	90	16.66	72.24
						Bhelwa	20	3.70	
						Char	20	3.70	
						Kakai	10	1.85	
						Aonla	10	1.85	

Contd...

Table 8.6—*Contd...*

Sl.No.	Range	Compartment No.	Grid/ Plot No.	Area of Compartment in ha.	Total No. of Trees per ha.	Species	No. of Trees	Percentage of Species (per cent)	Other Species (per cent)
3.	Machkot	33	175	272.761	650	Sal	240	39.92	37.04
						Bija	70	10.76	
						Harra	10	1.53	
						Kakai	40	6.15	
						Aonla	10	1.53	
						Char	20	3.07	
		36	148	271.196	860	Sal	50	5.81	77.93
						Bija	90	10.46	
						Mahua	30	3.48	
						Char	20	2.32	
		12	117	379.492	510	Sal	160	31.37	52.95
						Bija	50	9.80	
						Aonla	10	1.96	
						Harra	10	1.96	
						Bhelwa	10	1.96	
4.	Darbha	244	258	140.021	360	Sal	160	44.44	44.46
						Bija	20	5.55	
						Char	20	5.55	
		248	245	192.671	270	Sal	60	22.22	66.68
						Kusum	20	7.40	
						Bhelwa	10	3.70	

Contd...

Table 8.6—Contd...

Sl.No.	Range	Compartment No.	Grid/Plot No.	Area of Compartment in ha.	Total No. of Trees per ha.	Species	No. of Trees	Percentage of Species (per cent)	Other Species (per cent)
		179	216	416.546	480	Sal	90	18.75	72.92
						Char	30	6.25	
						Bija	10	2.08	
5.	Bakawand	314	448	245.9	280	Sal	20	7.14	78.58
						Baheda	20	7.14	
						Char	10	3.57	
						Kusum	10	3.57	
		290	476	306	380	Sal	250	65.78	24.00
						Char	10	2.63	
						Bhelwa	10	2.63	
						Harra	10	2.63	
						Baheda	10	2.63	
		324	435	264	280	Sal	140	50.00	42.86
						Char	20	7.14	
6.	Bhanpuri	314	418	230.8	420	Aonla	20	4.76	92.86
						Char	10	2.38	
		283	428	134	410	Char	10	2.43	43.93
						Sal	200	48.78	
						Aonla	10	2.43	
						Harra	10	2.43	
		352	376	114	650	Sal	490	75.38	9.25
						Char	80	12.30	
						Bhelwa	20	3.07	

Table 8.7: Ethno medicinal plants, collection season and their rates.

Sl.No.	Hindi Name	Gondi/Halbi Name	Nature	Botanical Name	Family	Parts Used	Rate/Kg. (Rs.)	Collection Season
1.	Adusa	Pedavali, Adusa	Shrub	*Adhatoda vasica*, Nees	Acanthaceae	Leaf	150-200	Nov.-Dec.
2.	Am	Marka	Tree	*Mangifera indica*, L.	Anacardiaceae	Kernel	5	April-May
						Amchur	25	
3.	Amaltas	Dhanbahar, Sonarli, Rela	Tree	*Cassia fistula*, L.	Caesalpiniaceae	Fruit	25	April-June
4.	Anantmul	Sugandhi, Phalurai	Creeper	*Hemidesmus indicus*, Linn.	Periplocaceae	Root	300-400	Nov.-Jan.
5.	Aonla	Aonra, Neli	Tree	*Emblica officinalis*, Gaerth.	Euphorbiaceae	Fruit (Dry)	20	Nov.-Jan.
6.	Arjun, Koha	Kahu, Mangi	Tree	*Terminalia arjuna*, W and A	Combretaceae	Bark	20	Nov.-Jan.
7.	Asgandha	—	Shrub	*Withania somnifera* L.	Solanaceae	Root	200	Nov.-Jan
8.	Bahera	Tahka, Mandu	Tree	*Terminalia bellerica*, Roxb.	Combretaceae	Fruit	20	Jan.-March
9.	Baibirang	Duli	Herb	*Embelia tsjerium*, Cottam.	Myisinaceae	Seed	150-120	Oct-Nov
10.	Baichandi	Kaliapapad	Herb	*Doscorea hispida*, Denn.st.	Dioscoreaceae	Tuber	100-150	Nov.-Dec.
11.	Bel	Mahak, Mandu	Tree	*Aegle marmelos*, Corr.	Rutaceae	Pulp	50	March-June
12.	Bhelva	Bhela	Tree	*Semicarpus anacardium*, L.	Anacardiaceae	Fruit (nut)	50	April-May
13.	Bhojraj	—	Herb	*Peucedanum nagpurense*, Prain	Apiaceae	Root	700-800	Nov.-Dec.
14.	Bhringraj	—	Herb	*Eclipta prostrata* L.	Asteraceae	Leaf	300-400	Oct.-Nov.
15.	Bhuileem	Kalmegha, Bhuiimb	Shrub	*Andrographis paniculata*, Nees	Acanthaceae	Whole Plant	100-150	Nov.-Jan.
16.	Bija	—	Tree	*Pterocarpus marsupium*, Roxb.	Fabaceae	Bark	50	Nov.-Jan.
17.	Bramhi	—	Herb	*Centella asiatica*	Apiaceae	Leaf	400-500	Nov.-Dec.
18.	Buch	—	Herb	*Acorus calamus* L. Bach. (H)	Araceae	Root	100-150	Dec.-Feb.

Contd...

Table 8.7–Contd...

Sl.No.	Hindi Name	Gondi/Halbi Name	Nature	Botanical Name	Family	Parts Used	Rate/Kg. (Rs.)	Collection Season
19.	Char	Char	Tree	*Buchanania lanzan*, Spreng.	Anacardiaceae	Chironji	300.400	April-May
20.	Charota	Charota	Herb	*Cassia tora*, Linn.	Caesalpiniaceae	Seed	5	Sept.-Oct.
21.	Chiraita	—	Herb	*Swertiad chirayita*, Roxb. Ex Flem	Gentianaceae	Whole plant	100-150	Oct.-June
22.	Vhiysest	—	Herb	*Plumbago zeylanica* L.	Plumbaginaceae	Root	400-500	Nov.-Jan.
23.	Chitrak	Chitrak	Herb	*Plumbago zeylanica* L.	Plumbaginaceae	Root	200-300	Dec.-Jan.
24.	Dhawai	Dhai, Dhaul, Hit	Shrub	*Woodfloria fruticosa*, L.Kur.	Lythraceae	Flower	20	April-May
25.	Gheekumar	Gwarpatha	Herb	*Aloe barbadensis* Mill.	Liliaceae	Leaf	50-100	Dec.-Jan.
26.	Giloe	Gurbel	Climber	*Tinospora cordifolia* Willd	Menispermaceae	Stem	100-150	Nov.-Jan.
27.	Gumchi	Gunchi	Shrub	*Abrus precatorius* L.	Fabaceae	Seed	200	Feb.-April
28.	Gundmar	Marasingi	Herb	*Gymnema sylvestre* Retz.R.Br.	Asclepiadaceae	Leaf	100-150	Oct.-Nov.
29.	Harjor	—	Climber	*Peristrophe bivalves*, Merill	Acanthaceae	Stem	100-200	Entire year
30.	Harra	Hirala, Karka	Tree	*Terminalia chebula*, Retz	Combretaceae	Fruit	5	Dec.-Jan.
31.	Hiranjutia	—		*Colchicum luteum*, Bakes.	Liliaceae	Root	600-800	Sept.-Oct.
32.	Imli	Ita, Imli	Tree	*Temarindus indica* L.	Caesalpiniaceae	Deseeded Fr.	25	Jan-April
33.	Indrayan	Citru	Herb	*Citrullus colocynthis* L.	Cucurbitaceae	Root	150-200	Oct.-Nov.
34.	Kakai	Banjogni, Banramtila	Shrub	*Flacourtia indica*, Merr.	Flacourtiaceae	Seed	50-100	Sept.-Oct.
35.	Kali Musli	Banjogni	Herb	*Curculigo orchioides* Gaertn.	Amaryllidaceae	Rhizome	200-300	Sept.-Oct.
36.	Kalihari	Jhagren	Climber	*Gloriosa superba*, Linn.	Liliaceae	Root	30-50	Sept.-Oct.
37.	Kamraj	Kakasmati, Kirinjmati	Climber	*Doscorea oppostifolia* Linn.	Dioscoreaceae	Roots	700-800	Oct.-Nov.
38.	Kewach	Kawich	Climber	*Mucunna prurita*, Hook.	Leguminaceae	Seed	100-150	Jan.-Feb.
39.	Kuchla	Duchla	Tree	*Strychnos nuxvomica*, Linn.	Loganiaceae	Seed	200-300	Feb.-March

Contd...

Table 8.7–Contd...

Sl.No.	Hindi Name	Gondi/Halbi Name	Nature	Botanical Name	Family	Parts Used	Rate/Kg. (Rs.)	Collection Season
40.	Kusum	Kusum	Tree	Schleichera oleosa, Lour.	Sapindaceae	Seed	25	April-May
41.	Mahua	Irpi, Irrhu, Mahu	Tree	Madhuca indica J.F. Gmel.	Sapotaceae	Flower part	15	March-May
42.	Malkangini	Pena, Pitto, Tonda	Climber	Celastrus painculatus, Willd.	Celastraceae	Seed	300-400	July-Aug.
43.	Marorphali	Telka	Shrub	Helicteres isora, Linn.	Sterculiaceae	Fruit	50	Feb.-March
44.	Musli (Safed)	Kawrakanda	Herb	Chlorophytum tuberosum Roxb.	Liliaceae	Root	400-600	Sept.-Oct.
45.	Nagrmotha	Mutha	Grass	Cyperus scariosus R.Br.	Cyperaceae	Root	400-500	Nov.-Dec.
46.	Neem	Neem	Tree	Azadirachta indica, Juss.	Miliaceae	Seed	25	March-April
47.	Nirgundi	—	Shrub	Vitex negundu L.	Verbenaceae	Root	100-200	Dec.-Jan.
48.	Palas Bel	Bodel Laha	Climber	Butea superb, Roxb.	Fabaceae	Bark	40	Nov.-Jan.
49.	Piperamul	Bhumia	Creeper	Piper longum Linn.	Piperaceae	Root	300-400	Nov.-Jan.
50.	Sal	Sargi, Mhang	Tree	Shorea robusta, Gaertn.	Dipterocarpaceae	Seed	5	May-June
						Dhup	50	Nov.-Feb.
51.	Sarpagandha	Kukdichend	Herb	Rauvolfia serpentina, L.	Apocynaceae	Root	100-200	Dec.-Jan.
52.	Satawar	Salavar, Deobadani	Climber	Asparagus racemous Wild.	Liliaceae	Root	150-200	Dec.-Jan.
53.	Shivlingi	—	Shrub	Bryonopsis laciniosa L.	Cucurbitaceae	Seed	500-600	Dec.-Jan.
54.	Tejraj	—	Herb	Peucedanum dhana, Buch-Ham	Apiaceae	Root	700-900	Nov.-Dec.
55.	Tikhur	Tikhur	Herb	Curcuma angustifolia Roxb.	Zingiberaceae	Rhizome	70-100	Oct.-Dec.
56.	Vajradanti	—	Herb	Dicoma tomentosa, Cass.	Asteraceae	Root	100	Sept.-Nov.

resources. We should not assume that forest resources are easily available national property. Forests are regular productive assets. Each individual has right to utilize it and duty for protection. Diversion of forestland into non-forestland should be done under supervision of ecological specialists. Plants and animal species, which are being depleting, should be conserving through tissue culture and biotechnology. The culture of the tribals who are living in the forest and surrounding the forest should be enhance, therefore forest and tribals and forest both will be benefited. Some possible suggesstions

☆ Thorough ecological, phytogeographical and ethno-anthropological studies are required to have complete knowledge of individual medicinal plant.

☆ The wide ranging commercial survey of the markets and buyers are required to study the market trend and price fluctuation.

☆ Identification of drawbacks in the traditional trade practices are required.

☆ Extensive survey of possibilities of ex-situ conservation of ethno medicinal plants.

Table 8.8(A): Production of ethnomedicinal plants/their products (Forest produce which are found in abundance).

Sl.No.	Plants/their Products	Quantity in quintals				
		1996-97	1997-98	1998-99	1999-2000	2000-2001
1.	Char (Chironji)	553	470	550	596	1098
2.	Am (Kernal)	690	530	450	600	580
	Amchur	270	280	290	275	300
3.	Aonla	2575	2122	1560	2055	1050
4.	Baheda	1000	1100	1000	1200	1212
5.	Harra	-	12	198	355	2223
6.	Imali	18400	18783	19000	18443	23003
7.	Mahua (Flowers)	4000	7350	6775	5500	13641
	Mahua Tora	1800	1850	1780	1670	1565
8.	Sal Seed	-	36827	109	28273	58255
9.	Chirayata	2000	1500	1200	1000	5900
10.	Dhawai phool	2850	3075	3175	1957	2215
11.	Charota seed	1900	1540	1680	1918	1116
12.	Malkangni	1200	1010	1100	1200	812
13.	Neem	850	700	1000	1200	900
14.	Satawar	600	500	450	400	300
15.	Bhelwa phal	1500	1220	1100	1000	200
16.	Nagarmotha	500	450	400	400	300
17.	Baichandi	500	660	400	300	200
18.	Baibidang	650	723	567	615	467

Table 8.8(B): Production of ethnomedicinal plants/their products (Forest produce which are not found in abundance)

Sl.No.	Plants/their Products	Quantity in quintals				
		1996-97	1997-98	1998-99	1999-2000	2000-2001
1.	Amaltas	20	26.6	21	24.5	18
2.	Arjun Bark	4.05	5.20	7.0	6.7	5.6
3.	Bel (Pulp)	2.90	3.8	5.10	4.7	3.6
4.	Bija (Bark)	70	110	95	125	100
5.	Kusum Seed	180	150	2020	150	100
6.	Bhuileem	14.5	12.3	15.8	14.2	16.53
7.	Pipermool	450	400	300	200	100
8.	Chitawar	0.60	0.85	1.05	1.35	0.75
9.	Maror Phally	40	60	35	84	97
10.	Anantmul	2.05	0.90	1.20	1.45	0.60
11.	Giloe	-	-	-	0.10	0.50
12.	Palasbel	4.30	5.50	6.10	4.10	2.25
13.	Adusa	0.25	0.65	0.50	0.75	0.80
14.	Indrayan	-	-	0.25	0.30	0.70
15.	Bhring raj	1.10	2.25	0.90	3.05	2.20
16.	Bramhi	2.50	3.00	1.75	2.80	1.05
17.	Harjor	0.40	0.20	0.70	0.55	0.40
18.	Hiranjutia	-	-	-	0.05	0.25
19.	Vajradanti	4.25	3.80	5.10	2.75	1.90
20.	Safed Musli	300	400	200	100	100
21.	Kali Musli	300	350	250	100	80
22.	Tikhur	300	250	200	100	50
23.	Gudmar	0.55	0.40	0.80	1.10	0.65
24.	Gumchi	0.10	0.35	0.50	0.40	0.44
25.	Kalihari	40	30	20	15	10
26.	Kakei	0.18	0.30	0.65	0.25	0.52
27.	Kuchla	0.85	1.10	1.5	1.75	2.05
28.	Kamraj	-	0.25	0.47	0.30	0.10
29.	Tejraj	0.30	0.45	0.77	0.56	0.35
30.	Bhojraj	0.30	0.55	0.80	0.60	0.30
31.	Gheekumar	-	-	-	0.10	0.25
32.	Chitrak	0.75	0.20	0.56	0.90	0.40
33.	Nirgundi	0.80	0.95	1.02	0.50	0.70
34.	Sarpagandha	305	275	250	200	100
35.	Buch	3.80	2.95	3.95	4.10	3.65
36.	Asagandha	0.25	0.40	0.60	0.35	0.45
37.	Shivlingi	1.20	1.35	1.00	1.10	1.05
38.	Kewach	0.90	1.00	0.80	0.90	0.75

☆ Role of natural forest in *in-situ* conservation when the forest protection involves peoples participation.

☆ Research and Extension programmes to be launched to promote cultivation of economically important ethno medicinal plants by small and large farmers, piece meal efforts will not suffice.

☆ Field level facilities of transport, improved collection techniques, processing, grading and storage to be increased.

☆ Strict fire protection measures need to be taken before and during summer.

☆ Special protection measures need to be adopted to prevent rivers and nalas of the district from chemical poisoning by fish catchers. This toxic water does not permit innumerable number of herbs/shrubs and climbers to grow on their banks. Beside this, toxic water is also harmful to all aquatic animals.

References

Bhowmick, P.K., 2001. Level of people's participation and involvement in tribal development programme in India: Need for policy intervention. *Indian Journal of Social Development*, Vol. 1.

Choubey, C.M.S., Negi and Pandey, Amit, 2001. Status of phytodiversity and population dynamics in preservation plots representing Moist Penisular Sal Forest in Chhattisgarh. *Indian Journal of Tropical Biodiversity*, 9(1–4): 18–30.

Dubey, A.K., 1988. Working Plan for Central Bastar division and Tongpal range of South Bastar forest division, Jagdalpur Circle. 1988–1989, Vol. 1–3.

Jain, S.K., 1991. *Dictionary of Indian Folk Medicine and Ethnobotany*. Deep Publications, New Delhi.

Khullar, Pankaj, 1992. Conservation of biodiversity in natural forests through preservation plots: A historical perspective. *Indian Forester*, 118(5): 327–337.

Ravishankar, T., 1996. Role of indigenous people in the conservation of plant genetic resources. In: *Ethnobiology in Human Welfare*, (Ed.) S.K. Jain. Deep Publications, New Delhi, p. 310–314.

Sahu, T.R., 1996. Life support promising food plant among aboriginal of Bastar M.P., India. In: *Ethnobiology in Human Welfare*, (Ed.) S.K. Jain. Deep Publications, New Delhi, p. 26–30.

Tewari, D.D., 1994. Developing and sustaining non-timber forests products: Policy issues and concerns with special reference to India. *Journal of World Forest Resource Management*, 7: 151–178.

2013, Biodiversity Conservation for Sustainable Management   Pages 77–83
Editor: Dr. K. Muthuchelian, Vice Chancellor, Periyar University, Salem
Published by: Daya Publishing House, NEW DELHI

Chapter 9

Social Forestry: A Tool for Carbon Sequestration, Sustainable Development and Conserving Biodiversity

Jayesh Pathak

ASPEE College of Horticulture and Forestry,
Navsari Agricultural University, Navsari, Gujarat

Introduction

Rising levels of greenhouse gases are already changing the climate. According to the Intergovernmental Panel on Climate Change (IPCC) Working Group I (WGI) Fourth Assessment Report, from 1850 to 2005, the average global temperature increased by about 0.76°C and global mean sea level rise by 12 to 22 cm during the last century. These changes are affecting the entire world, from low-lying islands in the tropics to the vast Polar regions. Climate change predictions are not encouraging; according to the IPCC WGI Fourth Assessment Report, a further increase in temperatures of 1.4°C to 5.8°C by 2100 is projected.

Current rates and magnitude of species extinction far exceed normal background rates. Human activities have already resulted in the loss of biodiversity and thus may have affected goods and services crucial for human well-being. The rate and magnitude of climate change induced by increased greenhouse gases emissions has and will continue to affect biodiversity either directly or in combination with other drivers of change.

Tropical deforestation provides a significant contribution to anthropogenic increases in atmospheric CO_2 concentration that may lead to global warming which is the most significant drivers of biodiversity loss by the end of the century.

Climate change lead to a sharp increase in rates of extinction. The rate of extinction increases 1000 times than earlier. According to one recent study focusing on five regions of the world, if the climate continues to warm it could dramatically increase the number of species going extinct. Predictions suggest that 15-37 per cent of species in these regions will be on their way to extinction by 2050 due to climate change. This study also indicates that for many species, climate change poses a greater threat to their survival than the destruction of their natural habitat. These estimates show the importance of rapid implementation of technologies to decrease greenhouse gas emissions and strategies for carbon sequestration.

Major options that sequester carbon are Forests, Oceans and Soil. About two-thirds of the globe terrestrial carbon, exclusive of that sequestered in rocks and sediments is sequestered in the standing forests. This include forest under storey plants, leaf and forest debris and forest soils. There are some non natural stocks in form of long lived wood products, waste dumps. This form of carbon stock is gaining important with increased global timber harvest and manufactured wood products over the past several decades.

Plant fixes atmospheric carbon in cell tissues as they grow thereby transforming carbon from the atmosphere to the biotic system. As a matter of fact drawing CO_2 out of the air and sequestering it into biomass is the only known practical way to remove large volumes of this green house gas form the atmosphere. Thus Forest is the best option to sequester carbon. In Forest mainly carbon sequester by standing biomass and soil. But due to increase in population anthropogenic pressure increases due to which pressure on Conventional forest increases hence there is need to conserve standing forest and to manage forest outside the conventional area of forest.

Forestation and other forest management options to sequester CO_2 in the tropical latitudes may fail unless they address local economic, social, environmental, and political needs of people. Social forestry systems are a better climate change mitigation option than oceanic, and other terrestrial options because of the secondary environmental benefits and helping to achieve Millenium Developments goals such as helping to attain food security and secure land tenure in developing countries, increasing farm income, maintaining watershed hydrology, and soil conservation. Social Forestry can play major role in restoring and maintaining above-ground and below-ground biodiversity as carbon sequestration can be pursued on agricultural lands and on lands with low productivity, *i.e.* on lands that are least suitable for agriculture or intensive forestry, and are compatible with the preservation of biodiversity over large areas.

Social Forestry defined as the practice of forestry on lands outside the conventional areas with application of forest technology for the benefit of the rural and urban population. Its main object is to bring under wood land and tree cover an adequate portion of the total land area of the country for the preservation of environment by moderating the climate and preventing soil erosion, improving the agro ecosystem and restoring the natural biodiversity to a large extent. Increasing in the area under wood land and tree cover results in adequate supply of fuel wood, fodder and small timber to rural people. It also provides other economic products and timber for the urban population and industries to some extent.

Social Forestry land use management includes:

1. Community based Forest management
2. Agroforestry and Agrisilviculture
3. Farm Forestry
4. Afforestation and reforestation on degraded land and waste lands

Community based Forest Management

☆ Forest management can contribute to carbon sequestration through promoting forest growth and biomass accumulation.

☆ Through proper management of world's forest the carbon sources could be increased to a greater extent.

☆ Effective management by implementing the J.F.M. micro plan prescriptions can further enhance the potential of carbon sequestration.

☆ In community managed teak forests of Harda forest division, M.P., carbon sequestered reported levels varying between 0.5 to 3.4 tonnes per ha per year.

☆ Negi *et al.* (2002)studied the carbon sequestration through community based forest management in two villages named Radhiapali and Kunjapali and found that carbon sequestrated more in Kunjapali village due to more standing biomass.

Table 8.1: Carbon sequestration through community based Forest management.

	Village	
	Radhiapali	*Kunjapali*
Control		
Unprotected area		
Basal area (m²/ha)	0.67	1.32
Standing woody biomass (t/ha)a	3.89	9.29
Protected area		
Basal area (m²/ha)	4.66	6.04
Standing woody biomass (t/ha)	37.08	48.56
Years of protection		
Standing biomass	10	6
Increment/ha/yr over control (MT)	3.32	6.55
Carbon sequestered*/ha/yr (MT)	1.53	3.01

Source: Negi *et al.* (2002)

A new approach for carbon estimation.

1 ton woody biomass = 0.46 tonnes of carbon.

Agroforestry

☆ Introducing trees in Agricultural farms may be useful technique to increase the soil carbon status because the presence of trees affects (dynamics, directly or indirectly).

☆ Various interacting factors which influences carbon stock in the soil under agroforestry are addition of:

1. Litter
2. Maintenance of higher soil moisture content
3. Reduced surface soil temperature
4. Proliferated root system.
5. Enhanced biological activities.
6. Decrease risk of soil erosion.

☆ In this system tree species are widely accepted by farmers which have minimum interaction with agriculture crops.

☆ Cultivation of fast growing trees with arable crops under agri-silviculture systems help in improving S.O.C. and sequestering carbon particularly in highly degraded barren lands.

☆ Karnal Das and Itnal *et al.* (1995) studied the SOC contents after six years of plantation with different land use-systems and found that agroforestry, agrihorticulture and agrisilviculture has more SOC than sole cropping systems

Table 8.2: SOC contents after six years of plantation with different landuse-systems.

Land Used Systems	Organic Carbon (g kg^{-1}) in	
	0.15cm	0.30cm
Sole cropping	4.2	3.9
Agroforestry	7.1	7.2
Agri-horticulture	7.3	7.3
Agri-silviculture	3.8	4.7

Central Soil Salinity Research Institute.

Karnal Das and Itnal *et al.* (1995)

Farm Forestry

☆ It involves the growth and management of trees on private farm lands for the purpose of supply of wood and non-wood products, provides a balanced mix of economic, social and environmental outcomes.

☆ Of the 80 per cent potential that the tropics possess to conserve and sequester carbon, forestation and agro forestry/farm forestry account for 50 per cent.

☆ Farm forestry compares favorably to reforestation or afforestation activities not only in terms of utilizing substantial amounts of atmospheric carbon but costs.

☆ Unlike traditional carbon sequestration techniques, farm forestry has the potential of being presented by policy makers as a primary instrument a people's welfare and industrial growth.

☆ Singh *et al.* (2000) studied the total carbon sequestered under farm Forestry in UP during the period 1979-94 and found that 1525.44 million trees sequestrated 1.9 billion tonnes of CO_2.

Table 8.3: Total carbon sequestered under farm Forestry in UP during the period 1979-94.

Year	Number of Tress Planted (in Million)	Numbers of Trees @ 80 per cent Survival (in million)	Trees Converted to ha. @ 1500/ha	Volume of Wood Produced per ha/Year (m^3)	Quantity of Carbon Sequestered @ 0.5 tonnes per m^3 of Wood (in tonnes)
1979-80	16.3	13.04	8693		
1980-81	66.7	53.36	35773	53889	26949
1981-82	146.8	117.44	78293	141904	70952
1982-83	254.0	203.20	135467	302064	151032
1983-84	252.3	201.84	134560	812859	406429.3
1984-85	355.0	284.00	189333	1701531	850765.3
1985-86	151.5	121.20	80800	2685872	1342936
1986-87	144.8	115.84	77227	4023888	2011944
1987-88	136.8	109.44	72960	4525920	2262960
1988-89	142.2	113.76	75840	4832432	2416216
1989-90	110.3	88.24	58827	4726923	2363461
1990-91	62.6	50.08	33387	4627520	2313760
1991-92	32.2	25.76	17173	4280773	2140387
1992-93	18.4	14.72	9813	3780784	1890392
1993-94	16.9	13.52	9013	3093413	1546707
	1906.8	1525.44	1016960	39589781	19794891

TERI, New Delhi.

Singh *et al.* (2000).

Afforestation and Reforestation

☆ Afforestation have tremendous potential for carbon sequestration not only in above ground carbon biomass but also root carbon biomass in deeper soil depths.

☆ Reforestation of 75 million ha most degraded lands with suitable trees and grasses/crops may sequester about 4Pg of carbon.

☆ If 10 million ha area reforested per year, 400 x 10[6] ha of new forest will be created by the year 2040. By the time the new forest matures (40-100 year after plantation) 25 to 50Gt of CO_2 will have been sequestered from the atmosphere.

☆ Raising trees on degraded lands in arid areas irrigating with saline water also may help in C sequestration to a greater extent.

Table 8.4: Chronosequential age, total system C and proportion of aboveground carbon in tropical forests and lands covered by slash and burn agriculture.

Land Use	No. of Observation	Age Years	Total System C (MgC ha⁻¹)	Above Ground Total C
Original forest	10	-	305	0.72
Managed forest	9	-	181	0.73
Burned and cropped	18	2	52	0.23
Bush fallow	17	5	85	0.22
Tree fallow	8	9	136	0.48
Secondary forest	8	19	219	0.61
Pasture	9	10	48	0.21
Imperata grassland	8	13	47	0.05
Young agroforest	1	5	65	0.28
Mature agroforest	19	23	130	0.58

Source: Woomer *et al.* (2000).

Social Forestry and Biodiversity

As much as 90 per cent of the biodiversity resources in the tropics are located in human-dominated or working landscapes. Social forestry impinges on biodiversity in working landscapes in at least three ways. First, the intensification of social forestry systems can reduce exploitation of nearby or even distant protected areas (Murniati *et al.*, 2001; Garrity *et al.*, 2003). Second, the expansion of Social forestry systems can increase biodiversity in working landscapes. And third, social forestry and agroforestry development may increase the species and within-species diversity of trees in farming systems.

Social Forestry and Sustainable Development

The well being of the land is directly tied to the well being of its habitants. Only when rural people and poor framers have a way to earn sustainable, stable livelihoods will the planet's biodiversity be safe. It is not futile to attempt to conserve tropical forests without addressing the needs of poor local people, nor is it desirable.

The world agroforestry Centre (ICRAF) has identified seven key challenges related to the Millennium Development Goals that social forestry and agroforestry science and practice can materially address:

1. Help eradicate hunger through basic, pro-poor food production systems in disadvantaged areas based on social forestry methods of soil fertility and land regeneration;

2. Lift more rural poor from poverty through market-driven, locally led tree cultivation systems that generate income and build assets;

3. Advance the health and nutrition of the rural poor through social forestry and agroforestry systems;

4. Conserve biodiversity through integrated conservation-development solutions based on social forestry and agroforestry technologies, innovative institutions and better policies;

5. Protect watershed services through social forestry based solutions that enable the poor to be rewarded for their provision of these services;

6. Assist the rural poor to better adapt to climate change, and to benefit from emerging carbon markets, through tree cultivation; and

7. Build human and institutional capacity in social and agroforestry research and development.

Conclusion

As in introduction it is discussed that major biodiversity loss is due to climate change which show the importance of rapid implementation of technologies to decrease greenhouse gas emissions and strategies for carbon sequestration. Hence in different options of carbon sequestration Social forestry can work as a tool for all the three objective simultaneously *i.e.,* carbon sequestration, sustainable development and conserving biological diversity. Carbon can be sequestrated in large amount by different landuse management in Social forestry which in turn conserve biodiversity of standing forest, creating new habitat and restoring natural biodiversity to large extent and play key role in achieving millenium development goals.

References

Cairns, Michael A. and Meganck, Richard A., 1994. *Environmental Management,* 18(1/January): 13–22.

Garrityk D.P., 2004. *Agroforestry Systems,* 61: 5–17.

IPCC, 2002. *IPCC Technical Paper* V.

Pandey and Narayan, Deep, 2002. *Climate Policy,* 2(4): 367–377.

2013, Biodiversity Conservation for Sustainable Management   *Pages 84–87*
Editor: Dr. K. Muthuchelian, *Vice Chancellor, Periyar University, Salem*
Published by: Daya Publishing House, NEW DELHI

Chapter 10

Role of Nurseries in Stable Source of Income for Rural People and Biodiversity Conservation

E. Mohan* and K. Rajendran

Post Graduate and Research Department of Botany, Thiagarajar College,
Madurai–625 009, Tamil Nadu

ABSTRACT

Forests are considered as main source of rural employment in Western Ghats regions of Southern parts of Tamil Nadu. The various activities such as development of plantations, nurseries, maintenance of forests, soil and water improvement works etc., are labour oriented and generate income for many people in rural areas. In addition to this, collection of fuel wood, fodder and medicinal plants etc., also generate sustainable income in rural sectors. Besides this, there is tremendous potential of gainful employment generations in rural areas by developing forest based activities such as establishment of nurseries to supplement the planting stock requirement to various agencies.

Survey on village and forest nurseries in Western Ghats regions of Southern parts of Tamil Nadu shows that there are different type of tree species being developed by the Government, NGOs and private sectors for the large scale cultivation. Among all the species *Casuarina equisetifolia, C. junghuhniana, Eucalyptus tereticornis, Acacia auriculiformis, A. nilotica, Terminalia arjuna, Mimusops elengi, Bauhinia variegata, Thespesia populnia, Pongamia pinnata,*

* Corresponding Author: E-mail: easmohan@yahoo.co.in

Cassia siamia, Jatropha curcas, Azadirachta indica, Syzygium cumini are predominantly developed in Western Ghats regions of Southern parts of Tamil Nadu. The above species are not only giving income for the rural people but also conserve the biodiversity of living being.

Keywords: *Nurseries, Income generation, Western Ghats, Bioconservation.*

Introduction

The survival and well being of a nation depends on sustainable development. It is the process of social and economic betterment that satisfies the needs and value of all interest groups without foreclosing future options. Plant nurseries are one of the essential to economic development and the maintenance of all forms of life. For the vast majority of our rural people, the income need is very urgent for their survival. Self employment is defined as an individual or family unit, which derives income from various sources among them, rearing collection, processing, maintenance and marketing of seedlings of different plants of both wild and domesticated. Since ancient time, mankind depended mainly on the plant kingdom to meet its need for medicine, fragrance and flavours. Indian subcontinent is blessed with varied and diverse soil and climatic conditions, which are suitable for the growth of almost every plant species.

Nursery farm is a major subsistence as well as economic activity of poor people. Rural people not only depend on wild plants as sources of food, medicine, fodder and fuel, but have also developed methods of resource management, which may be fundamental to the conservation of some of the world's important habitats (Gemedo-Dalle *et al.*, 2005).

The main aim of the present study is the survey of the nursery plants of both wild and to domesticated for sustainable utilization of plant resources and to provide data for conservation aspects. It is also aimed to encourage the farmers as well as peoples those who are an unemployment conditions to go for conservation of plants of suitable wild and local species in this agro-climatic region through the nursery management. As such survey and observations were carried out in Madurai and Coimbatore districts.

Materials and Methods

Description of the Study Area

Survey and information collected from various nurseries located in different altitude of Madurai and Coimbatore Districts. Madurai district lies between 9° 39'–10° 30' N latitude and 77° 00' E–78° 30' E longitude. The district receives an annual rainfall is about 600–850 mm. The maximum and minimum temperature varies between 18° and 40°. The Coimbatore district is located between 11° 02' N latitude and 76° 58' E longitude at an elevation of 409 m above the sea level and the site receives the annual precipitation averages 450-650 mm. The maximum and minimum temperature varies between 14° and 36°C.

Methodology

Survey was conducted and observation made to collect information regarding the multipurpose usable wild tree species which are being cultivated for various purposes in nurseries located in both rural and urban areas of the above mentioned districts. The nomenclature of the plants recorded was identified with the help of the Flora of Tamil Nadu Vol. I, (Nair and Henry, 1983), Flora of Tamil Nadu, Vol. II–III (Henry *et al.,* 1987; 1989) and An Excursion Flora of Central Tamil Nadu, India (Matthew, 1991).

Results and Discussion

The present survey revealed that there are 21 multipurpose usable plant species, belonging to 18 genera spreading 14 families and 17 species of 12 genera belonging to 11 families are being commonly cultivated in both rural and urban nurseries of Madurai and Coimbatore district respectively. 8 plant species (*Acacia nilotica* L., *Azadirachta indica* A.Juss., *Cassia siamea* Lam., *Ceiba pentandra* (L). Geartn., *Pongamia pinnata* L., *Tectona grandis* L. f., *Terminalia anjuna* Roxb. ex DC. and *Terminalia chebula* Retz). are being under cultivation in the nurseries of both the districts surveyed.

Table 10.1: List of plant species being cultivated in the nurseries of Madurai district.

Sl. No.	Botanical Name	Family
1.	*Acacia nilotica* L.	Mimosaceae
2.	*Azadirachta indica* A. Juss.	Meliaceae
3.	*Bauhinia racemosa* Lam.	Caesalpiniaceae
4.	*Butea monosperma* Lam.	Fabaceae
5.	*Caesalpinia coriaria* Jacq.	Caesalpiniaceae
6.	*Cassia siamea* Lam.	Caesalpiniaceae
7.	*Ceiba pentandra* (L). Geartn.	Bombacaceae
8.	*Crataeva religiosa* Dc	Capparidaceae
9.	*Ficus amplissima* J.E.Smith	Moraceae
10.	*Ficus recemosa* L.	Moraceae
11.	*Gmelina arborea* Roxb.	Verbenaceae
12.	*Hibiscus tiliaceus* L.	Malvaceae
13.	*Madhuca longifolia* Koen.	Sapotaceae
14.	*Mimusoaps elengi* L.	Sapotaceae
15.	*Pongamia pinnata* L.	Fabaceae
16.	*Syzygium cumini* L.	Myrtaceae
17.	*Tectona grandis* L. f.	Verbenaceae
18.	*Terminalia arjuna* Roxb. ex DC.	Combretaceae
19.	*Terminalia catappa* L.	Combretaceae
20.	*Terminalia chebula* Retz.	Combretaceae
21.	*Thespesia populnea* L.	Malvaceae

By cultivating these multipurpose usable plants through nurseries, the particular group of people those who are involving in this nursery development gets stable source of income to meet their daily life needs for generation to generation. Most of the recorded plants species are presently occur in wild condition. But nowadays they are being cultivated domestically through nurseries, by which the gene pool of a particular plant species being conserved, even though they may become extinct from their wild nature in future. Therefore, training programmes on proper management and financial support should be given to encourage the people to develop the nursery techniques which lead to the conservation of particular plant to be cultivated.

Table 10.2: List of plant species being cultivated in the nurseries of Coimbatore district.

Sl. No.	Botanical Name	Family
1.	*Acacia auriculiformis* A. Cunn.	Mimosaceae
2.	*Acacia leucophloea* (Roxb). Willd.	Mimosaceae
3.	*Acacia nilotica* L.	Mimosaceae
4.	*Azadirachta indica* A.Juss.	Meliaceae
5.	*Bambusa aurundinacea* (Retz). Roxb.	Poaceae
6.	*Cassia siamea* Lam.	Caesalpiniaceae
7.	*Casuarina equisetifolia* L. Diss.	Casurinaceae
8.	*Casuarina junghuniana*	Casuarinaceae
9.	*Ceiba pentandra* (L). Geartn.	Bombacaceae
10.	*Emblica officianalis* L.	Euphorbiaceae
11.	*Eucalyptus tereticornis* Sm.	Myrtaceae
12.	*Jatropha curcas* L.	Euphorbiaceae
13.	*Pongamia pinnata* L.	Fabaceae
14.	*Tectona grandis* L. f.	Verbenaceae
15.	*Terminalia arjuna* Roxb. ex DC.	Combretaceae
16.	*Terminalia bellirica* (Geartn). Roxb.	Combretaceae
17.	*Terminalia chebula* Retz.	Combretaceae

References

Gmedo-Dalle, T., Maass, B.L. and Isselsten, J., 2005. Plant biodiversity and ethnobotany of *Borana pastoralists* in Southern Oromia, Ethiopia. *Economic Botany*, 59: 43–65.

Henry, A.N., Chitra, V. and Balakrishnan, N.P., 1989. *Flora of Tamil Nadu, India*, Series–I, Analysis Volume III, Botanical Survey of India, Southern Circle, Coimbatore.

Henry, A.N., Kumari, G.R. and Chitra, V., 1987. *Flora of Tamil Nadu, India*, Series–I, Analysis Volume II, Botanical Survey of India, Southern Circle, Coimbatore.

Matthew, K.M., 1991. *An Excursion Flora of Central Tamil Nadu.* Oxford & IBH Publishing Co. Pvt. Ltd., New Delhi.

Nair, N.C. and Henry, A.N., 1983. *Flora of Tamil Nadu, India*, Series–I, Analysis volume I, Botanical Survey of India, Southern Circle, Coimbatore.

2013, Biodiversity Conservation for Sustainable Management   Pages 88–94
Editor: Dr. K. Muthuchelian, Vice Chancellor, Periyar University, Salem
Published by: Daya Publishing House, NEW DELHI

Chapter 11

Joint Forest Management: A Participatory Approach in Biodiversity Conservation– A Case Study from Tirunelveli District, Tamil Nadu

R. Sankaravadaiammal[1] and Kailash Paliwal[2]
[1]*Department of Botany, Sri Parasakthi College for Women,*
Courtallam, Tirunelveli District, Tamil Nadu
[2]*Indian Institute of Advance Research, Gandhi Nagar, Gujarat*

Introduction

Joint Forest Management (JFM) is a participatory forest management strategy adopted in India for restoration, development, protection and conservation of degraded forests. National Forest Policy, 1988, JFM resolution 1990 and JFM guidelines 2000 are oriented towards decentralized approach to natural resource management, marking a paradigm shift from 'Departmental Policing' to 'Social Fencing'. Initiation of JFM calls for basic institutional reforms in the forest bureaucracy and changes in the mindset of foresters, local communities and NGOs, the stakeholders in forest management. JFM operates on the concepts of institutionalization and capacity building, organizing the villagers into Village Forest Councils (VFCs) and training the forest dependents for self employment and extending financial assistance to select income generating activities. Hence, the success of JFM programmes depends on the effective functioning of VFCs.

This chapter analyses the role of the community in forest management, through a study on villages adopted under two JFM programmes, namely, Tamil Nadu Afforestation Project (TAP) and Eco Development Project (EDP).

Study Area

The study was carried out in Punnaiahpuram–Dharmapuram, Singilipatty and Karkudi–Kudiyiruppu adopted under TAP and in Forest Bungalow–Kudiyiruppu, Kalyanipuram and Sambangulam adopted under EDP.

Primary data were collected from members of VFCs of the study villages by personal interview by the investigator based on structured schedules. Secondary data were collected from JFM records maintained in the forest range offices. Sampling technique was applied for collecting primary data and the sample size was fixed as 40 households for each village, except Forest Bungalow Kudiyiruppu for which the sample size was 35, for three reasons, no record of VFC membership, the fluctuating number of VFC membership and the small size of the particular village.

The schedules had been designed so as to extract information on the socio economic background of the respondents and the 35 attributes identified as related to the functioning of VFC.

The schedules had been designed so as to extract information on the socio economic background of the respondents and the attributes related to the functioning of VFC. 35 attributes were identified. Data were analysed using Chi square test for correlation of attributes using SPSS–12.

Observations and Discussion

Functioning of VFCs was assessed in terms of:

☆ VFC membership
☆ Attendance in VFC meetings and participation in discussions
☆ Involvement of the local community
☆ Forest department–Local community relationship
☆ Transparency in the implementation of the programme
☆ Empowerment of VFC

VFC Membership

Social, economic, political, cultural and ecological variables influence JFM of all, the pecuniary benefits are the single most propelling factors to seek community collaboration in forest management. As evident from Table 11.1, only a small percentage of the village population has been enrolled as VFC members.

Moreover, certain activities of the forest department have made the programme project –oriented rather than people–oriented. Microplan prepared for Dharmapuram, a hamlet with a population of 500, almost all of them being forest dependent for their livelihood, extended to Punnaiahpuram without appropriate modifications in the microplan, treating Karaikudi and Kudiyiruppu, two villages geographically separated by a distance of 15 kms as a single unit are a few such activities.

Table 11.1

Village	VFC Membership	Per cent of Population
Forest Bungalow–Kudiyiruppu	180	43.27
Kalyanipuram	291	25.98
Sambangulam	392	24.08
Punnaiahpuram–Dharmapuram	156	5.0
Singilipatty	250	9.5
Karaikudi–Kudiyiruppu	566	25.4

Attendance in VFC Meetings and Participation in Discussions

Participation is democratic empowerment. In general the attendance in the VFC meetings was not encouraging. In the TAP villages only 22.5 per cent of the respondents attended the meetings regularly. Forest dependents were more regular than the forest non dependents. Caste factor to some extent decided the attendance of the respondents in the VFC meetings. Educational level of the male members enhanced their attendance while the attendance of the female members was decided by factors other than their educational level, probably socio cultural. Even the literate women did not take part in the discussions. In general the attendance pattern and the extent of participation show that the involvement of the local community in the implementation of the programme is not satisfactory. Regular participation of the forest officials in the meetings also seems to have impact to certain extent.

In the Eco villages, the attendance of women in the VFC meetings was satisfactory. Interestingly in Sambangulam, the membership and attendance of women in the VFC meetings was slightly higher than that of men. It may be because many of the men of this village are employed either outside the state or outside the country and they come to their village only on occasions.

The performance of the VFCs seemed to be unaffected by the caste of the members or the caste structure of the community. Invariably in all the villages studied the VFC members of the scheduled caste did not attend the meetings regularly. Probably the non-interfering nature of the caste factor is due to the local arrangements made by the local community in consultation with the forest department to keep the caste factor away from VFC functioning. This may seem to be an intelligent and effective strategy for successful implementation of the project in the village but it amounts to denial of justice to the poor and the downtrodden.

Involvement of the Local Community

The realization by the local people that forest protection is the joint responsibility of the villagers and the forest department and the mutual trust of these two stakeholders help the project to take off; but there are many more factors such as, awareness of the programme on the part of the local villagers, the attitude of the foresters towards the local community, livelihood security to the forest dependents and availability of alternative fuels that decide the success of the programme by making the villagers involve themselves in forest protection.

The continued pilferage of wood from the forest areas reported in the present study indicates the lack of involvement of the local community. This is the major problem in many JFM areas. Moreover lack of incentives to the EC chairman also was an impediment for VFC functioning. In one of the study villages, Karaikudi –Kudiyiruppu, the VFC chairman was getting remuneration for recovering the loan from the beneficiaries and majority of the respondents supported such an incentive to the EC members.

Forest Department: Local Community Relationship

Initiation and implementation of JFM is historical making the beginning of a revolutionary change in the forestry sector. It demands an attitudinal change in the foresters towards the local community and the former has to win the confidence of the latter and to develop a rapport with them. The changing institutional framework and the mindset of all the stakeholders is one of the factors to be considered in formulating strategy for JFM programme and earlier experience has shown that improved relationship between the foresters and the people guarantees the success of the programme.

Of the various variables studied, none had any impact on the relationship between the foresters and the local people. But in the opinion of the majority of the respondents of Eco villages and about 50 per cent of the respondents from TAP villages the relationship with the forest department had improved following the initiation of JFM in their villages. At the same time one can not ignore the finding of the reviews on JFM programmes that foresters are less convinced regarding the various methods of involving people and are of firm belief that more and more involvement of people in field like scientific forestry will become the cause for the ruin of forestry. Nevertheless, enormous efforts put in this direction will have its impact on the success of the programme. Posting of forest staff with right attitude and aptitude for a reasonable time, involvement of people's representatives such as MLAs and MPs and other local leaders and formation of coordination committees will help improve the relationship of the forest department with the local community.

Transparency in Implementation of the Programme

Micro plans are integral part of JFM and as per JFM guidelines they should be prepared by the forest officers and the Village Forest Protection Committees. Generally the foresters dominate the preparation and the implementation of the micro plans and micro planning exercise becomes a source of manipulation. The present study also supports this. Majority of t he VFC members from Eco villages as well as TAP villages had no idea on the micro plan. Moreover the local community was not involved in the construction of structures like check dams and percolation ponds, as measures of soil and water conservation. Such a lack of transparency has obviously resulted in the low level of awareness and the lack of feeling of ownership of the JFM initiatives and the institution on the part of the villagers. But there are cases quoted as success stories of open mindedness of the foresters from states of Gujarat, Andhra Pradesh, Central Himalayas Madhya Pradesh etc.

Empowerment of VFC

As per JFM guidelines, 2000 and order from Government of Tamil Nadu, all the VFCs have been registered under Tamil Nadu Societies Registration Act, 1975. Sections 4.10 and 4.11 of the Tamil Nadu JFM guidelines empower the Executive Committee to impose fines against individuals/hamlets for illicit removal of firewood or illicit grazing of cattle and to expel any of its members or member of VFC after providing an opportunity of defence. But in practice as the majority of the respondents feel the VFC/EC does not have the punitive power against forest offences. At the same time it interesting to know that the VFC of Forest Bungalow Kudiyiruppu imposed a penal interst of Rs.25 per day for the delayed repayment of loan amount. JFM experience across the country has shown that VFCs are only functional bodies without constitutional sanctity.

Even though the present study could identify a few positive impacts of the JFM programmes such as, realization of the responsibility in forest protection by the local community, a marked decline in the recorded forest offences a marked decline in the number of livestock in the study villages and identification of a good number of income generating activities by the forest department and extending financial assistance to the same, it identified some grey areas too. Attendance in VFC meetings, poor women participation in the project and the role of the caste of the members or the caste structure of the village are those areas identified. From the observations made during the study it quite obvious that the JFM programmes studied have to go a long way so as to achieve the target.

The following are a few recommendations to the forest department based on the observations made during the study.

☆ A coordination committee may be formed in every project village and the members of the same should be involved in every stage of implementation of the project.

☆ VFC meetings should be held regularly.

☆ The local arrangements to keep the caste factor away from the project should be avoided.

☆ The financial assistance to the forest dependents should be equitable and the income generating activities should be monitored by the forest department.

☆ Vocational training to the individuals as well as the Self Help Groups will help the forest dependents to take up new enterprises hopefully and fruitfully.

☆ The VFC or at least the EC should be empowered to inspect the target forest areas periodically.

☆ Federation of VFCs at the district level will expand and strengthen the concept of institutionalization of the local people and involve them to a greater extent on the targeted task of forest protection.

Acknowledgements

The authors wish to thank the Principal Chief Conservator of Forests, Chennai for the permission accorded to carryout research in the forest areas and villages adopted under JFM and the range officers of Kadayam, Tenkasi, Kadayanallur and Courtallam for the assistance extended during the study.

References

Bhojvaid and Pawar, Rajat, 2000. JFM: The joint financial munching. In: *Proc. International Workshop on 'A Decade of Joint Forest Management: Retrospection and Intraspection.* 19, 20, June, MoEF and Ford Foundation, p. 248–252.

Gupta, H.S., 1997. A note on forest management. *The Indian Forester,* 123(6): 565–567.

Hiremath, S.R., 2000. Empowerment of rural communities and joint forest management. In: *Proc. International Workshop on 'A Decade of Joint Forest Management: Retrospection and Intraspection.* 19-20 June, MoEF and Ford Foundation, p. 106–09.

Jain, A.K., 1998. Joint forest management in Cuddapah forest division: A successful example. *The Indian Forester,* 124(7): 524–530.

Johri Anil. (2000). JFM–The economic arousal. In: *Proc. International Workshop on 'A Decade of Joint Forest Management: Retrospection and Intraspection.* 19, 20, June, MoEF and Ford Foundation, p. 110–14.

Melkania, Uma and Bisht, N.S., 2000. Identifying indicators for successful implementation of joint forest management in Andhra Pradesh. *The Indian Forester,* 126(5): 537–544.

Mitra, Amit, 1997. Joint forest management: Case studies. *Yojana,* 41(8): 41–45.

Mukherji, S.D., 1997. Is handing over forests to local communities solution to deforestation? Experience in Andhra Pradesh. *The Indian Forester,* 123(6): 461–471.

Mukherji, S.D., 1998. Update on joint forest management programme in Andhra Pradesh. *The Indian Forester,* 124(6): 413–424.

Pande, I.D. and Pande, Sameer, 2000. Women's participation in joint forest management: A case study of village Parwara, Central Himalayan Region of Kumaon. In: *Proc. International Workshop on 'A Decade of Joint Forest Management: Retrospection and Intraspection.* 19, 20, June, MoEF and Ford Foundation, p. 236–239.

Ray, P.N., 2000. Status of joint forest management in Tripura. *The Indian Forester,* 126(5): 483–492.

Sankaravadaiammal, R. and Kailash Paliwal, 2008. Joint forest management– Decentralization and devolution: A case study from Kalakad Mundanthurai Tiger Reserve, Tirunelveli district, Tamil Nadu. *The Indian Forester,* 134(2): 177–189.

Shanker, Ajay, 1995. Forest resource management through people's participation in Barwani forest division, Madhya Pradesh. *The Indian Forester,* 121(5): 355–358.

Verma, D.P.S., 1989. Women and forestry (Tale of six Harijan women of Ganeshpura). *The Indian Forester,* 115(11): 780–788.

2013, Biodiversity Conservation for Sustainable Management   Pages *95–101*
Editor: **Dr. K. Muthuchelian,** Vice Chancellor, Periyar University, Salem
Published by: **Daya Publishing House, NEW DELHI**

Chapter 12

Importance of Wood Protection for Conservation of Tree Diversity in India

R. Sundararaj
Institute of Wood Science and Technology,
18th Cross Malleswaram, Bangalore – 560 003

Introduction

Wood is one of the most important renewable natural resources available to man. It is perhaps the first material used by man since time memorial when he was roaming the icy wilderness and is prevalent in our everyday lives and the economy (Bell and Rand, 2006). Until this century wood was the single greatest material aid and comfort in every century of our ancestors lives. It is the prime product of forests for diverse industrial and structural applications. Contrary to the common but incorrect perceptions, use of wood is in fact, ever increasing, more so in developing countries like India (Rao, 2002). On a worldwide scale, the industrial use of wood approximates that of cement and steel and far exceeds plastics (Schulz, 1991). Along with new uses, more and more applications of wood are being re-discovered and documented, especially its importance as an environmentally friendly material that is remarkably reusable, recyclable, biodegradable and more importantly, as a renewable natural resource. Because of the many advantages in its usage, it is put to a myriad of applications. Wood degradation is the destructive changes in the properties of wood caused by a range of biotic, physical or chemical agents. The biological reasons of wood deterioration both during the processing and in service are a matter of great concern to wood technologists, industrialists, wood users and environmentalists. The economic importance of wood losses due to decay fungi, wood-boring insects and marine borers demands the conservation of wood through better understanding of these biological agencies. Immeasurable losses of wood happen due to invasion of insects and microbes in the wood of living trees in the vast forests of the world. Other than this there are biological agencies, which cause inestimable amount of loss in the logs, timbers or wood in service. Various attempts have been made to estimate the losses in different countries. Much of the losses could be avoided by proper knowledge of the biological requirements of various deteriorating organisms, means of identifying the source of damage and employing simple precautions like proper seasoning and preservation.

Degradation of Wood: A Necessary Evil

In natural conditions since tree is a part of ecosystems, it must decompose for contributing to the supply of nutrients for other plants and animals. Wood is an organic material, which is a source of energy for wood feeding fungi and insects. But for the activities of fungi and wood eating borers and termites, the earth would have been dumped with the dead parts of trees, wood being the major junk, making the ecosystem functioning impossible and hampering the sustainable renewal of the living components of the ecosystem. However, when deterioration affects the objects, structures and products of wood made for human consumption, it is a matter of alarm and man is alerted to fight against this evil.

Biological Agents Causing Wood Biodegradation

Fungi

Fungi which are unable to produce their own food feed on organic materials including wood. They use the simple sugars obtained by the break down of wood. There are five groups of fungi harmful to wood. Mold and sapstain, decay fungi, brown rot, white rot and soft rot causes deterioration due to different group of fungi.

Insects Causing Wood Degradation

Various groups of insects, belonging to several orders, are classified as wood borers, confine their attack to various timber species under different conditions. Among these, the most important order is Coleoptera (or beetles). Wood wasps and ants belonging to Hymenoptera and termites (Isoptera) are the other major wood invaders. Timber in structures/houses is mainly attacked by insects belonging to the Coleopteran families, Platypodidae and Scolytidae (ambrosia beetles), Lyctidae, Bostrychidae, Cerambycidae and Anobiidae. Ghoon beetles/borers or bostrychids or shot hole borers particularly *Heterobostrychus aequalis, Schistoceros anoboides* and *Sinoxylon* spp. also cause extensive damage to timber in service and storage. Termites are one of the principle destroyers of wood in buildings in the tropics and the subtropics. The extent of financial loss due to termite damage has not been computed precisely due to the difficulties associated with such computations. Edwards and Mill (1986) estimated that US$ 1,920,000 was spent worldwide per annum by the pest control operations for the treatment of buildings against termites. In India itself, the cost of treatments as preventive and remedial measures comes to about Rs. 28 million annually (Rawat, 2004). Damage to constructional timbers and timber products in buildings can be caused either by drywood termites which do not maintain soil connection for moisture requirements or by the soil dwelling termites that obligatorily maintain soil connection for moisture requirements. However *Coptotermes* spp. are known to survive without soil connection provided moisture source is available near their colony sites. The soil dwelling termites cause the maximum infestation and damage in India and other tropical countries.

Marine Wood Biodegradation

Degradation of marine timber by biological agencies is an age old problem that has descended upon mankind from the very day man ventured out in the sea on a primitive wooden craft and a problem that has defied a solution ever since. Marine wood borers in general fall under two main classes, viz, Bivalvia of Mollusca and Crustacea of Arthropoda.

Wood Protection

Naturally durable timbers account for only 10 per cent of the total volume of wood used in the industrial sector (Purushotham, 1975). Presence of various chemical compounds and lignin are the factors which account for the natural resistance of timbers (Walcott, 1946 and 1947; Abushama and Abdel Nur, 1973 and Behr *et al*, 1972). Wood protection or preservation is an art through which wood can be protected against degrading agencies and its life increased 5 to 8 times depending upon the adversity of condition of use. Wood as a material is very durable and in sheltered environments it can last for thousands of years without sustainable changes. The saying "an ounce of prevention is better than a pound of cure" is most befitting in the case of management of wood biodeterioration. The methods of management are mainly preventive. The measures can be divided into two categories.

Natural or Non-chemical Measures

Safe Felling Period

The safest felling period is during winter(November to March) in North India and immediately after rains in South India when most of the coleopterans donot remain on wings or are in hibernation.

Debarking

Females of long horn borers oviposit in the bark. Debarking soon after felling along with the use of endcoating of antiseptic, antisplitting compositions with coal tar, paraffin wax etc., will prevent oviposition by these beetles.

Starch Depletion

Presence of starch is compulsory for the infestation by powder post beetles. Amount of starch in sap wood depends on the species of tree, felling season, and treatment of logs after felling. High girdling of trees before felling reduces the starch content. Immersion of logs in fresh water for varying periods or continued sprinkling of water on the logs results in the depletion of starch.

Storage

Storage in the open is preferable to storage under shade because it reduces pin hole borer attack. Proper hygiene in the storage areas is essential for the prevention of borer attacks.

Seasoning of Wood

The process of seasoning is essentially one of removing moisture from timber. Seasoning as applied to wood is primarily a drying process. Freshly felled timber contains large quantity of moisture, the major portion of which has to be removed before the timber is fit for use for most purposes. Seasoning reduces the likelihood of mould, stain or decay; reduces weight thereby reduces transportation and handling costs; as wood dries most of its strength properties improve; strength of joints made with nails and screws is greater in dried wood than in green; and also improves finishing of wood. There are two commercially important methods of seasoning wood practiced in our country, *viz.*, air seasoning (natural seasoning) and kiln drying (artificial seasoning).

Prophylactic Chemical Treatments

Prophylactic treatment can be given either by spraying contact insecticides or by dipping in wood protecting solutions. In rainy seasons it is advisable to incorporate a compatible fixative to prevent washing off of the chemical. Microemulsions of pyrethroid insecticides give protection from a wide range of borers and termites. Tree poisoning is a method where the tree to be felled is treated *insitu* with water soluble insecticides and fungicides using simple pressure injection techniques. Fumigation is resorted to eradicate the existing infestation and is no cover against future infestations. Borates have low toxicity to man and are approved for interior uses of wood treatment. They protect attacks by both fungi and insects. They are the preservatives of choice for remedial treatments of wood in service.

Wood Preservatives

These are chemical substances, which are applied to wood to make it resistant to attack against decaying agents. Coar or Creosote, Copper and Zince Naphthenates/ Abietates, Pentachlorophenol, Benzene Hexachloride, DDT, Copper-Chrome-Arsenic (CCA), Acid Cupric-Chromate (ACC), Chromated-Zinc Chloride, Copper-Chrome-Bornon are some of the important preservatives. Treatment methods include both Non-Pressure process and Pressure process. Non-Pressure process in dry condition are by surface application and soaking treatment and in dry condition are by sap displacement method, Boucherie Process and Diffusion Process. In Pressure process Full cell process and Empty cell process are used. The economic value of wood preservation is convincingly demonstrated by the extended service life that pressure treatment with creosote or multisalts can impart to the original timber. Untreated railway sleepers may last 5 to 6 years; treated sleepers, 20 to 25 years; untreated transmission poles may stand for about 4 to 6 years; properly impregnated poles for 30 years; untreated mining timber may last 2 to 3 years; treated timber up to 20 years or more; untreated timber for cooling towers may last for an average of 10 years; suitably pressure treated timber for 25 to 30 years (Swiderski, 1968).

Surface Protection of Wood

This involves chemical modification in the surface of the wood for improving dimensional stability and photostability in wood. Modified wood has superior resistance to UV degradation, improved dimensional stability, biological resistance and paint retention. It also helps to enhance the life of coated wood by chemical pretreatment

Role of Wood Protection for Conservation of Tree Diversity

India is rich in biodiversity and is one of the 12 mega biodiversity centres of the world. It occupies 2.5 per cent of the world's land area and 1.8 per cent of global forest area. It supports more than 15 per cent of the human and 14 per cent of the cattle population and therefore forests in India are under immense biotic pressure. The present forest cover is 63.7 m ha which is 19.4 per cent) of the total land area. India is essentially a biodiversity dependent country and the wood industry fulfils several key needs of the society (Rao, 2002). Its forests have been exploited for millennia and intensively during the past two centuries (Meena *et al.*, 2007). The NFAP (National Forestry Action Programme, India, Ministry of Environment and Forests, Government of India, 1998) projects the annual requirement of timber in India for housing, furniture, agriculture implements, and industrial uses to be about 64 million m^3 against a supply of 43 million m^3 (12 million m^3 from natural forests + 31 million m^3 from farm forestry and other sources). For fuel wood, which continues to be the major source of energy for rural India the projected requirement is around 201 million tonnes against a supply of only 95 million tonnes (17 million tonnes from forests + 78 million tonnes from farm forestry and other sources). Despite being bestowed with over 4000 woody species, the country is a "timber deficient country" and the gap between supply and demand is ever widening (Shashidhar and Aggarwal, 2006). There are acute shortages and phenomenal hikes in prices of conventionally preferred species of timbers like Teak, Sal, Deodar, Rosewood, Red sanders, Sandal etc. The scenario has now become quite alarming as the productivity gap in wood biomass is far beyond the current productivity in natural forests and the expected productivity from the plantation forestry. Adoption of wood preservation technologies has the potential to save approximately 2 m^3 of wood for every 3 m^3 and because of increase in average life of timber from 5 to 15 years would saving at least 5.6 m ha of forests raising the forest cover by 2 per cent in India (Kumar, 1991). The advancements in wood protection encourage the use of secondary/fast growing/plantation timbers and these industrial plantations gradually displacing wood from natural forests although the main purpose of growing such species is meant earlier to meet pulpwood demands (Kumar *et al.*, 2004). The tree species like *Acacia auriculiformis, A. mangium, Maesopsis eminii, Eucalyptus camaldulensis, E. tereticornis, Leucaena leucocephala, Gyrocarpus americanus* and *Swietenia mahagoni* are being utilized in furniture industry. Rubber wood is an example which is traditionally used as fuel wood finds its application in furniture industry. Prophylactic treatments with scientific storage, conversion, seasoning and preservative treatment of rubber wood alone meets India's timber requirement to as high as 2 per cent. It has the potential to conserve more than 20000 ha of rain forests. The recent development of wood polymer composites which are produced by mixing wood flour or fibre and plastics in extruders to produce a material that can be processed like a conventional plastic and has the best features of wood and plastic. Wood polymer composites find application in exterior building components, railings, window profile, shutters, building profiles, automobile interiors, door panel, dashboard, trims, etc., and also in moulded products. Hence production and proper use of wood form a very important part of the global level effort to conserve the tree diversity of the earth, while achieving their most economic utilization.

Conclusion

The destruction of the world's forests is a major concern in our age. A whole range of environmental problems is associated with deforestation. (Perlin, 1989). In view of the dwindling natural forests and the logging ban in force, many countries including India have to depend more and more on "Agro-Wood" and imported wood to meet its requirements. India is importing wood and forest products ton the tune of approximately US$1,028 million every year (FAO, 2000). Further more and more species, hitherto under-utilized, have now become necessary to be put to use and upgraded to suit different end users. The adoption of a new "National Wood Use Policy" is being advocated at many forums (Rao, 2002). To achieve this, wood science/ technology interventions have now come to become very necessary. The Wood Industry as well will also be confronted with this challenge. We owe a great responsibility to our future generations and have to address this problem with a tangible and longstanding solution. As part of our commitment to intergenerational equity and Conservation of Biodiversity, we have to give a serious thought for an amicable solution to these problems. The World Commission on Forests and Sustainable Development in its report (1999) concluded, "forests can no longer be used in the same way as they have been in the past, forest products and services must be assured through new political choices and policy decisions that ensures the survival of forests". A report of IUCN Working Group states that the challenge in the 21st Century is forest resource conservation and sustainable use (Graham Bennett, 2004). And hence more conservation oriented progressive forest policies for greening India is required for stabilization of carbon stocks in our forests and their tree diversity. These approaches need to adept and change with the prevailing social and economic realities of our world on local, national and global levels.

References

Abushama, F.T. and Nur, H.O. Abdel, 1973. Damage inflicted on wood by the termite *Psammotermes hybostoma* Desneux in Khartoum District, Sudan, and measures against them. *Z. Angew. Entomol.*, 73: 216–223.

Behr, E.A., Behr, L.F. and Wilson, L.F., 1972. Influence of wood hardness on feeding by the eastern subterranean termite, *R. flavipes* (Isoptera: Rhinotermitidae). *Ann. Entomol. Soc. Erica*, 65: 457–460.

Bell, V.B. and Rand, P., 2006. *Materials for Design.* Princeton Architectural Press, New York, 270 pp.

Edwards, R. and Mill, A.E., 1986. *Termites in Buildings: Their Biology and Control.* Rentokil Limited, East Grinstead, United Kingdom.

FAO, 2000. *FAOSTAT Database.* Food and Agricultural Organization of the United Nations. Website http://apps.fao.org.

Graham, Bennett, 2004. *Integrating Biodiversity Conservation and Sustainable Use: Lessons Learned from Ecological Networks.* IUCN, Gland, Switzerland, and Cambridge, UK, 55 pp.

Kumar, P., Rao, R.V. and Sudheendra, R., 2004. Strength properties of 6 year old plantation grown *Acacia crassicarpa* and *A. mangium* under irrigation. *J. Indian Acad. Wood Sci. (N.S)*, 1(1 and 2): 10–17.

Kumar, S., 1991. Preservative: An economical way to extend wood resource. *Wood News*, 2: 15–16.

Meena, P., Nehara, S. and Trivedi, P.C., 2007. Biodiversity profile of India. In: *Global Biodiversity: Status and Conservation,* (Ed.) P.C Tivedi. Pointer Publishers, Jaipur, India, p. 42–127.

Perlin, John, 1989. A forest journey. In: *The Role of Wood in the Development of Civilization.* Cambridge, MA, London, p. 15.

Purushotham, A., 1975. *Biological Deterioration of Wood in Storage and Use.* Report No. FOR–IUFRO/DI/75/6–15.

Rao, K.S., 2002. Current trends of wood use: An Indian perspective. *Wood News*, 12(2): 12–15.

Rawat, B.S., 2004. Termite control in buildings: Indian scenario. *Pestology*, 28(4): 11–23.

Schultz, H., 1991. The development of wood utilization in the 19th, 20th and 21st centuries. *Forestry Chronicle*, 69(4): 413–418.

Shashidhar, K.S. and Aggarwal, P.K., 2006. Utilization of plantation grown timbers: Future needs. *Plywood digests*, July–August: 5–8.

Swiderski, J., 1968. The importance of wood protection in tropical countries. *Unasylva*, 22(3): 12.

Walcott, G.N., 1946. Factors in the natural resistance of wood to termites attack. *Caribbean Forest*, 7: 121–134.

Walcott, G.N., 1947. The permanence of termite repellents. *J. Econ. Ent., Menasha*, 40(1): 124–129.

World Commission on Forests and Sustainable Development, 1999. Final report. IISD Publication centre, 37 pp.

2013, Biodiversity Conservation for Sustainable Management   Pages *102–108*

Editor: Dr. K. Muthuchelian, *Vice Chancellor, Periyar University, Salem*

Published by: Daya Publishing House, NEW DELHI

Chapter 13

Strategies for Conservation and Management of Sacred Groves in Kerala

U.M. Chandrashekara, B.S. Corrie, R.S. Neethu, Bincy K. Jose and E.C. Baiju

Kerala Forest Research Institute Sub Centre,
Nilambur – 679 342, Kerala

ABSTRACT

Community Conserved Biodiverse Areas (CCBAs) are those areas where the biological diversity is conserved with the involvement of local communities, in decision–making. In India, sacred groves are prominent among different forms of CCBAs. These sacred groves represent a tradition of nature worship by dedicating patches of forest to ancestral spirits/deities and providing protection to such forest patches. As a result, these sacred groves continue to be home to a wide range of rare, endemic and threatened flora and fauna. Apart from supporting rich biological diversity, the sacred groves provide ecological services in the rural landscape. However, due to global change, many sacred groves are now threatened and altered in terms of size, vegetation structure and species composition. When the ecological and socio-cultural dimensions of sacred groves are considered it is important to identify and adopt sacred grove management practices. In this context, a study has been conducted in 28 selected sacred groves in Kerala belonging to the Devaswom Board and Temple Trusts. Field visits and stakeholders' meetings conducted in these sacred groves led to the identification of 12 major threats to sacred groves; among that dumpage of solid waste materials, trespassing, illegal collection and removal of small fallen timbers and other forest products are prominent. Altogether 26 management strategies were recognized for the conservation and protection of these sacred groves. Even though the social barrier is more appropriate, the study revealed that in the present day socio-cultural context, physical barriers such as fencing and compound wall are needed to protect sacred groves till the attitude of stakeholders towards sacred groves becomes positive. Other major management strategies identified for the sacred groves include enrichment planting with species characteristic to sacred groves, management of exotics weeds, and protection of water bodies and development of soil and water retention structures. In the present chapter, ways and means to adopt these management strategies are highlighted.

Introduction

Sacred groves represent patches of forests protected by assigning them as the abode of Gods and Goddesses. In India, in spite of very high land to man ratio, sacred groves have survived under a variety of ecological situations. Sacred groves are considered as one of the land use systems with ecological and socio-cultural importance in the region. Well conserved sacred groves may be compared with the regional natural forests for various ecological attributes. Very often, in a given region, the stand quality of sacred groves is better than that of several natural forests. Sacred groves are also regarded as the treasure houses of rare and endemic species. For example, analysis of the phytogeographical elements of sacred groves of Kerala indicated that out of 721 species (including *Gnetum ula*) recorded from sacred groves, 154 are endemic to Western Ghats and 33 per cent of them are trees. Sacred groves often serve as the last refuge for many wild plants and animals. The role of well managed sacred groves as the gene-pool gardens for *in-situ* conservation of genetic resources has been appreciated. In fact, sacred groves can also be considered as gene-banks of several economically important plants.

Being the landscape unit in a rural landscape, the sacred grove performs several ecological functions, and the preservation of these unique landscape units is of key importance which directly or indirectly can help to maintain the biodiversity as well as comprehensive ecosystem health of all other interacting landscape units.

At present in Kerala the erosion of sacred groves takes place in different ways, due to the socio-economic conditions and land use system, many sacred groves are now threatened and altered both in terms of number, size, vegetation, structure and species composition. In this context, the Kerala Forest Department, as a part of the centrally sponsored scheme '*Intensification of Forest management*' is implementing a programme for protection and conservation of sacred groves of Kerala. In this chapter we discuss the management strategies for the conservation of sacred groves in Kerala.

Methodology

The Chief Conservator of Forests of Biodiversity Cell [CCF (BDC)], Kerala Forest Department has requested the Assistant Conservator of Forests (Social Forestry) of the particular division to prepare a detailed proposal for the Management of the Sacred Groves. And KFRI has also been asked to prepare the management plans as a participatory programme through PRA (Participatory Appraisal) exercises with full involvement of local community, committee members and Forest Department officials.

Thus, 28 Sacred Groves were selected in different parts of Kerala which is coming under the Devaswoms/Temple Trustees and presented Management proposals before the expert committee constituted by the CCF of Biodiversity Cell.

Study Area

In this map we can see the selected twenty eight sacred groves which have been distributed all around the fourteen districts of Kerala.

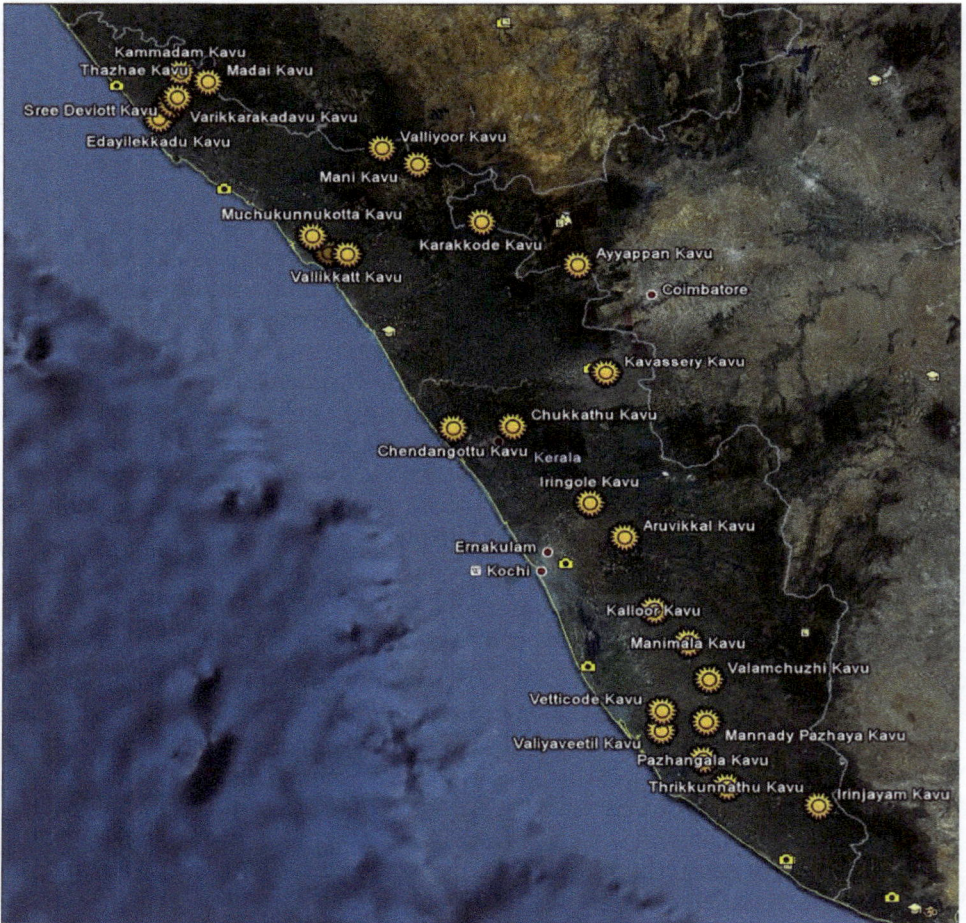

Vegetation

Mainly, four major types of forests have been found in the twenty eight sacred groves of kerala, Evergreen forest, Semi-evergreen forest, Moist deciduous forest, and Mangrove forest.

Good, thickly wooded with high land evergreen forest have been found in the kavus like Kammadam kavu, Sree Deviottu kavu, Poyilkavu, Vallikkattu kavu, Iringole kavu *etc.* At the same time highly degraded status of forest with light demanding evergreen forest and also with high infestation of exotic weeds and climbers has seen in the kavus like Karimanal kavu, Varikkarakadavu, Edayilekkavu *etc* Semi-evergreen forests with good vegetation and a variety of woody climbers have been found in the kavus like Thazhe kavu, Iringole kavu, Aruvikkal kavu, Kalloor kavu, Manimala kavu, Valiya veetil kavu, Pazhangala kavu. At the same time highly degraded semi-evergreen forests having poor regeneration and light tolerant species with high infestation of exotic weeds and climbers have seen in the kavus like Muchukunnukotta and Karakkode Kavu Moist Deciduous Forest with Good vegetation and rich in species diversity has seen in the kavus like Mani kavu, Valliyoor, Mannady, Pazhangala, Valamchuzhi, Vetticode, Thrikkunnathu and Irinjayam. And degraded with sparsely distributed vegetation have found in Madai, Kavassery, Chendangottu, Chukkath and Ayyappan Kavus. Good mangrove vegetation along with semi-evergreen has found in Thazhe kavu of Kannur District and Poor regeneration of Mangrove have seen in Kammadam Kavu and Poyilkavu.

Among the whole unique features is the presence of *Anophyllum wightii,* a tree species endemic to Western Ghats has been found in Iringyam kavu of Thiruvanthapuram district, and thazhe kavu of Kannur district could possibly the only area in the world giving home to both Mangrove and Semi-evergreen vegetation, and which also could possibly be the only sacred grove closest to the Arabian sea possessing remnants of species characteristic to the lowland semi-evergreen forests of Kerala. In this Kavu, endangered mangrove species such as *Avicennia marina, Acanthus ilicifolius* and *Cerbera manghas,* and vulnerable mangrove species such as *Rhizophora mucronata, Excoecaria agallocha,* and *Acrostichum aureum* can be seen. A fine patch of '*Myristica swamp*' is another feature of some of the groves like Kammadam kavu, Sree deviottu kavu, Poyil kavu, Vallikkattu kavu *etc.* Another remarkable feature is the luxuriant growth of pteredophytes and rare endangered ferns like *Crepidomanes christii, Crepidomanes intramarginale* and *Alsophila gigantean* have found in Kammadam kavu of Kasargod district. A large number of medicinal plants including *Nervilea infundibulifolia* and *Aporusa lindleyana* are also identified in some kavus.

Species commonly found in the sacred groves are *Hopea parviflora, Hopea ponga, Memecylon umbellatum, Hoya wightii, Strychnos vanpruki, Gymnostachium febrifugum, Myristica malabarica, Knema attenuata, Curcuma oligantha, Ixora brachiata, Plumeria rubra* Large trees such as *Lagerstroemia microcarpa, Mesua ferrea, Xylia xylocarpa, Antiaris toxicaria, Aporusa lindleyana, Cinnamomum malabathrum,* Mallotus *tetracoccus, Syzygium cumini, Vitex altissima, Pterocarpus marsupium, Persea macrantha, Holigarna arnottiana, Alstonia scholaris, Holigarna arnottiana, Ficus tsjehela, Grewia tiliaefolia, Terminalia panicualta* Some of the dominant species among are *Dalbergia latifolia, Terminalia bellerica, Terminalia paniculata, Terminalaia crenulata, Mimusops elanji, Mangifera indica, Artocarpus hirsutus, Hydnocapus pentandra, Ficus religiosa, Ficus benghalensis, Michelia chempaca, Swietenia macrophylla, Cassia fistula, Macaranga peltata, Cycas circinalis* and *Ficus exasperata.* And a variety of woody climbers like *Calicopteris floribunda, Strychnos involucra* etc are also found.

Adjoining Landscape Description

Sacred groves are surrounded by farming and non-farming units. Among the twenty eight sacred groves, most of them (15 kavu) are surrounded by farming units such as paddy fields, rubber estates plantations of coconut and arecanut and homesteads. Rest of the kavus are surrounded by non-farming units like rivers and back waters. Among that Ayyappan kavu of Palakkad district is surrounded by both the units like southern side with *Shokanashini puzha* and the other sides with paddy fields and coconut plantations. Thazhekavu of Kannur district is located in an island and is surrounded by Valapattanam River on one side and Pazhayangadi River and back waters of Arabian Sea on the other sides. Valamchuzhi kavu of Pathanamthitta district is surrounded by Achen Kovil River. In the case of Aruvikkal kavu one side is covered by coffee and rubber plantations and the adjacent patches highly degraded thus dominated by exotic weeds and light demanding native species.

Water Sources

Generally one or two water bodies are associated with almost all sacred groves; however, they are not serving the purpose. Mainly three types of water bodies are present like ponds, perennial streams/rivers and wells. The ponds found are Vadukunda pond in Madai kavu and a natural fresh water pond '*Thirukuzhi*' is present in Poyil kavu, dried recently and unable to store water for longer period. Mostly ponds are not serving its purpose due to silt accumulation and inflow of brackish water, also very pathetic due to dumpage of solid waste materials. However, by removing silt and deepening the pond, it can become functional. In some kavus like Edayilekkadu kavu the kavu committee dug out a pond for the storage of water. Presence of Streams/River is another source of water in some kavus, Ayyappan kavu, '*Shokanashini puzha*' a tributary of Chittur Puzha being a perennial rivulet provides water throughout the year; Manimala Kavu of Kottayam district, itself is situated near the bank of Manimala River. In mani kavu a spring originates from the down stream area falls on the idol of God Shiva throughout the year acts as source of water and a perennial stream in Varikkarakadavu Kavu had dried completely due to intensive harvest of trees and shrubs. Open well is another source of water and have been found in three kavus, an open well in Ayyappan kavu, two in Valliyoor kavu and a well near the temple is the source of water but it needs to be deepened and repaired in Chukkath kavu. Among the twenty eight groves water source is absent in three sacred groves like Kalloor Kavu, Mannady Pazhayakavu and Valamchuzhi kavu.

Conspicuous Changes in the Habitat since Inception

Major changes have been occurred due to direct and indirect anthropogenic interference. Indirect Anthropogenic interference includes, a large number of tall trees have been fallen down and the area has been utilized for developing temple complex/vehicle parking area. Direct anthropogenic interference causes drastic changes in the habitat due to several physical interventions, includes tremendous change in vegetation structure and composition due to trespassing, illegal collection of fallen trees and leaf litter as well as modification in land use structure. And also the utilization of land under the grove for non forestry purpose are increasing, leading to decrease in land under vegetation and even the degradation of remaining forest land. A large extent of Kavu has been transformed into other landuse, and was also under grazing pressure. Presence of exotic invasive species becoming dominant around the temple and inside the forest land also changes the habitat caused naturally. Apart from those, in fifteen kavus species characteristic to Semi-evergreen forests are established well due to the better regeneration.

Threats

Associated with faiths, taboos and beliefs, over years, local people have developed a strong affinity towards temple and the forest of Kavu. The local people in general also believe that their livelihood, security and cultural existence are complementary to the blessings of the deity of the Kavu. However, due to changes in economic and socio-cultural scenario in the region, the forested lands of the groves are facing several threats. Among the twenty eight kavus of Kerala, twelve major threats have been found. The threats have been categorized under four main groups such as Threats due to lack of awareness, lack of physical barriers, lack of legal support, and also due to natural calamities.

Four major threats have been identified due to lack of awareness, Dumpage of solid waste materials (16 kavus) as second major threat among the twenty eight kavus. Encroachment of sacred grove area (3 kavus) Settlement solid waste from the river and Collection of Pneumatophore have identified in Thazhe Kavu of Kannur district. Threats like Incidence of pouching wild animals (Kammadam), Grazing (Karakkode and Valamchuzhi Kavus), Trespassing and damaging established seedlings and poles have been found due to lack of physical barriers. Among twenty eight kavus trespassing is identified as the major threat (18 Kavus). Threats due to lack of legal support includes Incidence of illegal activities by anti-social elements (Kavasseri and Iringole Kavu) Increase in tourism activity (Iringole Kavu) Illegal collection and removal of small fallen timbers and firewood as third major threat (10Kavus) among the twenty eight. Erosion of forest fringes (Thazhe kavu and Ayappan Kavu) Premature fall of trees due to lack of wind break (Iringole Kavu) are identified as the threats due to natural calamities.

Management Proposals

From the forgoing description and analysis of sacred groves in terms of their present condition it is evident that the preservation/conservation of these unique landscape units of key importance, should get maintained for the biodiversity as well as comprehensive ecosystem health of given landscape. Twenty seven management options have been proposed for the twenty eight sacred Groves of Kerala. And are categorized under four major groups, Construction/Repairing physical barriers, Conservation of water conservation structures, Weed management measures and enrichment planting and other management proposals.

In order to avoid trespassing and encroachment of the land belonging to the respective kavu, it is proposed to construct boundary demarcation structures all around the kavu, includes Construction of boundary demarcation posts (Kammadam Kavu), Construction of mud wall/lateritic brick wall (Edayilekkadu Kavu), Chain-link fence around the kavu (Valliyoor and Kavassery Kavus), Construction of truck path from the forest fringe to temple complex (Vallikkattu and Karakkode Kavu), Construction of soil retention wall (Karakkode), Barbed wire fencing to prevent trespassing, litter and biomass removal *etc.*(Iringole Kavu) and Repairing barbed wire fence (Valliyoor Kavu). In order to avoid soil erosion and also to conserve water, it is proposed to construct soil and water retention structures include De-silting and repairing/cleaning the existing ponds/open wells (8 Kavus) Digging a well and undertaking soil and water conservation measures (Varikkarakadavu) Repairing a tank situated in front of the temple for water conservation (Vallikkattu Kavu Redevelopment of natural fresh water pond *Thirukuzhi* (Poyil Kavu) Construction of low-cost pond for rain water harvesting (Karakkode Kavu) Construction of check dams in order to check run-off and store water (Aruvikkal Kavu) and Conservation of soil and rain water conservation structures (Valamchuzhi Kavu).

Management proposal for weed management measures and enrichment planting includes Management of weeds and climbers and augmentation of natural regeneration (Kammadam and Karimanal Kavu), Management of weeds and climbers and enrichment planting (6 Kavus), Management of exotic weeds and climbers (6 Kavus) Enrichment planting alone (5 Kavus), Establishment of *Nakshathra vanam* (Manikavu), Enrichment planting in the forest patch around the temple (Valliyoor, Chendangottu and Chukkathu Kavus) Planting along the river bank (Vallitoor Kavu), and Planting on either sides of the temple (Vetticode Kavu).

Other management proposals include Engaging protection watchers (Kammadam and Iringole Kavus), Fire control activities (Manikavu), Labeling important species and signages (all the 28 Kavus), Printing and distributing information brochures (except 5 Kavus) have been proposed by the management community for the twenty eight sacred groves of Kerala.

Conclusion

Management Plans for twenty eight Sacred Groves have been prepared by considering all the foregoing aspects. The BDC of Kerala Forest Department has submitted the proposal to the Government of India for financial support to manage the Sacred Groves as per the Management Plans. And is also recognized that similar attempts need to be extended to other poorly managed Sacred Groves too.

2013, Biodiversity Conservation for Sustainable Management   Pages *109–113*
Editor: **Dr. K. Muthuchelian,** *Vice Chancellor, Periyar University, Salem*
Published by: **Daya Publishing House, NEW DELHI**

Chapter 14

People's Participation in the Conservation of Sacred Groves

M. Kumar[1] and Lalitha Sundari[2]
[1]Saraswathi Narayanan College, Madurai
[2]Kalasalingam University, Krishnankoil, Tamil Nadu

Introduction

The ancient people loved the nature and allotted a patch of land as the property of god. These patches of lands associated with religious rituals are called sacred groves. The sacred groves are seen in all districts of Tamil Nadu.

Many traditional societies all over the world revered and worshipped nature and considered certain plants and animals as sacred. Some communities also followed the practice of setting aside certain patches of land or forest as sacred groves dedicated to a deity or village god and protected and worshipped. In India sacred groves are found all over the country and abundantly along the Western Ghats and the West coast and in several parts of Kerala, Karnataka, Tamil Nadu, Maharashtra, Madhya Pradesh, Rajasthan, Orissa and Himachal Pradesh. In Tamil Nadu still there are several sacred groves being maintained by the rural people for several generations. In recent times the sacred groves offer excellent avenues of research.

Sacred groves in India are known under different names in different parts of the country as *'Dev'* in Madya Pradesh. *'Deorais/Deovani'* in Maharashtra, *' Sarnas'* in Bihar, *'orari'* in Rajasthan, *'Devarakadu'* in Karnataka, *koil kadu/kattu koil/swamy shola* in Tamil Nadu.

Nandhavana and Sthala vruksham

Historical evidences prove that the ancient land lords, Zamindars and kings of Tamil Nadu liberally donated land and money for the establishment of sacred groves and their management. The protected gardens within the campus of large temples are called "*Nandhavanas*". The plants grown in these gardens are considered as sacred. The flowers derived from such temple gardens are generally used for the decoration of the idols. Certain species are of medicinal values. The practice of maintaining the temple gardens is still in vogue in several temples of Tamil Nadu. Worship of sacred trees, sacred animals (cows, bulls, monkeys) snakes, birds (eagle, crow, peafowl) and weapons of God have been practiced in several temples and the sacred groves of Tamil Nadu.

The trees are considered as the abode of ancestors. Though tree worship has been a common practice of Hindus in India, the worship and conservation of specific tree or plant species in each temple and sanctity attributed to these trees are unique to the temples of Tamil Nadu. Such sacred trees are grown in the temple as '*sthalavrukshas*'. In Tamil Nadu every Siva temple possesses a sacred tree inside the temple premise. In some sacred groves the sacred trees meant for ritual purposes. In Allinagaram sacred grove the mango tree adjoining the temple is considered as the sacred tree for ritual purposes. The devotees tie the brass bells, bangles and wooden cradles on these sacred trees for fulfillment of their prayers.

The first Inspector General of Forest Brandis wondered and recorded the presence of several sacred groves in Shervaroy hills. The exact number of sacred groves in India is not known. Almost every village had a sacred grove. In course of time several sacred groves were replaced by temple structure in the middle of the villages. The area of the sacred grove ranges from a few trees to a few hectares. The sacred groves located in the Western Ghats and the Himalayas may range to several hundred hectares. The sacred groves form a part of the entire forest range. The total number of sacred groves in Tamil Nadu is 448 with and area of about 21003.90 ha.

Deities in Sacred Groves

In Kerala, the sacred groves owned collectively by the villagers are mostly dedicated to Lord Ayyappa and are called '*Ayyappan kavu*' or '*Sasthan kavu*' and to Goddess '*Bhagavathi kavu*' or '*Amman kavu*'. Sacred groves owned by tribal communities are dedicated to '*Vanadevatha*', the Goddess of the forest, or to natural spirits or demons or ancestral spirits. There are some sacred groves exclusively meant for snake gods. Such sacred groves are called *sarpakavus*. In *sarpakavus,* the snake-gods are worshipped in the form of stone images. A serpent *kavu* is an indispensable adjunct to well-to-do Nair and Nambudiri families of Kerala and Kanyakumari district of Tamil Nadu. They allot a portion of their land for snakes and keep them undisturbed. Many sacred groves in Tamil Nadu have more than one deity, the main deity and two or more sub deities. Ayyanar is the main deity in several sacred groves of Tamil Nadu.

The ancient Tamil people were intelligent in selecting the Aiyanaar (Aiyappan) as their God in the sacred groves. Mythological story narrates that Aiyanaar was born between Lord Siva and Lord Vishnu. So there was a fusion between two major religious parties namely the Saivaites and Vaishnavaites.

Management of the Sacred Groves

The management of the sacred groves varies from place to place. In Tamil Nadu the smaller groves are generally in private ownership while the larger groves are controlled by a trust. Trusteeship generally goes on as hereditary privilege. In many smaller groves owned by orthodox families all types of biomass extraction are not permitted. Traditions have been distorted in large groves managed by temple trusts. In such cases biomass is often utilized to derive income for maintenance/expansion of the temple complex. For example in Kandanoor sacred grove (Sivaganga district) dead wood on the forest floor is used for cooking food during annual celebration.

Taboos and Belief System and Rituals

In several sacred groves strict taboos exist. These taboos are associated with mystic folklores and beliefs. Entering the sacred groves wearing chapels, ladies during menstrual period, removal of dead wood and killing of animals are all forbidden. People believe that such acts may provoke the wrath of God and Goddesses. People associated with the sacred groves celebrate rituals in different forms. They fulfill their vows by offering the boiled rice, by animal sacrifice, by tonsuring etc. The devotees pray to the God and Goddess of the sacred grove for marriage, childbirth, good harvest, curing of disease, victory in competitions etc. For this they fulfill their vow in the form of rituals. They believe that the God of the sacred grove will ward off the evil spirits and look after the welfare of the people.

Festivals

Sacred groves have always been part of cultural life of people of Tamil Nadu. Many sacred groves in Tamil Nadu are associated with annual festivals. The days of the festival are decided by the trustees/sacred grove management. The festival days may be one day to one week. On the day of the festival, the deities are decorated with flowers. All the local people along with their friends and relatives take part and offer boiled rice to the deities. Animal sacrifice is a part of the rituals during festivals. The offerings are shared among the family members. In some sacred groves all the participants share a common feast. During the festivals temporary shops are erected. In the evening folk-arts are performed. The folk arts include street play, (locally called *therukoothu*) or stage plays are enacted. Mostly the dramas are related to the Hindu epic or religious stories which have been popular for several years. The villagers in Pudukottai, Tanjore, and Sivaganga districts arrange the Jallikattu (taming the Bull by brave youths) in connection with the village festivals. Thus the festivals develop communal harmony among the village people and give an opportunity to stay with their relatives and friends. Offering of terracotta images (horse/elephant) is one of the important features in the festivals of several sacred groves of Tamil Nadu. In some groves one or two terracotta images are offered to the deity on behalf of the entire village. In some other sacred groves devotees are allowed to offer the terracotta images. The size of the image may be from 1-20 feet. In several sacred groves permanent horse/horses of various sizes are kept. Offering such terracotta images to the deity represents that the God may use them as vehicles to protect the villagers from evils and diseases. In Ilangudipatti of Pudukottai district hundreds of terracotta images are kept on either side of the pathway inside the grove. In Virasilai sacred grove the terracotta images were repainted and arranged around the newly renovated temple.

Water source is an important component in many sacred groves of Tamil Nadu. The water source may be a river or lake or a pond. The ponds are locally called *ooranies*. In certain sacred groves for *e.g.* Sadayaarkoil in Tanjore district there are two types of ponds, one that is adjacent to the temple is exclusively meant for religious purposes. The other pond is meant for public use (domestic and irrigation). The water source in the sacred groves is used by the visiting devotees for bathing before worshipping the God. Those sacred groves which are located in the catchments serve as a source of springs and streams. Thus, the water source in the sacred groves serves a variety of functions such as irrigation, fish culture and for the percolation for nearby wells.

Biodiversity and Conservation

The sacred groves protect several plant and animal species, valuable for food, medicine and other uses. Some sacred groves harbour rare species of plants. The sacred groves of Tamil Nadu harbour several endemic plants. The sacred groves of Kanyakumari district in Tamil Nadu harbour several endemic plants. Continuous protection of the tree species for several generations can be seen in some sacred groves of Tamil Nadu. The Allinagaram sacred grove of Theni district harbours huge *Terminalia arjuna* trees (locally called *Marudhamaram*) measuring more than 15 meters girth. It indicates that this sacred grove has been maintained by the people for several generations. Several wild plant varieties are still found in some sacred groves. Wild mangoes in Allinagaram, wild tamarind in Illangudipatti, and wild jambolina (naval fruits) in Kandanoor are some of the examples. Poisonous plants such as *Abrus precatorius, Strychnos nux-vomica, Dolichandrone falcata, Gloriosa superba, Cleistanthus collinus* are found in the sacred groves.

Several sacred groves offer a comfortable residence for a variety of animals including birds. Ilangudipatti sacred grove in Pudukottai district harbour a variety of reptiles found among the old terracotta images. Pea-fowls, and monkeys are common in several sacred groves. Thousands of fruit eating bats are perching on the branches of tall trees in Allinagaram sacred groves. Migratory birds often visit the sacred groves in search of food.

Livelihood for Several People

The sacred groves offer livelihood for people living around them. They provide fuel wood, fruits, fodder to the cattle and medicinal plants for the rural traditional physicians. During festivals several shop owners such as candies, children toys, groceries, people of folk arts and the priests are benefited. Thus the sacred groves serve as an excellent medium for communal harmony.

Conclusion

Thousands of huge trees have been sacrificed for broadening the National Highways. This is quite inevitable for the sake of development of the country. Now the time has come for the media, environmental biologists, voluntary organizations, educational institutions and village panchayats should come forward to protect the existing sacred groves form degradation. The sacred groves are still reflecting the glorious ancient Tamil culture. Our ancestors cleverly planned and established the sacred groves for us. We have to protect these groves and hand over to our future generations.

2013, Biodiversity Conservation for Sustainable Management Pages *114–127*
Editor: Dr. K. Muthuchelian, *Vice Chancellor, Periyar University, Salem*
Published by: Daya Publishing House, NEW DELHI

Chapter 15

Vegetation Structure and Species Composition of Manjal Nathi Amman Sacred Grove, Erasakkanayakanoor Theni District

S.M. Sundarapandian[1]*, M. Muthukumaran[1], M. Mari[2] and R. Jayakumararaj[3]

[1]*Department of Ecology and Environmental Sciences, School of life sciences, Pondicherry University, Pducherry – 605 014*
[2]*Saraswathi Narayanan College, Madurai – 625 022*
[3]*Department of Botany, RD Government Arts College, Sivagangai – 630 561*

ABSTRACT

Vegetation structure and species composition were studied in the Manjal Matha (Manjal Nathi Amman) sacred grove near Erasakkanayakanoor, Uttamapalayam taluk, Theni district of Tamil Nadu. A Total of 71 plants species belonging to 61 genera and 30 families were recorded. Twenty five adult tree (>10 cm DBH) species were recorded in 1 hectare plot of the sacred grove. Sixteen shrub species, nine climbers and seventeen herbaceous species were recorded in the sacred grove during the study period. A total of 122 stem (tree density) occurred in 1 hectare plot of the study area. The diversity index value of

* Corresponding Author: E-mail: smspandi@yahoo.co.in

tree community is 2.52. A total of 30 families were represented in the sacred grove. Exotic plants such as *Lantana camara* and *Eupatorium odaratum* were dominated in the shrub community. The invasion of alien plants inhibits the growth of natural plant biodiversity. These sacred groves harbour a sizable proportion of the region's characteristic flora but also reflect the cultural tradition associated with them. They cannot be conserved based only on spiritual belief system and social taboos alone. Therefore, it is suggested that the area adjacent to the groves may be developed as buffer zone from where people can satisfy their basic minimum requirements for their domestic consumption (normally expected from the sacred groves). This in turn, would also reduce anthropogenic pressure on these sacred groves.

Introduction

Sacred groves were established with a view to preserve, share and save natural resources from the region where they existed. The concept of sacred groves grew over time when some of the important ecological and economic species of plants or of animals were conserved (or protected) in a grove. Sacred grove was one way of expressing the gratitude of man towards the vegetation which sustained and supported life under respective agro-ecological condition. However, studies on socio-economic aspects as well a complete inventory of sacred groves representing different agro-ecological regions are still lacking (Gadgil and Vartak 1976; Chandran and Gadgil, 2003; Ramakrishnan *et al.,* 1998; Christopher Mcled 2007). The values of the sacred groves are manifold: from the materials available, they serve aesthetic, ecological, and socio-cultural functions. Any sacred grove, anywhere, despite being various stages of decline and degradation shares one or more of these features (Gadgil and Vartak 1975; 1976; Ramakrishnan *et al.*1998; Chandrashekara and Sankar 1998; Swamy *et al.,* 2003).

The sacred groves invariably occur throughout India (Ramakrishnan *et al.,* 1998). In Tamil Nadu, the sacred groves are popularly called '*Koil kadugal' or kavus. Swamy sholas* (Brandis 1897). Most of the sacred groves in Tamil Nadu are associated with the village deities (Whitehead 1921). The literature on the study of sacred groves in Tamil Nadu is limited. A comprehensive list of sacred groves in Tamil Nadu with their area, deities and their location in each district was recorded by Amirthalingam (1998). The seacred groves selected in the presesent study is not included in the above-mentioned list. Tree diversity was studied in various sacred groves of Tamil Nadu (Parthasarathy and Karthikeyan 1997; Swamy 1997; Swamy *et al.,* 1998; Britto *et al.,* 2001; Swamy *et al.,* 2003; Kumar, 2006; Mani and Parthasarathy, N. 2009). Vegetation studies were conducted in sacred groves in many parts of India such as Kerala (Chandrashekara and Sankar 1998), Pondicherry (Ramanujam and Kadamban 2001; Ramanujam and Cyril 2003;) Manipur (Khumbongmayum *et al.,* 2005a), Mehghalaya (Mishra *et al.,* 2004), Uttarakhand (Bisht and Childiyal, 2007), West Bengal (Bhakat, 2009) and few sacred groves in Tamil Nadu (Swamy *et al.,* 2003; Kumar, 2006). It is therefore imperative to survey the sacred groves and properly assess their role in nature conservation so that these forests may continue to be preserved even if the religious beliefs associated with them weaken and may disappear

(Gadgil and Vartak 1975; Swamy *et al.,* 2003). Most of the sacred groves are rich in biodiversity, source of rare, endemic and endangered plants and also wild crop relatives (Gadgil and Vartak 1976; Swamy *et al.,* 2003). Therefore, an analysis of the vegetation structure, plant diversity and species composition of the existing sacred groves is important to record plant resources and also understand ecosystem structure and function. Hence, the present study is intended to evaluate the vegation structure and species composition of Manjal Matha (Manjal Nathi Amman) sacred grove near Erasakkanayakanoor, Uttamapalayam taluk, Theni district of Tamil Nadu.

Study Area

Manjal Matha (Manjal Nathi Amman) sacred grove near Erasakkanayakanoor, Uttamapalayam taluk, Theni district of Tamil Nadu was selected for the present study. It is located 20 km from Uttamapalayam. Total area of of the sacred grove is about 200 acres and it was considered as sacred grove and maintained more than 200 years. It comes under Forest Department. However, Mango and coconet trees in the sacred grove are maintained and used by family trustee. Manjal Nathi River originated from Megamalai is run through it. Communities involved in the sacred gove are Maravar, Naiyakars, Chettiar, Saliyar and the inhabitants of local and neighbouring villages. Principal Deity is Manjal Matha Swamy (Kali). Sthalavriksha of the sacred gorve is *Schefflera Stellata* (Gaertn). Harms.

Methods

Phytosociological studies were carried out in the selected sacred grove. The density, frequency and basal area were estimated using 1 ha plot for trees [individuals with girth at breast height (GBH) more than 30 cm]. Twenty quadras (1x1 m^2) were studied for herbs (Kershaw 1973, Misra 1968). Similarly shrubs and lianas (climbers of all sizes) whose base fell inside the quardrats (5x5 m^2) were enumerated. Importance value index were calculated by the summation of relative density, relative basal area and relative frequency. The plant samples were identified in the field with the help of Gamble's (1925) and Matthew (1988) floras.

Species diversity (Margalef, 1968), index of dominance (Simpson, 1949), species richness (Menhinick, 1964) and evenness index (Pielou, 1966) were calculated.

Results

A Total of 71 plants species belonging to 61 genera and 30 families were recorded from Majal Matha sacred grove near Erasakkanayakanoor, Uttamapalayam Taluk, Theni District of Tamil Nadu (Table 15.1). Twenty five adult tree (>10 cm DBH) species were recorded in 1 hectare plot of the sacred grove. Sixteen shrub species, nine climbers and seventeen herbaceous species were recorded in the sacred grove during the study period. A total of 122 stem (tree density) occurred in 1 hectare plot of the study area. Shrub contribution was grater when compared to other life forms in terms of density in the sacred grove. A similar trend was observed in the case of the basal area also. The diversity index value of tree community is 2.52. The herbaceous community showed greater diversity index value compared to other life forms. In contrast, shrubs showed greater dominance index value when compared to other life

Table 15.1: Consolidated details of phytosociological analysis of the Manjal Matha sacred grove.

Criteria	Values
Number of species	
Trees (>30 cm DBH; No./ha)	25
Juveniles (>10-30 cm DBH; No./0.1 ha)	13
Shrubs (No./0.1ha)	16
Climbers (No./0.1ha)	9
Herbs (No./0.001ha)	17
Density	
Trees (>30 cm DBH; No./ha)	122
Juveniles (>10-30 cm DBH; No./0.1 ha)	93
Shrubs (No./0.1ha)	252
Climbers (No./0.1ha)	92
Herbs (No./0.001ha)	218
Basal area	
Trees (>30 cm DBH; cm^2/ha)	76907.7
Juveniles (>10-30 cm DBH; cm^2/0.1 ha)	2953.5
Shrubs (mm^2/0.1ha)	27902
Climbers (mm^2/0.1ha)	4775.94
Herbs (mm^2/0.001ha)	2564.6
Diversity index	
Trees	2.522
Juveniles	2.36
Shrubs	2.375
Climbers	2.171
Herbs	2.614
Dominance index	
Trees	0.163
Juveniles	0.112
Shrubs	0.134
Climbers	0.117
Herbs	0.092
Species richness	
Trees	2.263
Juveniles	1.348
Shrubs	1.008
Climbers	0.938
Herbs	1.151
Evenness index	
Trees	1.804
Juveniles	2.119
Shrubs	1.972
Climbers	2.275
Herbs	2.124

forms except for adult trees. Species richness showed greater value in trees followed by herbs. Least was observed in climbers while climbers showed greater value in evenness index.

A total of 30 families were represented in Manjal Matha sacred grove (Table 15.2). Papilionoideae was the dominant families in terms of genus, species and density also. There are seventeen families represented only one genera. Among these families, fourteen genera represented only one species alone. Ten families represented less than ten individuals (density) in the study area.

Table 15.2: Family wise distribution of plant communities in the Manjal Matha sacred grove.

Families	Genus	Species	Density
Acanthaceae	2	2	123
Amarantaceae	1	1	4
Anacardiaceae	1	1	14
Annonaceae	1	1	6
Arecaceae	2	2	8
Asclepiadaceae	1	1	11
Asteraceae	3	3	31
Bignoniaceae	1	1	20
Bombacaceae	1	1	12
Boraginaceae	1	1	7
Caesalpinioideae	3	5	46
Combretaceae	1	1	10
Cyperaceae	2	2	35
Ebenaceae	1	2	7
Euphorbiaceae	4	4	30
Graminieae	5	6	47
Liliaceae	3	3	30
Melastomaceae	1	1	7
Meliaceae	1	1	2
Mimosoideae	3	5	30
Moraceae	1	3	13
Papilionoideae	7	8	82
Piperaceae	1	1	9
Pteridophyte	1	1	80
Rhamnaceae	1	3	18
Rubiaceae	3	3	24
Rutaceae	2	2	13
Simarubaceae	1	1	2
Tiliaceae	1	1	6
Verbinaceae	2	2	32

Density, basal area and importance value indices of various life forms of plant community are presented in Tables 15.3a–15.3e. *Mangifera indica* was the dominant tree (>10 cm DBH) species compared to others in terms of density, basal area and importance value index (Table 15.3a). *Ceiba pentandra* and *Terminalia chebula* are co-dominant species in terms of density while *Ficus bengalensis* is the co-dominant species in terms of basal area and importance value index. However, *Morinda tinctoria, Melia azedarach, Cordia wallichii, Caryota urens* and *Ailanthus excelsa* showed poor (<2-density) representation in adult forms.

Table 15.3a: Density (D: No./ha), basal area (BA: cm²/ha) and importance value indices of tree community in the Manjal Matha Sacred grove.

Name of the Species	D	BA	IVI
Acacia nilotica (L). Brenan.	3	425.8	3.013
Ailanthus excelsa, Roxb.	2	722.6	2.579
Albizzia amara, Roxb.	5	496.6	4.744
Albizzia labbeck, L.	4	673	4.154
Annona squamosa, L.	3	224.9	2.751
Atalantia monophylla, Corr.	4	428.7	3.836
Bauhinia racemosa, Lam.	6	2293.3	7.900
Bridelia crenulata, Roxb.	6	724.2	5.860
Caryota urens, L.	2	696.2	2.545
Ceiba pentandra, (L). Gaertner	12	2673.8	13.313
Cocos nucifera, L.	6	2607.5	8.308
Cordia wallichii, G. Don.	2	308.4	2.040
Dalberja latifolia, Roxb.	5	1047.9	5.461
Diospyros ebenum, J.Koening ex Rotz,	3	359.8	2.927
Ficus bengalensis, L.	3	10155.7	15.664
Ficus religiosa, L.	6	741.1	5.882
Ficus tomentosa, Roxb.	4	445.2	3.858
Ixora brachiata, Roxb.	3	224.9	2.751
Mangifera indica, L.	14	48529.3	74.576
Melia azedarach, L.	2	71.7	1.733
Millingtonia hortensis, L.	3	245.2	2.778
Morinda tinctoria, Roxb.	1	71.7	0.913
Pongamia pinnata, (L) Pierre.	9	1102.2	8.810
Tamarindus indica, L.	4	454.2	3.869
Terminalia chebula, Retz.	10	1183.8	9.736

Strobilanthes kunthianus was the dominant shrub species in terms of density in the sacred grove (Table 15.3b). However, the shrub community was dominated by exotic species such as *Lantana camara* and *Eupatorium odaratum* in the sacred grove. *Euphorbia antiquorum* showed the greater value of basal area, which reflects in the importance value index also.

Table 15.3b: Density (D: No./ha), basal area (BA: cm²/ha) and importance value indices of juvenile population of tree community in the Manjal Matha Sacred grove.

Name of the Species	D	BA	IVI
Albizzia amara, Roxb.	3	139.9	7.963
Annona squamosa, L.	3	123.2	7.397
Atalantia monophylla, Corr.	4	180.5	10.412
Cascabela thevitia, L.	14	290.4	24.886
Cordia wallichii, G. Don.	5	146.5	10.337
Diospyros montana, Roxb.	4	11.3	4.684
Ixora brachiata, Roxb.	9	377.1	22.445
Millingtonia hortensis, L.	17	668.6	40.917
Morinda tinctoria, Roxb.	5	225.3	13.005
Pongamia pinnata, (L) Pierre.	9	166.8	15.325
Tamarindus indica, L.	12	467.2	28.722
Tectona grandis, L.	4	116.5	8.246
Ziziphus mauritiana, Lam.	4	40.2	5.662

Table 15.3c. Density (D: No./0.1ha), basal area (BA: mm²/0.1ha) and importance value indices of Shrub community in the Manjal Matha Sacred grove.

Name of the Species	D	BA	IVI
Bauhinia sp.	6	471.00	9.866
Cassia auriculata, L.	10	384.65	11.144
Cassia sp.	8	401.92	13.311
Clausena denata, (Willd). Roem.	5	192.33	8.471
Eupatorium odaratum, L.	17	854.08	15.604
Euphorbia antiquorum, L.	9	17663	72.670
Grewia wightiana, Drumm.ex Dunn	6	169.56	10.235
Jatropha curcas, L.	6	923.16	8.588
Lantana camara, L.	28	2198.00	27.684
Mallotus tetracoccus, (*Roxb*). *Kurz*.	9	346.19	10.609
Memecylon malabaricum, (Cl). Cogn.	7	269.26	10.989
Micranthus oppositifolius, Wendl	11	310.86	12.726
Pavetta indica, L.	4	200.96	6.655
Strobilanthes kunthianus, T.And.	112	3165.1	67.382
Ziziphus jujuba, Lam.	9	254.34	8.831
Ziziphus rugosa, Lam.	5	98.125	5.234

Grasses were the dominant species in the herbaceous community in the sacred grove (Table 15.3c). Based on density, *Adiantum* was the dominant species. In grasses, *Cyperus sp.* Contribution was more followed by other grasses.

In climbers, *Acacia pennata* showed maximum importance value index followed by *Abrus precatorius*, *Derris brevipes* and *Gloriosa suparba* (Table 15.3d). However, in terms of density, *Abrus precatorius* was the dominant climber followed by *Clitoria ternatea*, *Gloriosa suparba* and *Tylophora macranthus*.

Table 15.3d: Density (D: No./0.1ha), basal area (BA: mm²/0.1ha) and importance value indices of climbers in the Manjal Matha Sacred grove.

Name of the Species	D	BA	IVI
Abrus precatorius, L,	14	703.36	39.56
Acacia pennata, willd.	9	1384.7	50.315
Clitoria ternatea, L.	13	163.28	29.088
Derris brevipes, Bent.	8	904.32	37.246
Gloriosa superba, L.	12	339.12	31.683
Piper argyrophyllum, (Miq). Sysf.	9	113.04	23.688
Smilax zeylanica, L.	7	269.26	24.785
Tylophora macranthus, Hook.f.	11	552.64	35.066
Vigna umbellata, Thumb.	9	346.19	28.57

The juvenile population of tree community indicated that the dominant species showed poor regeneration status (Table 15.3e). Only nine adult tree species are represented in juvenile stages. No adult tree was observed in the four trees. They occurred only in the junile stage. *Millingtonia hortensis* and *Tamarindus indica* showed a moderate level of regeneration status among trees.

With increasing tree size classes, species richness [no of species per hectare] and density decreases (Table 15.4a). In contrast, reverse trend was observed in the case of juvenile population (Table 15.4b) of tree community.

Discussion

In the present study, totally 71 plant species were recorded from the Manjal Matha (Manjal Nathi Amman) sacred grove. Which is lower than the reported studies *i.e.* a total of 189 species were recorded in 6 selected sacred groves of Tamil Nadu (Kumar 2006). Ramanuajam and Kadamban (2001) reported 58 families, 121 genera and 136 species in Olagapuram sacred grove (Pondicherry). Khumbonmayum *et al.* (2005a) recorded 70 families with 145 genera and 173 species in four sacred groves of Manipur. However, the value in the present study lies closer to the Oorani sacred grove, Pondicherry value (41 families with 71 genera and 74 species; Ramanuajam and Kadamban 2001).

Dicot families were the dominant component in the present study is in agreement with the results reported (where the dicot families contributed more than 90 per cent of angiosperm families) by Mishra *et al.* (2004) for the sacred groves of Meghalaya and Khumbongmayum *et al.* (2005a) for the sacred groves of Manipur.

Table 15.3e: Density (D: No./0.001ha), basal area (BA: mm^2/0.001ha) and importance value indices of herbaceous community in the Manjal Matha Sacred grove.

Name of the Species	D	BA	IVI
Abrus pulchellus, Wall.	9	28.26	13.043
Adiantum sp.	80	251.2	49.617
Aerva lanata, Juss.	4	1256	53.934
Ageratum conyzoides, L.	4	3.14	6.6448
Asparagus racemosus, Willd.	11	34.54	11.08
Centrocema pubescens, Benth.	6	75.36	10.378
Cymbopogan martini, Wats.	8	226.08	21.86
Cymbopogon citratus, Stapf.	5	141.3	12.491
Cynodon dactylon, Pers.	19	14.915	13.985
Cyperus sp.	28	21.98	19.951
Desmodium styracifolium, (Osbeck) Merr.	6	169.56	17.176
Imperata cylindrica, Beauv.	3	84.78	7.8069
Kyllinga brevifolia, Rottb.	7	5.495	12.8
Mimosa pudica, L.	6	18.84	11.299
Panicum sp.	5	3.925	5.5716
Themeda cymbaria, Hack.	7	197.82	20.3
Tridax procumbens, L.	10	31.4	12.062

Table 15.4a: Diameter class-wise (DBH) distribution of species richness (no. of species) and density (No./ha) of trees (>10 cm DBH) in the Manjal Matha Sacred grove.

Diameter Class (cm)	Number of Species	Density
10-20	19	82
20-30	6	23
30-40	0	0
40-50	0	0
50-60	0	0
60-70	2	12
70-80	1	1
80-90	2	4

The number of tree species (species richness) >30 cm GBH in the sacred gove is 25. the value obtained in the present study is closer the reported values (20-23 in Kerala -Chandrashekara and Sankar 1998; in six sacred groves of Tamil nadu 11-17

Kumar 2006). However, this value is lower than the several reported values *i.e.* the species richness in Thirumanikuzhi sacred grove was 38, in Kuzhanthaikuppam sacred grove 52 (Parthasarathy and Karthikeyan 1997), in Puthupet sacred grove 52 (Parthasarathy and Sethi 1997) and in four sacred groves of Manipur 81 (Khumbongmayum *et al.,* 2005a). The low value of tree species richness in the present study may be attributed to anthropogenic pressures such as lopping, extraction of minor forest produce (fruits, seeds etc). and cattle grazing. These attributes may also be some of the reasons that might have resulted in poor tree regeneration through seedling recruitment and also stunted growth in lopped trees, thus leading to small openings in the canopy of sacred groves studied. Similar observations were reported by Mishra *et al.* (2004) for the sacred groves of Meghalaya and Britto *et al.* (2001) for the Malliganatham sacred grove of Pudukottai district.

Table 15.4b. Diameter class-wise (DBH) distribution of species richness (no. of species) and density (No./ha) of juvenile population of tree communities (<10 cm DBH) in the Manjal Matha Sacred grove.

Diameter Class (cm)	Number of Species	Density
<1	0	0
1-2	1	2
2-3	4	6
3-4	3	6
4-5	3	10
5-6	3	3
6-7	9	29
7-8	6	8
8-9	9	23
9-10	5	8

Tree density in the sacred grove is 122/ha which is lower than the values (367-657/ha) recorded in the sacred groves of Kerala (Chandrashekara and Sankar 1998) while, it is comparable to the density of Marakanam reserve forest near Pondicherry (Visalakshi 1995). However, the tree density range recorded in the present study is at lower range when compared to the sacred groves of Thirumanikuzhi sacred grove and Kuzhanthaikuppam sacred grove (Parthasarathy and Karthikeyan 1997) and the Puthupet sacred grove (Parthasarathy and Sethi 1997). The density range of trees (>30 cm gbh) in tropical forests is 245-859/ha (Richards 1952; Aston 1964; Campbell *et al.,* 1992). Such low density of tree species in the present study is governed by a complex array of environmental factors besides human interferences as also suggested by Visalakshi (1995). Density of plant population may be governed by the circular steady state in suitable niche loci, survival of seedlings to reproductive stages as affected by predation and environmental fluctuation in relation to relative favorableness of loci and the density of the seeds that the population of mature plants produce (Whittaker 1969). Ground clearing and ground fires occurred during occasional rituals and annual festivals (by the visiting devotees) may influence the tree density of the sacred groves. Man made disturbances such as cattle grazing, criss-crossing foot path, lopping of small branches for fodder may also account for the lower tree density. The canopy gaps were invaded by exotic weed like *Lantana camara and Eupatorium odaratum*. Thus influencing the course of natural regeneration of sacred groves (Ramakrishnan *et al.*, 1998). Menace of invasion of alien weeds was also reported in many sacred groves in India (Parthasarathy and Karthikeyan 1997; Ramakrishnan *et al.*, 1998; Ramanujam and Kadamban 2001; Kumar 2006).

In the present study the basal area of trees is 7.7 m²/ha. This range of basal area is comparatively lower than Thirumanikuzhi sacred grove Kuzhanthaikuppam sacred grove (Parthasarathy and Karthikeyan 1997), Puthupet sacred grove (Visalakshi 1995; Parthasarthy and Sethi 1997) and the three sacred groves of Kerala (Chandrashekara and Sankar 1998).

The tree diversity index (Shannon index) in the present study is 2.5, which is comparable to Thirumanikuzhi sacred grove (Parthasarathy and Karthikeyan 1997) and Kuzhanthaikuppam sacred grove (Parthasarathy and Karthikeyan 1997). However, the tree species diversity is higher than Puthupet sacred grove (Visalashi 1995; Parthasarathy and Sethi 1997). The tree diversity index is lower than the sacred groves of Kerala (Chandrashekara and Sankar 1998) and four sacred groves of Manipur (Khumbongmayum *et al.*, 2005a). The dominance index value of trees in the present study is 0.163, which is comparatively lesser than the dominance index recorded in Kuzhanthaikuppam sacred grove (Parthasarathy and Karthikeyan 1997), Puthupet sacred grove (Parthasarathy and Sethi 1997) and the sacred groves of Kerala (Chandrashekara and Sankar 1998). However, the dominance index value is comparable to that of Thirumanikuzhi sacred grove (Parthasarathy and Karthikeyan 1997). The higher dominance value in the present study is due to the dominance of single species in the sacred groves.

The stem density decreased with increased size classes of trees observed in the present study is in agreement with others (Chandrashekara and Ramakrishnan 1994; Parthasarathy and Karthikeyan 1997: Sundarapandian and Swamy 2000; Swamy *et al.*, 2000). This shows that all the sacred groves of the present study have been preserved by the belief systems and taboo.

Nair and Sastri (1990) published the rare, endemic and threatened flowering plants of India. Daniels *et al.* (1995) reported the rare, endemic and threatened flowering plants of South India. Parthasarthy and Karthikeyan (1997), Chandrashekara and Sankar (1998) reported the endemic and rare plants in the sacred groves. Khumbongmayum *et al.* (2005b) observed rare and threatened plants in the sacred groves of Manipur. In the present study the *Ficus benghalensis and Mangifera indica* was found to be the keystone species in the sacred groves.

The sacred grove harbour a sizable proportion of the region's characteristic flora but also the rich cultural tradition associated with them. They cannot be conserved based only on spiritual belief system and social taboos alone. Therefore, it is suggested that the area adjacent to the groves may be developed as buffer zone from where people can satisfy their basic minimum requirements for their domestic consumption (normally expected from the sacred groves). This in turn, would also reduce anthropogenic pressure on these sacred groves.

References

Amirthalingam, M., 1998. *Sacred Groves of Tamil Nadu.* C.P.R. Environmental Education Centre, Chennai, 190 pages.

Ashton, P.S., 1964. A quantitative phytosociological technique applied to tropical mixed rain forest vegetation. *Malaysian Forester,* 27: 304–317.

Bhakat, R.K., 2009. Chilkigarh Kanak Durga Sacred Grove, West Bengal. *Current Science,* 96(2): 185.

Bisht, S. and Childiyal, J.C., 2007. Sacred groves for biodiversity conservation in Uttarakhand Himalaya. *Current Science,* 92(6): 711–712.

Brandis, D., 1897. *Indian Forestry.* India Oriental Institute, Poona.

Britto, S. John, Balaguru, B., Natarajan, D. and Arokiasamy, D.I., 2001. Comparative analysis of tree diversity and its population density in a sacred grove at Malliganatham, Pudukottai district of Tamil Nadu. *Advances in Plant Science,* 14(2): 327–330.

Campbell, D.G., Stone, J.L. and Rosas, A.Jr., 1992. A comparison of the phytosociology and dynamics of the three floodplain (Verzea) forest of known ages, Rio Jurua, Western Brazilian and Amazon. *Botanical Journal of Linnaean Society,* 108: 213–237.

Chandran, M.D.S. and Gadgil, M., 1993. Sacred groves and Sacred trees of Uttara Kannada: A pilot study. Report submitted to the Indira Gandhi National Center for the Arts. New Delhi.

Chandrashekara, U.M. and Sankar, S., 1998. Ecology and management of sacred groves in Kerala, India. *Forest Ecology and Management,* 112: 165–177.

Christopher, Mcleod, 2007. *Sacred Groves of India.* http://www.sacredland.org/world_sites_pages/Sacred_Groves.html

Daniels, R.J.R., Anilkumar, N. and Jayanthi, M., 1995. Endemic, rare and threatened flowering plants of South India. *Current Science,* 68: 493–495.

Gadgil, M. and Vartak,V.D., 1975. Sacred groves in India: A plea for continued conservation. *Journal of Bombay Natural History Society,* 72: 314–320.

Gadgil, M. and Vartak,V.D., 1976. The sacred groves of Western Ghats in India. *Economic Botany,* 30: 152–160.

Gamble, J.S., 1925. *Flora of Presidency of Madras.* Adlard and Sons, London, Vols. 1–3: 2017.

Kershaw, K.A., 1973. *Quantitative and Dynamic Plant Ecology.* Edward Arnold, London, 308 pp.

Khumbongmayum, A.D., Khan, M.L. and Tripathi, R.S., 2005a. Sacred groves of Manipur, north-east India: Biodiversity value, status and strategies for their conservation. *Biodiversity and Conservation,* 14: 1541–1582.

Khumbongmayum, A.D., Khan, M.L. and Tripathi, R.S., 2005b. Survival and growth of seedlings of few tree species in the four sacred groves of Manipur, Northeast India. *Current Science*, 88(11): 1781–1788.

Kumar, M., 2006. Ecological studies on the selected sacred groves of Tamil Nadu. *Ph.D. Thesis*, Madurai Kamaraj University, Madurai, 140 pp.

Mani and Parthasarathy, N., 2009. Tree population and aboveground biomass changes in two disturbed tropical dry evergreen forests of peninsular India. *Tropical Ecology*, 50: 249–258.

Margalef, R., 1968. *Perspective in Ecological Theory*. University of Chicago Press. Chicago, 111 pp.

Matthew, K.M., 1988. *Flora of Tamil Nadu Carnatic*. St. Joseph College, Thiruchirapalli, 4: 915.

Menhinick, E.F., 1964. A comparison of some species diversity indices applied to samples of field insects. *Ecology*, 45: 859–861.

Mishra, B.P., Tripathi, R.S., Tripathi, O.P. and Pandey, H.N., 2004. Effect of anthropogenic disturbance on plant diversity and community structure of a sacred grove in Meghalaya. *Biodiversity and Conservation*, 13: 421–436.

Misra, R., 1968. *Ecology Work Book*. Oxford and IBH Publications, New Delhi, 244 pp.

Nair, M.P. and Sastri, R.S., 1990. *Red Data Book of Indian Plants*. Botanical Survey of India, Calcutta, Vols. 1–3.

Parthasarathy, N. and Karthikeyan, R., 1997. Plant biodiversity inventory and conservation of two tropical dry evergreen forest on the Coromandel coast, South India. *Biodiversity and Conservation*, 6: 1063–1083.

Parthasarathy, N. and Sethi, P., 1997. Trees and liana species diversity and population structure in a tropical dry evergreen forest in South India. *Tropical Ecology*, 38: 19–30.

Pielou, E.C., 1966. The measurement of diversity in different types of biological collections. *Journal of Theoretical Biology*, 13: 131–144.

Ramakrishnan, P.S., Saxena, K.G. and Chandrashekara, U.M. (Eds.) 1998. *Conserving the Sacred for Biodiversity Management*. Oxford and IBH Publications, New Delhi, 480 pp.

Ramanujam, M.P. and Kadamban, D., 2001. Plant biodiversity of two tropical dry ever green forests in the Pondicherry region of South India and the role of belief system in their conservation. *Biodiversity and Conservation*, 10: 1203–1217.

Ramanujam, M.P. and Cyril, P.K., 2003. Woody species diversity of four sacred groves in the Pondicherry region of South India. *Biodiversity and Conservation*, 12: 289–299.

Richards, P.W., 1952. *The Tropical Rain Forest*. Cambridge University Press, London.

Simpson, E.H., 1949. Measurement of diversity. *Nature*, London, p. 163–168.

Sukumaran, S. and Raj, A.D.S., 1999. Sacred groves as a symbol of sustainable environment: A case study. In: *Sustainable Environment*, (Ed.) N. Sukumaran. SPCES. M.S. University, Alwarkurichi, India, pp. 67–74.

Sundarapandian, S.M. and Swamy, P.S., 2000. Forest ecosystem structure and composition along an altitudinal gradient in the Western Ghats, South India. *Journal of Tropical Forest Science*, 12(1): 104–123.

Swamy, P.S., 1997. Ecological studies of sacred groves in Sivaganga district. UNESCO project.

Swamy, P.S., Sudarapandian, S.M. and Chandrasekaran, S., 1998. Sacred groves of Tamil Nadu. In: *Conserving the Sacred for Biodiversity Management*, (Eds.) P.S. Ramakrishnan, K.G. Saxena and U.M. Chandrashekara. Oxford and IBH Publications, New Delhi, 480 pp.

Swamy, P.S., Kumar, M. and Sundarapandian, S.M., 2003. Spirituality and ecology of sacred groves of Tamil Nadu. *Unasylva*, FAO of United Nations, 54: 53–58.

Swamy, P.S., Sundarapandian, S.M., Chandrasekar, P. and Chandrasekaran, S., 2000. Plant species diversity and tree population structure of a humid tropical forest in Tamil nadu, India. *Biodiversity and Conservation*, 9: 1643–1669.

Visalakshi, N., 1995. Vegetation analysis of two tropical evergreen forests in Southern India. *Tropical Ecology*, 36: 117–127.

Whittakar, R.H., 1969. Evolution and diversity of plant communities. In: *Diversity and Stability in Ecological System,* (Eds.) G.M. Woodwell and H.M. Smith. Brookhavan Symposium on Biology No. 22 U.S. Department of Commerce, Springfield, Virginia, pp. 179–93.

Whitehead, Henry, 1921. *The Village Gods of South India.* Asian Educational Service, Madras, 175 pp.

2013, Biodiversity Conservation for Sustainable Management   Pages *128–134*
Editor: Dr. K. Muthuchelian, *Vice Chancellor, Periyar University, Salem*
Published by: Daya Publishing House, NEW DELHI

Chapter 16

Causes and Consequences of Biodiversity Loss with Reference to the Forests of Kanyakumari in Tamil Nadu

S. Davidson Sargunam[1] and A. Selvin Samuel[2]
[1]*Environmental Educationist,*
23, Cave Street, Nagercoil – 629 001, Tamil Nadu
[2]*Department of Botany,*
St. John's College, Palayamkottai – 627 002, Tamil Nadu

ABSTRACT

Kanyakumari which is at the southern extremity of the land is a biodiversity Hot Spot, located at the tail end of the Western Ghats. It receives two monsoons and maintains a mediocre climate of mean 26°C. It has 14 types of forest ecosystems ranging from tropical ever green forest to thorn forest, with an area of 50,486 hectares of reserve forests, which amounts to 30 per cent of the total area of the district having 1672 sq.kms. Some areas of the forests are leased to other purposes meant for rubber, clove, hydel project and space research. Kanyakumari has a high population density, above the national and Tamil Nadu level. The forests are the rich repositories of biodiversity and the animals listed in the IUCN list as endangered, threatened are seen in the district. Biodiversity loss occurs due to lease of forest for other activities as monoculture rubber, rubber processing units that release obnoxious acid pollutants, introduction of exotic species, creation of infrastructure as roads, housing for forest dwellers, cattle grazing and goat browsing, population stress by people in periphery and forest inhabitants, unscientific herbal extraction, hunting, poaching and therapeutic uses of animals, illegal timber logging, stone quarrying and sand mining, private plantations, use of insecticides and pesticides, unplanned ecotourism, lack of eco-awareness and lack of environmental literacy. Monoculture rubber and private plantations drastically reduce green cover while rubber processing units damage soil and water. They destroy the habitats of birds, and dwindle bird population. Hunting, poaching and therapeutic uses of animals reduce faunal population. In the conflict between man and animal, animals are killed or poisoned, reducing the latter's population. Profuse use of fertilizers, insecticides and pesticides kill insects which are useful to ecology and man as butterflies and honey bees. Toddy palm inflorescences and cycas leaves used for rituals check regeneration of the palms.Forest fires cause loss of greenery and habitat destruction and result in soil erosion and landslides. The 'hunter gatherer instinct' of the forest dwellers and those in the forest periphery hunt minor animals and collect mushroom, edible seeds, nuts, tubers that check regeneration of species. Loss of food chain occurs due to loss of species that create eco-imbalance. Currently, macaques and wild boars have invaded areas beyond forest periphery, due to eco-imbalance. Pythons are caught every fortnight, in non-forest localities, even in town outskirts, signaling the impact of habitat destruction. Cattle grazing and goat browsing destroy herbal medicinal plants, which may result in extinction. The present impact of global warming is felt severely by the forest tribal inhabitants and biodiversity loss is predicted by climate change. To check biodiversity loss, scientific management of forests is highly essential. Reduction in forest resource dependence, regulation of ecotourism, drastic check on population pressure on forests, provision of more teeth to forest officials, genuine implementation of forest laws, sophistication in forest crime detection, sensitization to the cross section of the society and sustainable alternatives for forest dependent people would help in biodiversity conservation.

Profile of Kanyakumari

Kanyakumari, which is part of the Western Ghats, lies within 77p 7'–77p 35' in E longitude and 8 p5' -8p 35'N latitude, with an annual mean temperature of 26p C. Situated at the southern extremity of the Western Ghats, Kanyakumari forests, a biodiversity Hot Spot, cover an area of about 30 percent of the total area of 1672 sq. kms. The district receives two monsoons, South west and the North East, which make the district fertile. It is surrounded by sea on three sides, by the Bay of Bengal, the Arabian Sea and the Indian Ocean.

Reserve Forest Area

The Reserve Forests in Kanyakumari Forest Division is given in Table 16.1.

Forest Ecosystems of Kanyakumari Forests

Kanyakumari forests occupy an area of 50,186 hectares having 14 types of ecosystems ranging from luxuriant tropical evergreen to tropical thorn forests due to diverse factors.

Forest Types in Kanyakumari

1. Southern Hilltop Tropical Evergreen Forests
2. West Coast Tropical Evergreen Forests

Table 16.1: The reserve forests in Kanyakumari forest division.

Sl.No	Reserve Forests	Area in ha.
1.	Therkumalai East and West	1741
2.	Thadagaimalai	797
3.	Poigaimalai	1243
4.	Mahendragiri	4360
5.	Veerapuli	28109
6.	Velimalai	1126
7.	Old Kulasekaram	694
8.	Kilamalai	8106
9.	Asambu	4310
	Total	**50,486**

Source: Forest Department, Kanyakumari Division.

3. West Coast Semi Evergreen Forests
4. Moist Teak Forests
5. Slightly Moist Teak Forests
6. Southern Moist Mixed Deciduous Forests
7. Dry Teak Forests
8. Southern Dry Mixed Deciduous Forests
9. Dry Savannah Forests
10. Carnatic Umbrella Thorn Forests
11. Southern Thorn Forests
12. Southern Thorn Scrub
13. Southern Sub-tropical Hill Forests
14. Ochlandra Reed Brakes

Source: Forest Department, Kanyakumari.

Usage of Forest Land for Other Purposes

Kanyakumari forests are utilized for other purposes as listed below.

Table 16.2: Areas Leased for other activities (In hectares).

1.	Area leased to Arasu Rubber Corporation Ltd. for raising Rubber	14785.00
2.	Area leased to Arasu Rubber Corporation Ltd. for raising Clove	110.00
3.	Area leased for Space Research work to I.S.R.O.	1199.20
4.	Kodayar Hydro Electric Project T.N.E.B	133.24

Source: Forest Department, Kanyakumari Division.

Pressure on Forests by Population

The population density in Kanyakumari district is at a very high ratio when compared to the State ratio and at the national level. This density has deleterious impacts of population pressure on the forests, wielded by the forest dwellers and those residing in the forest fringes, as forests contribute to people's livelihoods providing food, medicine, shelter, energy, fodder and a multiple of other products and services.

Table 16.3: Population density in Kanyakumari district.

Place	Population	Population Density per sq.km
India	–	324
Tamil Nadu State	6,24,05,679	479.3
Kanyakumari	16,76,034	999

Source: Census of India 2001.

Forest Biodiversity of Kanyakumari

Biodiversity sustains the web of life and humanity fully depends upon it to meet food, health care and other needs. It provides innumerable services to humanity by direct and indirect ways. It sustains medicinal plants, clothing, shelter, spiritual, recreational and other needs of humankind. The wellbeing and survival of human populations are dependent on millions of species of flora, fauna and microbes.

Considering the vital importance of biodiversity, the Convention on Biological Diversity held at Rio de Janeiro in 1992, emphasised the ecological, economic and social aspects of it. The union ministry of Environment and Forests in India, developed a strategy for biodiversity conservation at macrolevel in 1999 and enacted the Biological Diversity Act in 2002. The National Environment Policy 2006, seeks to achieve balance and harmony between conservation of natural resources and development processes and also forms the basic framework for the National Biodiversity Action Plan.

Some of the animals seen in the forests of Kanyakumari are Indian elephant, tiger, panther, sloth bear, gaur, sambar, barking deer, black buck, spotted deer, mouse deer, wild boar, pangolin, bonnet macaque, common langur, Nilgiri langur, slender loris, jungle cat, small Indian civet cat, mangoose, jackal, Indian fox, Indian giant squirrel, Indian wild dog, large brown flying squirrel, Indian porcupine, Indian hare, Nilgiri tahr, and Indian monitor lizard.

Quail, tailor bird, woodpecker, hornbill, kingfisher, bee-eater, cuckoo, parakeet, swift, owl, night jar, dove, pigeon, kite, falcon, egret, heron, flycatcher, drongo, myna, swallow, bulbul, flower pecker, sun bird and munia are some of the birds that highlight the biodiversity of Kanyakumari forests.

In the Red Data Book, the Botanical Survey of India has categorized the present status of 427 Indian plants. Out of this, 123 species occur in Tamil Nadu, and as many as 62 species are endemic to Tamil Nadu, while 39 of these occur in the Western Ghats, where lies Kanyakumari district. The forests are endowed with a rich variety of herbal biodiversity.

Table 16.4: Animals listed in the Red List of IUCN as endangered, vulnerable, threatened, and near threatened are found in Kanyakumari forests.

Sl. No.	Name of the Wild Animal	Status in IUCN List
1.	Indian elephant	Endangered
2.	Gaur	Vulnerable
3.	Tiger	Endangered
4.	Travancore Flying squirrel	Near threatened
5.	Indian Rock Python	Near threatened
6.	Cobra	Endangered
7.	Travancore Tortoise	Endangered
8.	Sloth bear	Vulnerable
9.	Nilgiri tahr	Endangered
10.	Nilgiri langur	Threatened

Source: IUCN Red List–viewed 2010.

Causes of Biodiversity Loss

Biodiversity loss occurs due to lease of forests for monoculture plantations (Dulip, 1999), effluents from the rubber plantation units (Davidson, 1999), introduction of exotic species etc. The creation of infrastructure facilities for the forest dwellers, plantations, residential areas of the work force in plantations and the hydel power station fragment the forest areas.

Cattle grazing and goat browsing, illegal timber logging, exploitation of forest resources by forest dwellers and those in the peripheries result in habitat destruction. The pressing needs for food, fibre, shelter fuel and fodder exert enormous pressure on natural resources.

Demand for traditional medicines is on the increase with a rise in commercial gatherers, who have utter disregard for conservation practices (Davidson, 2004). The diversity of medicinal plants constitute critical resources for health care of rural communities and for the growth of Indian herbal industry.

Table 16.5: Owing to the traditional belief that some animals and birds have therapeutic effects, many animals are hunted for medicinal preparations.

Sl. No.	Name of the Animal	Preparation (in vernacular Tamil)
1.	Monitor Lizard	Udumbu Legium
2.	Lion tailed macaque	Karunkurangu Legium
3.	Sloth bear	Karadi Nei
4.	Small Indian civet	Punugu
5.	Tiger	Bone soup
6.	Python	Paampu Nei

Sand quarrying and sand mining destroy the green cover and result in habitat loss. Poaching, hunting and therapeutic uses of animals, heavy use of insecticides and pesticides in plantation and agricultural crops, unplanned ecotourism, lack of eco-awareness and lack of environmental literacy are some of the major causes for biodiversity loss.

Consequences of Biodiversity Loss

Monoculture rubber plantation and private plantations drastically reduce green cover. Large scale clear felling led to the disappearance of diversified tree species and introduction of plantations accelerated the speed of deforestation and other anthropogenic activities (P. Samraj, 1999). The chemical effluents from the rubber processing units reach the forest soil and in turn reach the water bodies polluting the soil and water. The rubber trees accelerate habitat destruction of birds and consequently bird population dwindles. The forest dwellers and those residing on the forest fringes still retain the hunter-gatherer instinct by hunting small animals by traditional methods using nets, nooses and traps and use of animals and birds for therapeutic uses, decrease the population of the respective species. In the man-animal conflict often animals as elephants, sloth bear, tigers, panthers, monkeys, wild boar, porcupines, bats are poisoned or killed by crude methods. Loss of habitats and over-exploitation have led to depletion of genetic diversity of several wild animals and cultivated plants.

In the plantations and agriculture sector, insecticides and pesticides are used, that have detrimental effects on insects that are useful to humans and agriculture as honey bees and butter flies. The forest inhabitants maintain the food culture of consuming seeds, nuts, mushrooms, tubers that checks the regeneration of species. The seeds of cycas palm are used to make flour and the inflorescence of Caryota urens (fish tail palm) is used for rituals with the traditional belief attached with regeneration.

Loss of food chain occurs due to loss of species that consequently creates eco-imbalance. Presently, macaques invade residential areas. Elephants visit agricultural fields for food and water. Wild boars destroy agricultural crops. Sloth bears visit forest peripheries for jack fruit and domesticated honey bee colonies. Pythons are caught every fortnight even in town outskirts. Cattle grazing and goat browsing destroy herbal medicinal plants, that may result in extinction of species.

Forest fires cause widespread damage and are responsible for large emissions of carbon into the atmosphere. They cause loss of greenery and habitat destruction and result in soil erosion and land slide. Forest fires eventually lead to heavy erosion of top soil from the hill slopes, enhances peak discharge and reduces infiltration (P. Samraj, 1999). Fires in the understorey of humid rainforests can cause tree mortality and canopy openness.(Christian Lambrechts-UNEP, 2009).

The present environmental situation—heavily influenced by climate change—could lead to massive destruction of forests and the extinction of countless species. The direct physical effects on forests by climate change, such as droughts, storms, fires and insect infestations, could also hurt the productivity of managed forests (UNEP, 2009).

Action Plan to Save Biodiversity

Efforts to conserve plants and animals in gene banks are vital, but an even more important task is to maintain biodiversity on farms and in natural habitats where it can to evolve and adapt to changing conditions. Provision of drinking water, health care, irrigation facilities are needed to improve the quality of life of people living inside the Reserve Forests (RFs) and fringe areas.

Provisions of supplementary and alternative livelihood support will reduce forest resource dependence. Regulation of ecotourism is essential to avoid nondegradable solid litter in forests, to save flora and fauna.

The Biological Diversity Act 2002 provides creation of a National Biodiversity Authority (NBA) at national, State Biodiversity Boards (SBBs) at state and Biodiversity Management Committees (BMCs) at local levels. The Act also stipulates preparation of People's Biodiversity Registers (PBRs) involving local people. The programmes should be implemented in a scientific manner.

Providing more teeth to forest officials with more sophistication of gadgets, intelligence gathering, regular monitoring and strict enforcement are effective ways to curtail illicit logging and poaching activities. There is a dire need to promote people's participation and solicit their cooperation to save the rich biodiversity.

References

Christian Lambrechts *et al.*, 2008. *Vital Forest Graphics*. UNEP, Nairobi, Kenya.

Davidson, S.S., 1999. Some factors responsible for the ecological degradation of the forest in Kanyakumari district. In: *Forest of Kanyakumari District*, (Ed.) R.S. Lal Mohan, Nagercoil.

Davidson, S.S., 2004. Environmental and social issues of the Kaani tribe of Western Ghats. In: *Save the Earth*, Research Cell, Women's Christian College, Nagercoil-1.

Dulip Daniels, A.E., 1999. Ecological impact of rubber and teak plantations on the forest of Kanyakumari district. In: *Forest of Kanyakumari District*, (Ed.) R.S. Lal Mohan, Nagercoil.

National Biodiversity Action Plan, Government of India, Ministry of Environment and Forests, New Delhi, 2008.

Samraj, P., 1999. Soil and water conservation measures for protecting the forests of Kanyakumari district. In: *Forest of Kanyakumari District*, (Ed.) R.S. Lal Mohan, Nagercoil.

Websites Visited

http://envfor.nic.in

www.iucnredlist.org

www.kanyakumari.tn.nic.in/forest.html

www.tn.nic.in/wlidbiodiversity/ws_kws.html

www.unep.org

2013, Biodiversity Conservation for Sustainable Management   *Pages 135–146*
Editor: **Dr. K. Muthuchelian,** *Vice Chancellor, Periyar University, Salem*
Published by: **Daya Publishing House, NEW DELHI**

Chapter 17

Biodiversity Conservation through Environmental Education for Sustainable Development: A Case Study from Puducherry, India

R. Alexandar* and G. Poyyamoli
*Department of Ecology and Environmental Sciences,
Pondicherry University, Puducherry – 605 014*

ABSTRACT

Promoting students commitment to protect local biodiversity is an important goal of Education for Sustainable Development (ESD) in India. The main focus of this Biodiversity education was to expose the complexity of ecosystems and interrelationships between organisms and their environment at local level. Students need to understand and develop skills to solve various biodiversity problems with reference to local context. In order to develop the Biodiversity consciousness, developing attitudes, values and skills, and promote participation among students to conserve their local Biodiversity. Activity based biodiversity education methods such as field trips, hands-on-activities, experiential education, debates, autobiography, games, and practical is vital to achieve sustainable biodiversity at local level in present and future. Local environment such as lakes, ponds, vegetation, animals, water, air and soil is the richest resource base for environmental studies for students. We developed

* Corresponding Author: E-mail: enviroalexandar@gmail.com

a comprehensive framework to assess the efficacy of biodiversity education modules in enhancing teaching and training in biodiversity conservation. Since the pre-test indicated little lesser than average interest in the relevance of biodiversity, the change in post-test was increased and significant to the high starting value.

Keywords: *Biodiversity conservation, Environmental education, Skills, Knowledge, Confidence.*

Introduction

Everyone in the world depends on natural ecosystem to provide the resources for a, healthy, and secured life (Millennium Development Goal-2010). Humans have made unprecedented changes in ecosystems in recent decades to meet their expanding populations and booming economy. Human activities have taken the planet to the edge of a substantial wave of species extinctions, further threatening our own well-being. The pressures on water, air, and natural ecosystems will increase globally in coming decades unless human attitudes and actions change (MDG-2010).

The world is facing a biodiversity crisis (Wilson 2002). In response, schools, teachers and parents are being urged to prepare students to face the real life issues they will routinely encounter in efforts to sustainably manage the biosphere and integrate biodiversity conservation with other societal goals (Noss 1997, Colker 2004, European Platform for Biodiversity Research Strategy 2006).

Status of India's Biodiversity

India occupies only 2.4 per cent of the worlds land area but its contribution to the world's biodiversity is approximately 8 per cent of the total number of species (Khoshoo 1995). From the biodiversity standpoint, India has some 59,353 insect species, 2,546 fish species, 240 amphibian species, 460 reptile species, 1,232 bird species and 397 mammal species, of which18.4 per cent are endemic and 10.8 per cent are threatened. The country is home to at least 18,664 species of vascular plants, of which 26.8 per cent are endemic. With only 2.4 per cent of the total land area of the world, the known biological diversity of India contributes 8 per cent to the known global biological diversity. It has been estimated that at least 10 per cent of the country's recorded wild flora, and possibly the same percentage of its wild fauna, are on the threatened list, many of them on the verge of extinction.

Understand and communicate how man's cultural activities (*e.g.*, religious, economic, political, and social and others) influence the environment from an ecological perspective. Understand and communicate how an individual's behaviors impact on the environment from an ecological perspective. Identify a wide variety of local, regional, national and international environmental issues and the ecological and cultural implications of these issues. Identify and communicate the viable alternative solutions available for remediation crucial environmental issues as well as the ecological and cultural implications of these various solutions. Understand the need for environmental issue investigation and evaluation as prerequisite to sound decision making. Understand the roles played by differing human beliefs and values in environmental issues and the need for personal values clarification as an important part of environmental decision making. Understand the need for responsible citizenship action in the solution of environmental issues. Identify and describe a wide variety of successful local, regional, national, and international sustainable development scenarios.

Status of Biodiversity in Puducherry

There is no appreciable forest cover in the U.T. The only existing patch of forest in Pondicherry is the vegetation available in Swadeshi Cotton Mills Campus, a sizable portion of which was cleared for the construction of District Court building. Mangrove vegetation is seen to some extent in the estuaries and along the sides of Ariyankuppam River (in Pondicherry region).

Wildlife population in Pondicherry comprises of small mammals, birds, reptiles and fishes. A large number of birds are sighted in the botanical garden and the two large water bodies namely, Ossudu and Bahour tanks. Similarly, the backwaters of Karaikal also attract a number of migratory aquatic birds including Ducks,Teals, Poczhards etc. The wild animals recorded in the U.T. are Jackal, Black Napped Hare, Bonnet Macaque, Jungle Cat, Civet Cat, Mongoose, Monitor Lizard, Olive Ridley Turtle and Leather Backed Turtle. Among these, the Olive Ridley Turtle and Leather Backed Turtle have been declared as endangered.

Traditional and substantial dependence on biodiversity resources for fodder, fuel wood, timber and minor forest produce has been an accepted way of life for the rural population that accounts for nearly 74 per cent of India's population. With radical demographic changes, the land to man ratio and forest to man ratio has rapidly declined. The lifestyles and the biomass resource needs having remained unchanged, the remnant forests have come under relentless pressure of encroachment for cultivation, and unsustainable resource extraction rendering the very resource base unproductive and depleted of its biodiversity.

Biodiversity Education

The evolution from nature conservation education to environmental education to education for sustainable development is one that can be characterized by an increasing awareness of the need for self determination, democratic processes, a sense of ownership and empowerment, and, finally, of the intricate linkages between environmental and social equity(Jensen and Schnack 1994, 1997, Hesselink *et al.,* 2000).

Several authors have shown that academic coverage of environmental topics and ecological principles increases student awareness, and positively affects attitudes, behaviors, and values regarding conservation issues (Leeming *et al.,* 1993; Zelezny 1999; Rickinson 2001; Humston and Ortiz-Barney 2005; Anderson *et al.,* 2007). It has been more difficult to create reliable instruments that correlate specific course teaching methods and learning objectives with changes in attitudes and values (Humston and Ortiz-Barney 2005).

Teaching biodiversity has been taught some hundred years ago, but due to low baseline level knowledge (Leather and Quicke, 2009) had become a challenging educational task at least since the conference of Rio in 1992 (Gaston and Spicer, 2004; van Weelie and Wals, 2002), and it has been emphasized again at the Conference of Bonn in 2008. From an educational point of view, however, biodiversity is a rather „ill-defined abstract and complex construct (van Weelie and Wals, 2002) which has to be transformed into small entities to enhance a sustained learning and understanding, especially during teaching at school. The most common entity used by conservation groups are species (van Weelie and Wals, 2002). Therefore, basic knowledge about animal species, their identification and life history has been targeted as a fundamental aspect for learning and understanding biodiversity (Gaston and Spicer, 2004; Lindemann and Matthies, 2005; Randler and Bogner, 2002), but baseline knowledge seemed to have declined significantly in recent decades (Leather and Quicke, 2009; but see Randler 2008).

Teaching about animals and about biodiversity in general should give a preference to outdoor ecological settings (Killermann, 1998; Lock, 1998; Prokop, Tuncer, and Kvasnièák, 2007a; Tilling, 2004). Previously, a lot of outdoor educational lessons often dealt with more or less immobile taxonomic groups such as plants or some invertebrates (Killermann, 1998). Within the context of ecology, many educational researchers emphasised measuring psycho-logical constructs such as attitude, perception and other personality factors rather than knowledge (Bogner, 2002; Randler and Bogner, 2002) but assessing cognitive learning outcome should sup-port the possible benefits of outdoor ecology education. Outdoor education must be enhanced and should be supported by previous learning within the classroom. This prepares the pupils for issues and task during outdoor field work and prevents pupils from novelty effects (Falk, 1983, 2005; Falk, Martin, and Balling, 1978).

In this paper, we report on the results of developing and piloting a activity based environmental education for sustainable development that measures and assesses learning gains in biodiversity education. We use this framework to evaluate the effectiveness of content learning gains, along with changes in students' interest in biodiversity, student self-perceptions of changes in process skills, and shifts in ecological worldview.

Objectives of the Program

☆ To discover/investigate varieties of plants and animals.

☆ To name some types of plants and animals found in the pupils respective villages and in nearby bushes or forests.

☆ To discover the value of animals and plants in different ecological groups.

☆ To recognize that other living things have a right to share the environment with us/attaching value to other living things in our environment.

☆ To discover the interdependence of several forms of living organisms.

☆ To discover how disturbance in the food web can lead to loss of biodiversity.

☆ To recognize how plants sustain environmental health.

☆ Learning to care for biodiversity at home and school.

☆ To recognize the diversity of natural resources around the school compound.

☆ To explain trees are vital resources.

Methods

Assessment Framework

We developed a comprehensive outcomes framework to assess the efficacy of biodiversity education modules in enhancing teaching and training in biodiversity conservation. The framework measured changes in conceptual understanding, improvements in self-perceptions of process skills, confidence in biodiversity knowledge, interest in biodiversity topics, and changes in environmental orientation. The methodology adapted and integrated three types of evaluation instruments in a pre-module exposure test/post-module exposure test format: To assess student learning outcomes; A self-reporting instrument measures changes in student confidence, interests, and process skills.

Biodiversity modules have prepared to expose the definition, importance of biodiversity and Threats to Biodiversity. Each module includes an interactive PowerPoint lecture slides with notes and discussion questions, a detailed topical synthesis paper, and a series of hands-on exercises in which students collect, in order to analyze, and synthesize biodiversity data from multiple sources. Each module component contains specific learning objectives to assist faculty teaching the material.

We have used presentations to introduce and discuss topics and applied the exercises as complements to lectures. Introduced the activity and answered questions at the end of the lecture, allowed students to work on the problems and then discussed the results in the following class, used an entire activity during one class session and another modified an exercise to last the whole semester. Variability in use and adaptation was allowed in this study since we were testing the proposed assessment framework rather than applying a quasi-experimental design.

Content Knowledge Tests

Content knowledge assessments measure student learning from the module component used. These assessments include true/false questions, multiple choice, matching, short answer, problem sets, and short essays. In addition to measuring knowledge recall, assessments focus on higher-order learning, including comprehension and application of material and problem solving in new situations. The biodiversity module used a written content knowledge test, consisting of twenty multiple-choice, true/false, and matching questions that were selected from the three modules, to measure changes in students' knowledge of biodiversity. Pre-tests were given prior to classroom use of the modules. The post-test was administered either immediately after teaching the modules.

Student Assessment of Learning

The Student Assessment of Learning is a self-reporting survey instrument that measures students' perceptions of their knowledge, attitudes, and skills The questionnaire was created to assess changes in students' confidence, interest and involvement in scientific modes of inquiry (Seymour and Hewitt 1997). Instrument to measure the perceptions of students in five areas: 1) confidence in knowledge and understanding of biodiversity conservation; 2) interest in the field of conservation biology; 3) confidence in process skills; and 4) preferred learning styles. Questions used a standard five-point Likert scale ranging from 1 (not at all confident) to 5 (extremely confident).

Categories of Questions

Biodiversity Knowledge Confidence–Assessed the Student's Confidence
1. Defining biodiversity
2. Identifying threats to biodiversity
3. Providing examples of the importance of biodiversity
4. Describing methods and strategies used in conservation
5. Identifying issues in a conservation controversy
6. Analyzing/synthesizing information on an issue

Biodiversity Interest–Assessed the Student's Interest
1. Understanding the relevance of biodiversity to real world issues
2. Taking additional courses related to biodiversity and conservation
3. Majoring in a related subject
4. Exploring career opportunities
5. Considering changes in lifestyle choices

Biodiversity Process Skills–Assessed the Student's Confidence
1. Oral communication
2. Written communication
3. Identifying underlying conservation problems
4. Gathering credible information to support a thesis
5. Sorting and filtering diverse sources of information
6. Predicting potential outcomes
7. Applying critical thinking
8. Collecting data and managing information
9. Working collaboratively with and in a group

With regard to confidence in process skills, biodiversity module identified thirteen skills that are important within the conservation biology profession, including: professional oral and written communication; public communication and outreach; problem and question definition; information gathering, critical inquiry, and research skills; sorting and filtering diverse sources of information; predicting potential outcomes and consequences; critical thinking for decision-making; data collection and management; data analysis and interpretation; graphical expression and interpretation; collaborative working skills; and project coordination and management skills. Biodiversity module exercise emphasizes at least one of these process skills. Because biodiversity module emphasizes active-learning approaches, we developed the questionnaire to allow students to rank their preferred learning styles. Choices ranged from traditional lectures to hands-on activities and outdoor field experiences. The standard Likert scale ratings ranged from Strongly Disagree to Strongly Agree. Demographic information, including gender, ethnicity, class standing and major, as well as reason for enrolling in the course were also collected.

Results and Discussion

Respondents indicated that interactive, hands-on learning exercises in pre test (36.7 per cent) in post test (43.3 per cent) were the preferred learning content, followed by lectures in pre test (28.3 per cent) in post test (36.7 per cent), outdoor field activities in pre test (41.7 per cent) in post test (52 per cent), class room discussions led by the students in pre test (21.7 per cent) in post test (28.3 per cent) group projects (40.2 per cent), and student presentations (57.1 per cent). Statistical analysis revealed there are significant differences with respect to overall changes in content knowledge tests and the students learning's (reflecting changes in confidence in biodiversity knowledge, interest in biodiversity conservation, and confidence in biodiversity skills, with demographic and motivational variables such as school, which has been compared with control group (non exposure group) there is no significant differences in control group pre-test and in post-test.

Changes in Content Knowledge

Students significantly increased their content knowledge of biodiversity conservation program differences in pre–and post–testing, in control group there is no changes in pre-test and post-test. Students gained significantly in their confidence in biodiversity conservation knowledge (Table 17.1), showed a significant increase in interest in biodiversity conservation (Table 17.2), and an increase in overall confidence in their knowledge (Table 17.3). Student confidence in biodiversity conservation knowledge increased uniformly for all questions including those relating to: defining biodiversity, identifying principal threats, providing examples of how biodiversity is important to human society, describing methods and strategies used in conservation, identifying underlying issues in a conservation controversy, analyzing/synthesizing information on an issue, and critically reviewing the content quality of researched material. There was change between before and after biodiversity education program in content knowledge and change in confidence in content knowledge in individual students. Since the pre-test indicated little lesser than average interest in the relevance of biodiversity, the change in post-test was increased and insignificant relative to the high starting value. In terms of biodiversity process skills, students reported significant gains in confidence in their skills in identifying conservation issues and evaluating diverse sources of information. However, students showed a significant decline in overall confidence in biodiversity related process skills.

Table 17.1: Biodiversity knowledge confidence-Assessed the student's confidence in

Define and Understand the Term Biological Diversity or Biodiversity	*CSS*		*JNV*	
	Pre-test	*Post-test*	*Pre-test*	*Post-test*
Not confident	16.7	13.8	3.3	1.7
A little confident	41.4	52.7	6.7	11.7
Somewhat confident	23.3	37.9	60	63.3
Highly confident	3.3	6.9	5	23.3

Identify the Principal Threats to biodiversity	*CSS*		*JNV*	
	Pre-test	*Post-test*	*Pre-test*	*Post-test*
Not applicable	13.8	3.3	13.3	10
Not confident	41.4	33.3	30	30
A little confident	20	37.9	35	36.7
Somewhat confident	43.3	6.9	16.7	18.3

Provide Specific Examples of the Importance of Biological Diversity to Human Societies	*CSS*		*JNV*	
	Pre-test	*Post-test*	*Pre-test*	*Post-test*
Not applicable	3.3	3.4	6.7	5
Not confident	23.3	10.3	23.3	28.3
A little confident	43.3	24.1	31.7	33.3
Somewhat confident	23.3	48.3	31.7	36.7
Highly confident	6.7	13.8	1.7	1.7

Describe some of the Methods and Strategies that Biologist Use to Address Conservation Questions	*CSS*		*JNV*	
	Pre-test	*Post-test*	*Pre-test*	*Post-test*
Not applicable	3.3	3.4	3.3	3.3
Not confident	13.3	10.3	13.3	13.3
A little confident	23.7	27.3	40	40
Somewhat confident	53.3	44.8	28.3	28.3
Highly confident	6.7	20.7	11.7	11.7

Identify the Underlying Issues and Perspective Regarding Controversy in Conservation	*CSS*		*JNV*	
	Pre-test	*Post-test*	*Pre-test*	*Post-test*
Not applicable	10.3	6.7	3.3	3.3
Not confident	36.7	27.6	10	6.7
A little confident	23.3	31	26.7	31.7
Somewhat confident	26.7	31	33.3	36.7
Highly confident	6.7	1.1	16.7	21.7

Table 17.2: Biodiversity interest assessed the student's interest in

Underlying the Relevance of Biodiversity to Real World Issues	CSS		JNV	
	Pre-test	*Post-test*	*Pre-test*	*Post-test*
Not at all interested	17.2	3.3	5	5
A little interested	26.7	37.9	15	15
Somewhat interested	33.3	41.4	20	21.7
Highly interested	3.4	36.7	43.3	45

Taking Additional Courses/ Course Work Related to Biodiversity and Conservation	CSS		JNV	
	Pre-test	*Post-test*	*Pre-test*	*Post-test*
A little interested	17.2	30	1.7	5
Somewhat interested	36.7	48.3	3.3	11.7
Highly interested	20.7	30	13.3	23.3
Extremely interested	3.3	13.8	25	36.7

Majoring in a Subject Area Related to Biodiversity and Conservation	CSS		JNV	
	Pre-test	*Post-test*	*Pre-test*	*Post-test*
Not applicable	3.3	6.9	1.7	3.3
Not at all interested	13.3	37.9	1.7	11.7
A little interested	13.3	34.5	11.7	30
Somewhat interested	30	20.7	31.7	33.3
Highly interested	36.7	13.2	31.7	21.7
Extremely interested	3.3	7.7	21.7	26

Considering Changes in Lifestyle Choices in Activities Related to Biodiversity and Conservation	CSS		JNV	
	Pre-test	*Post-test*	*Pre-test*	*Post-test*
Not at all interested	6.7	6.9	1.7	10
A little interested30	6.9	10	26.7	
Somewhat interested	33.3	24.1	25	35
Highly interested26.7	37.9	35	28.3	
Extremely interested	3.3	24.1	28.3	25

Table 17.3: Biodiversity process skills-Assessed the student's confidence in

Oral Communication and Public Speaking	CSS		JNV	
	Pre-test	*Post-test*	*Pre-test*	*Post-test*
Not at all	10	27.6	1.7	1.7
Just a little	30	48.3	1.7	1.7
Somewhat	33.3	24.1	8.3	8.3
A lot	26.7	1.1	25	25

Contd...

Table 17.3–Contd...

Written Communication	CSS		JNV	
	Pre-test	Post-test	Pre-test	Post-test
Not applicable	3.3	3.4	1.7	2.9
Not at all	13.3	10.3	13.3	3.2
Just a little	13.3	34.5	31.7	36.1
Somewhat	33.3	31	33.3	33.6
A lot	30	13.8	20	24
A great deal	6.7	6.9	32	44

Identifying Underlying Conservation Problems or Questions	CSS		JNV	
	Pre-test	Post-test	Pre-test	Post-test
Not at all	16.7	13.8	5	13.3
Just a little	23.3	20.7	16.7	28.3
Somewhat	40	34.5	31.7	36.7
A lot	16.7	17.2	28.3	21.7
A great deal	3.3	13.8	18.3	27

Sorting Potential Outcomes or Consequence of an Action or Situation	CSS		JNV	
	Pre-test	Post-test	Pre-test	Post-test
Not at all	10	3.4	16.7	1.7
Just a little	23.3	17.2	30	16.7
Somewhat	36.7	55.2	41.7	30
A lot	30	24.1	11.7	30

Predicting Potential Outcomes or Consequence of an Action or Situation	CSS		JNV	
	Pre-test	Post-test	Pre-test	Post-test
Not at all	20	37.9	5	18.3
Just a little	20	31	20	31.7
Somewhat	26.7	31	26.7	35
A lot	26.7	10.2	20	15
A great deal	6.7	3.7	28.3	38

Applying Critical Thinking to make Good Decisions	CSS		JNV	
	Pre-test	Post-test	Pre-test	Post-test
Not at all	16.7	3.4	1.7	5
Just a little	30	13.8	5	18.3
Somewhat	30	24.1	16.7	23.3
A lot	16.7	37.9	48.3	23.3
A great deal	6.7	20.7	18.3	30

Contd...

Table 17.3–*Contd...*

Collection Data and Managing Information	CSS		JNV	
	Pre-test	*Post-test*	*Pre-test*	*Post-test*
Not at all	20	3.4	5	2.1
Just a little	16.7	3.4	10	17
Somewhat	33.3	20.7	20	15
A lot	16.7	48.3	36.7	45
A great deal	13.3	24.1	28.3	29

Conclusion

This activity based EESD methods for biodiversity education for development encompasses comprehensive aspects of students cognitive, affective-and behavioral-development related to the perception and understanding about local biological diversity. These types of experiments can make learning about their local biodiversity practical and meaningful potentially having long term impacts on student's attitudes to conserve local biodiversity and also in shaping their future life concerning to their natural world. As a result of these experiments reflects the student's experiences and actions in their homes, school and community as this will get them pondering about everyday habits and occurrences in biodiversity dimension. These programs will help them to acquaint with the local biodiversity problems, and create an interest in them to identify more environmental problems and conserve and protect local biodiversity.

References

Bogner, F.X., 2002. The influence of a residential outdoor education programme to pupil's environ-mental perception. *European Journal of Psychology of Education*, 18: 19–34.

Colker, R.M., and Day, R.D., 2004. Issues and recommendations–A conference summary: Conference on personnel trends, education policy and evolving roles of federal and state natural resources agencies. *Renewable Resources Journal*, 21: 6–32.

Gaston, K.J. and Spicer, J.I., 2004. *Biodiversity.* Oxford.

Hesselink, F. van, Kempen, P.P. and Wals, A.E.J. (Eds.), 2000. *ESDebate: International Online Debate on Education for Sustainable Development* (Geneva: IUCN).Available online: http://www.xs4all.nl/¹esdebate.

Humston, R. and Ortiz-Barney, E., 2007. Evaluating course impact on student environmental values in undergraduate ecology with a novel survey instrument. *Teaching Issues and Experiments in Ecology*, Vol. 5.

Jensen, B.B. and Schnack, K., 1994. Action competence as an educational challenge. In: *Action and Action Competence as Key Concepts in Critical Pedagogy,* (Eds.) B.B. Jensen and K. Schnack. Studies in Educational Theory and Curriculum, Vol. 12, Copenhagen: Royal Danish School of Educational Studies, p. 5–19.

Khoshoo, T.N., 1991a. Census of India's biodiversity: tasks ahead. *Current Science,* 69: 14–17.

Killermann, W., 1998. Research into biology teaching methods. *Journal of Biological Education,* 33: 4–9.

Leather, S.R. and Quicke, D.J.L., 2009. Do shifting baselines in natural history knowledge threaten the environment? *Environmentalist,* 30: 1–2.

Leeming, F.C., Dwyer, W.O., Porter, B.E. and Cobern, M.K., 1993. Outcome research in environmental education: A critical review. *Journal of Environmental Education,* 29: 28–34.

Lindemann-Matthies, P., 2005. Loveable mammals and lifeless plants: how children's interests in common local organisms can be enhanced through observation of nature. *International Journal of Science Education,* 27: 655–677.

Lock, R., 1998. Fieldwork in the life sciences. *Int. Journal of Science Education,* 20: 633–642.

The Millennium Development Goal Report, 2010, United Nations New York

Noss, R.F., 1997. The failure of universities to produce conservation biologists. *Conservation Biology,* 11: 1267–1269.

Prokop, P., Tuncer, G. and Kvasnièák, R., 2007a. Short-term effects of field programme on students knowledge and attitude toward biology: A Slovak experience. *Journal of Science Education and Technology,* 16: 247–255.

Randler, C. and Bogner, F.X., 2006. Cognitive achievements in identification skills. *Journal of Biological Education,* 40: 161–165.

Rickinson, M., 2001. Learners and learning in environmental education: A critical review of the evidence. *Environmental Education Research,* 7: 207–230.

Tilling, S., 2004. Fieldwork in UK secondary schools: Influences and provision. *Journal of Biological Education,* 38: 54–58.

UNCED (United Nations Conference on Environment and Development), 1992a. *Agenda 21.* Rio de Janeiro, UNCED.

UNCED (United Nations Conference on Environment and Development), 1992b. *Convention on Biological Diversity (CBD).* Rio de Janeiro, UNCED.

Van Weelie, D. and Wals, A., 2002. Making biodiversity meaningful through environmental education. *International Journal of Science Education,* 24: 1143–1156.

Wilson, E.O., 2002. *The Future of Life.* Alfred A. Knopf, New York, NY.

Zelezny, L.C., 1999. Educational interventions that improve environmental behaviors: A meta-analysis. *The Journal of Environmental Education,* 31: 5–14.

2013, Biodiversity Conservation for Sustainable Management   Pages *147–153*
Editor: **Dr. K. Muthuchelian,** *Vice Chancellor, Periyar University, Salem*
Published by: **Daya Publishing House, NEW DELHI**

Chapter 18

Soil Carbon Sequestration through Management Practices in Agriculture

M.P. Kavitha and K. Sujatha
Agricultural Research Station (TNAU),
Vaigai Dam – 625 562

ABSTRACT

The increase in greenhouse gases (GHG) in the atmosphere and the resulting climatic change will have major effects in the 21st century. So, it is essential that a number of actions be undertaken in order to reduce GHG emissions and to increase their sequestration in soils and biomass. In this connection, new strategies and appropriate policies for agricultural and forestry management must be developed. One option concerns carbon sequestration in soils or in terrestrial biomass, especially on lands used for agriculture or forestry. Carbon sequestration in the agriculture sector refers to the capacity of agriculture lands and forests to remove carbon dioxide from the atmosphere. Carbon dioxide is absorbed by trees, plants and crops through photosynthesis and stored as carbon in biomass in tree trunks, branches, foliage, roots and soils.

Practically, the three areas of farm management that can affect soil carbon sequestration are the tillage, cropping intensity and fertilization. Tillage and soil carbon are negatively related. The greater the tillage, the less will be the soil carbon. No-till systems build soil organic matter, which is about 58 percent carbon. Reducing tillage reduces soil disturbance and helps mitigate the release of soil carbon in to the atmosphere. Cultivation of legume cover crops, inclusion of legumes in crop rotation, application of farm yard manure, compost, vermicompost, and incorporation of crop residues as mulch not only increase the organic carbon content in the soil but also increase the physical and microbial properties of the soil.

Keywords: Climate change, Soil carbon sequestration, No-till, Legumes, Cover crops, Crop rotation.

Introduction

SINCE the beginning of the industrial revolution, carbon dioxide concentration in the atmosphere has been rising alarmingly. Prior to the industrial revolution carbon concentration was around 270 ppm, but today it is around 372 ppm. If the pace of increase in carbon concentration remains constant and efforts are not made to reduce it, carbon concentration in the atmosphere would go up to 800–1000 ppm by the turn of the current century, which may create havoc for all living creatures on earth. Increasing concentration of atmospheric carbon dioxide (CO_2) is considered the predominant cause of global climatic change. Hence, Carbon management is a serious concern confronting the world today. A number of summits have been organized on this subject ranging from the Stockholm to Kyoto protocol.

Soil may be an important sink for the carbon storage in the form of soil organic carbon. Increase in greenhouse gases (GHG) in the atmosphere and the resulting climatic change will have major effects in the 21st century. In this connection, new strategies and appropriate policies for agricultural and forestry management must be developed. One option concerns carbon sequestration in soils or in terrestrial biomass, especially on lands used for agriculture or forestry.

Carbon Sequestration

Carbon sequestration in the agriculture sector refers to the capacity of agriculture lands and forests to remove carbon dioxide from the atmosphere and storing in the soil in the form of soil organic matter. Carbon sequestration can be defined as the capture and secure storage of carbon that would otherwise be emitted to or remain in the atmosphere. The idea is to (1) prevent carbon emissions produced by human activities from reaching the atmosphere by capturing and diverting them to secure storage, or (2) remove carbon from the atmosphere by various means and stores it. Carbon dioxide is absorbed by trees, plants and crops through photosynthesis and stored as carbon in biomass in tree trunks, branches, foliage and roots and soils. The ability of agriculture lands to store or sequester carbon depends on several factors, including climate, soil type, type of crop or vegetation cover and management practices. The amount of carbon stored in soil organic matter is influenced by the addition of carbon from dead plant material and carbon losses from respiration, the decomposition process and both natural and human disturbance of the soil. Several farming practices and technologies can reduce greenhouse gas emissions and prevent climate change by enhancing carbon storage in soils; preserving existing soil carbon; and reducing carbon dioxide, methane and nitrous oxide emissions.

Soil Carbon Sequestration

Soils are the largest carbon reservoirs of the terrestrial carbon cycle. Soils contain about three times more Carbon than vegetation and twice as much as that present in the atmosphere. Soils contain much more C (1500 Pg of C to 1 m depth and 2500 Pg of C to 2 m; 1 Pg = 1 × 1015 g) than is contained in vegetation (650 Pg of C) and twice as much C as the atmosphere (750 Pg of C). Carbon in the form of organic matter is a key element to healthy soil. It is estimated that each tonne of soil organic matter releases 3.667 tonnes of CO_2, which is lost into the atmosphere. Similarly, the build-up of each tonne of soil organic matter removes 3.667 tonnes of CO_2 from the atmosphere. (Rajeew Kumar, et al, 2006). The conversion of natural habitats to cropland and pasture, and unsustainable land practices such as excessive tillage frees carbon from organic matter, releasing it to the atmosphere as CO_2. Depleted of organic carbon, soils develop a carbon deficit. Soils can regain lost carbon by reabsorbing it from the atmosphere. This process is called carbon sequestration. Through photosynthesis, plants convert CO_2 into organic forms of carbon, viz. sugars, starch and cellulose, also known as carbohydrates. In natural habitats, carbon from plants is deposited in the soil through roots and plant residues, such as fallen leaves.

Agricultural Land's Ability to Store or Sequester Carbon Depends on many Factors including the Following

The rate of soil organic carbon sequestration with adoption of recommended technologies depends on soil texture and structure, rainfall, temperature, farming system, and soil management. Strategies to increase the soil carbon pool include soil restoration and woodland regeneration, no-till farming, cover crops, nutrient management, manuring and sludge application, improved grazing, water conservation and harvesting, efficient irrigation, agroforestry practices, and growing energy crops on spare lands (Lal, 2004).

Climate

In cooler climates, decomposition happens more slowly, so the plant residue has a greater chance of becoming humus, which is a stable part of soil with high organic carbon content.

Soil Type

Poorly-drained soil types have the capacity to store carbon more readily than others.

Type of Crop or Vegetation Cover

Crops like wheat and corn produce higher amounts of residue, *i.e.*, more organic matter (and more carbon) is left after harvest. Warm season grasses in conservation buffers are more effective at storing carbon than cool season grasses. Making certain plant choices can help capture more carbon from the atmosphere and make it available to processes that may lead to longer-term storage.

Management Practices

Most of the management practices that favor carbon sequestration also improve soil health, reduce erosion, and have other environmental benefits.

Management Practices for Increasing Soil Carbon Sequestration

Several farming practices and technologies can reduce greenhouse gas emissions and prevent climate change by enhancing carbon storage in soils; preserving existing soil carbon; and reducing carbon dioxide, methane and nitrous oxide emissions.

Conservation Tillage

Conservation tillage refers to a number of strategies and techniques for establishing crops in the residue of previous crops, which are purposely left on the soil surface. It reduces the mixing of oxygen and soil, slowing the process by which carbon in the soil is oxidized and released as CO_2. Reducing tillage reduces soil disturbance and helps mitigate the release of soil carbon into the atmosphere. (Baker, 2007). Conservation tillage also improves the carbon sequestration capacity of the soil. Additional benefits of conservation tillage include improved water conservation, reduced soil erosion, reduced fuel consumption, reduced compaction, increased planting and harvesting flexibility, reduced labor requirements and improved soil tilth. No tillage or conservation tillage includes crop residue management on site, which ensures the input of organic matter, and direct seeding of planting through the residue cover (FAO, 2001). Tillage practices often induce soil aerobic conditions that are favourable to microbial activity and may lead to a degradation of soil structure. As a result, mineralisation of soil organic matter increases in the long term.

Conservation tillage practices can minimize the rapid breakdown of plant residues, reduce CO_2 emission, and reduce the production of inorganic dissolved nitrogen (*i.e.*, nitrate and ammonium) in soil. When conventional tillage is converted to conservation tillage, both CO_2 emission from soil and N-uptake by crops are reduced. Reduction in CO_2 emission from soils enhances soil organic carbon (SOC) content, but reduction in N-uptake decreases residue production and hence, organic C storage in soils. Also, it was found that reducing tillage significantly decreases SOC loss from soils with high organic matter content. Principal mechanisms of carbon sequestration with conservation tillage are increase in micro-aggregation and deep placement of SOC in the sub-soil horizons. Pretty *et al.* (2002) consider tillage to be one of the major factors responsible for decreasing carbon stocks in agricultural soils.

Planting Cover Crops

Cover crops offer many benefits for agriculture that include erosion control; reduced compaction and nutrient leaching; increased water infiltration; improved soil biodiversity; weed control and disease suppression; increased carbon sequestration and maximum nutrient recycling; improved air, soil, and water quality; and wildlife enhancement. Legume cover crops are commonly used for nitrogen contribution because of their inherent capacity to fix atmospheric N (inert gas) into usable form to be used by succeeding crops. The carbon: nitrogen ratio of soil organic matter results in stable organic matter typically within a range of about 8-10:1. If insufficient nitrogen is present to permit stable formation of soil organic matter via soil microbial degradation of crop residues, then little carbon may be sequestered.

Planting cover crops maximizes the amount of carbon pulled into plant matter per unit of land as well as reduces erosion. Eliminating summer fallow also assists with this. Planting legumes as a cover crop is also a good way to replenish nitrogen levels in the soil as well as adding carbon to the soil. The Rodale Institute Farming Systems Trial has shown organic wheat/maize/soybean cropping systems with cover crops sequestered ~1,000 kg C/ha/yr, (range 667 to 1,381). Green manures and cover crops can provide important contributions to soil carbon. (Paul Reed Hepperly et al, 2008).

Improved Cropping and Organic Systems

Organic systems of production increase soil organic matter levels through the use of composted animal manures and cover crops. Organic cropping systems also eliminate the emissions from the production and transportation of synthetic fertilizers. Components of organic agriculture could be implemented with other sustainable farming systems, such as conservation tillage, to further increase climate change mitigation potential. The adoption of no-tillage systems and the maintenance of permanent vegetation cover using direct seeding Mulch-based Cropping system, may increase carbon levels in the topsoil (Martial Bernoux *et al.*2006). Using higher-yielding crops or varieties and maximizing yield potential can also increase soil carbon. The cropping sequence of Rice-Rice under wet season with early incorporation of crop residues resulted in 11-12 per cent more carbon sequestration and 5-12 per cent more N accumulation than the Rice –Maize cropping system with late incorporation of crop residues under dry condition.(Witt *et al.,* 2000) Soils under legume-based rotations tend to preserve residue C. A positive effect on SOC (an increase of 2-4 tonnes/ha) was also found with legumes and alternate cattle grazing in semi-arid Argentina. The effectiveness of rotations for sequestering C is likely to be greatest where they are combined with conservation tillage practices.

Land Restoration and Land Use Changes

Land restoration and land use changes that encourage the conservation and improvement of soil, water and air quality typically reduce greenhouse gas emissions. Modifications to grazing practices, such as implementing sustainable stocking rates, rotational grazing and seasonal use of rangeland, can lead to greenhouse gas reductions. Converting marginal crop land to trees or grass maximizes carbon storage on land that is less suitable for crops.

Irrigation and Water Management

Improvements in water use efficiency, through measures such as irrigation system mechanical improvements coupled with a reduction in operating hours; drip irrigation technologies; and center-pivot irrigation systems, can significantly reduce the amount of water and nitrogen applied to the cropping system. This reduces greenhouse emissions of nitrous oxide and water withdrawals.

Proper Amounts of Nutrients

It will help ensure optimal plant growth, resulting in better yields and more plant residue available to become soil organic carbon. However, using nitrogen in excess of crop needs not only wastes a producer's resources, but also results in greater GHGs emissions–CO_2 from the manufacture of fertilizers and NO_2 from the fertilizer itself. Scientists are exploring the effect of using different kinds of fertilizers to see if certain types enhance yields and carbon sequestration without high NO_2 emission rates.

Composting/Crop Residue Management

Composting or manuring are traditionally used in agriculture with proven beneficial effects on the soil. A problem in many countries is the decreasing source of such amendments, linked to animal husbandry. There is competition for plant residues or plant cover–to be used for feeding animals or for returning to the soil. But careful management associating cropping with livestock production can permit reintroduction of new sources of manure or composted manure. Crop residue management is another important method of sequestering C in soil and increasing the soil organic matter content. Residue-burning has negative consequences, even if they are sometimes mitigated by the great stability of the mineral carbon which is formed.

Conclusion

Soil C sequestration is a strategy to achieve food security through improvement in soil quality. It is a by-product of the inevitable necessity of adopting recommended management practices for enhancing crop yields on a global scale. While reducing the rate of enrichment of atmospheric concentration of CO_2, soil C sequestration improves and sustains biomass/agronomic productivity. It has the potential to offset fossil-fuel emissions by 0.4 to 1.2 Gt C/year, or 5 to 15 per cent of the global emissions. Soil organic carbon is an extremely valuable natural resource. Irrespective of the climate debate, the soil organic carbon stock must be restored, enhanced, and improved. A widespread adoption of recommended management practices by resource poor farmers of the tropics is urgently warranted. The soil C sequestration potential of this win-win strategy is finite and realizable over a short period of 20 to 50 years. Yet, the close link between soil C sequestration and world food security on the one hand and climate change on the other can neither be overemphasized nor ignored.

References

Baker, J.M., 2007. Tillage and soil carbon sequestration: What do we really know? *Agriculture, Ecosystems and Environment,* 118: 1–5.

Carlos, Martial Bernoux, Cerri, C., and Milne, Eleanor, 2006. Cropping systems, carbon sequestration and erosion in Brazil: A review. *Agron. Sustain. Dev.*, 26: 1–8.

FAO, 2001. Meeting on verification of country-level carbon stocks and exchanges in non-annex I countries. FAO, Rome.

Hepperly, Paul Reed, Moyer, Jeff, Pimentel, David, Douds, David Jr., Nichols, Kristine and Seide, Rita, 2008. Carbon sequestration in organic maize/soybean cropping systems. *16th IFOAM Organic World Congress*, Modena, Italy, June 16–20.

Kumar, Rajeev, Pandey, Sharad and Pandey, Apurv, 2006. Plant roots and carbon sequestration. *Current Science* 91(7): 885–890.

Lal, R., 2004. Soil carbon sequestration impacts on global climate change and food security. *Science,* 304 (5677): 1623–1627.

Pretty J. and Ball, A., 2002. Agricultural influences on carbon emissions and sequestration: A review of evidence and the emerging trade options. Centre for Environment and Society. *Occasional Paper 2001–03*.

Witt, K.G., Cassman, D.C., Olk, U., Biker, S.P., Liboon, M.I. Samson and Ottow, J.C.G., 2000. Crop rotation and residue management effects on carbon sequestration, nitrogen cycling and productivity of irrigated rice systems. *Plant and Soil,* 225: 263–278.

2013, Biodiversity Conservation for Sustainable Management   Pages *154–162*
Editor: Dr. K. Muthuchelian, *Vice Chancellor, Periyar University, Salem*
Published by: Daya Publishing House, NEW DELHI

Chapter 19

Biodiversity: Valuation of Ecosystem Services

M. Ravichandran*

Department of Environmental Management,
Bharathidasan Univeristy, Tiruchirappalli

ABSTRACT

Biodiversity is the base upon which ecosystem services exist. Conservation of biodiversity is imperative for the sustainability of services, which in turn facilitate the survival of all species on earth. Efforts at the global level began during 1992 Earth Summit in the form of Biodiversity Convention. However reaching targets set out in the convention is abysmally low by countries concerned. In this milieu, valuation of ecosystem services assumes importance, which is an exercise meant for environmenal managers. All countries may realize its biodiversity potential through learning the monetary value of services provided. Ecosystem services were construed as free goods in the past and which is why ecosystems components were misused and abused. Upon realizing the economic potential, valuation of the services began and has been made use from time to time. Robert Costanza made a poineering work in valuing ecosystem services in a more systematic manner. He estimated the value to be US$ 33 trillion per year and this creates an indelible impression in our mind regarding the importance of biodiversity.

Keywords: Biodiversity convention, Ecosystem services, Valuation.

---

* Corresponding Author: E-mail: muruguravi@yahoo.co.in

Introduction

Earth is unique amongst planets by virtue of its life sustaining capability. The annals of earth's history records its genesis to be 4.5 billion years, though life as such appeared 2 billion years ago. From the inception of unicellular organisms to the present, where complex multi cellular organisms like human evolved, earth has been providing all amenities conducive to life. A dynamic steady state, full of flux is the mechanism that prevails earth and its inhabitants. Homeostasis can be achieved through cyclic changes that occur in all three components viz. land, water and atmosphere, wherein all constituents, be it small or large, have a definite role to play. Man is also one of the components of earth's ecosystem, with right to use his/her share of space and resources. Human living as apart of this ecological web, utilising only his due and contributing his role in the cyclic reactions would pose no problem, as all other components are always functioning to the optimum. Problems arise when man tries to conquer nature, over exploit its resources and throw waste indiscriminately just to satisfy his own needs and greed, totally ignoring the rights of other organisms on earth. Even if we take the example of evolution, the constant interaction of environment and the genetic materials has lead to development of new species and each and every entity on earth had a role to play in the origin of new species. Similarly newer species would originate, down the line, in a few hundred years and some old species could face extinction. These processes have been existing over millions of years and the basis for this is the biodiversity—variability of life forms—of our planet. Entire earth had been rich in biodiversity when man lived with nature as an integral part it. Once civilisation and later industrial revolution exploded, man started his exploitation of nature at the expense of other inhabitants leading to disastrous consequences. Biodiversity has shrunk that now we have only a handful of biodiversity hot spots and that too many are threatened, facing extinction. The homeostasis of earth at present is skewed, imbalanced and disrupted that the mere existence of life on earth is in peril. Conservation of biodiversity or at least not to endanger it further, is the crux of the matter.

Analysing of humanity's ruthless exploitation of nature reveals the truth behind such carelessness, that all basic amenities on earth are provided free, as nature does not come under the realm of market. When no cost is incurred there is a tendency to over use, waste, hoard for future and exploit the share of the vulnerable. In such a context, is it possible to fix a cost for the oxygen we breathe, photosynthesis that produce food, nutrient cycle, water conservation, assimilation of waste done by nature, fertility of soil by micro organisms, climatic changes, monsoon, etc. If all these come under market what would be the cost of living, no human can exist if he needs to pay for all he gets from nature. Experiments have been conducted to create Biosphere II, artificial earth, with all non natural inputs and was found to be so expensive even to produce earth like conditions in a very small area. The main point to be reckoned is that whatever may be the unnatural inputs or synthetic inputs, the raw materials have to come from nature. Pricing of such non-market goods to some extent, with whatever methodology available, would create awareness among people regarding the importance of nature. Environmentalists have developed many methodologies for the same, ecological footprint measurement, ecosystem services valuation, environmental impact assessment are to list out a few. Of these ecosystem services valuation is central to this paper.

Valuation

Biodiversity sustains a significant population by means of providing all amenities and livelihood support. Primarily subsistence goods like food, clothing, building material, medicines and fodder for domesticated animals. Secondarily environmental services like maintenance of physcio-chemical properties of air, water and land, water shed, nutrient cycling, waste assimilation etc. Third is livelihood support by provision of tradable goods like wood, fish, food, medicines, wild life and genetic resources. Finally education and aesthetics like ecotourism, genetic stock for research to find new varieties of plants and animals. Natural capital stocks or biodiversity, provides the ground for all ecosystem services which are essential for life support systems of earth. If these services are stopped the earth would cease to exist but its importance is less understood, as most services are provided free of cost. The exercise of valuation of all the services, though an exact estimation is untenable, would reveal the types of services existing, bring out the approximate value of global services, inculcate the necessity to protect and conserve such systems and finally set a frame work for further research and new methodologies. Though estimation has been done by many groups of scientists, viz. A static general equilibrium input-output model of globe, estimation of maximum sustainable surplus value of ecosystem services, the valuation of the worlds ecosystem services and natural capital by Robert Costanza and his team in the year 1997 gives a holistic view of earth's value.

For the entire biosphere, the value is estimated to be in the range of US$ 16-54 trillion per year, with an average of US$ 33 trillion per year, with due respect to uncertainties this must be considered as a minimum estimate (Robert Costanza 1997). Ecosystem services will encompass all the natural cycles occurring on earth to maintain homeostasis and the capital natural stock available excluding the limited stocks like minerals and oil. For the purpose of valuation earth has been divided into 16 biomes.

A) Marine
1. Open ocean

      Coastal   2. Estuaries
                       3. Seagrass/algae
                       4. Coral reefs
                       5. Shelf

B) Terrestrial
      Forest   6. Tropical
                       7. Temperate
                       8. Grasslands

Wetlands   9. Tidal marsh/mangroves

                  10. Swamps/floodplains

                  11. Lakes/rivers

                  12. Desert

                  13. Tundra

                  14. Ice rocks

                  15. Croplands

                  16. Urban land

Ecosystem services maintain a flow of materials and energy which with human capital and manufactured goods provide the welfare of humanity. The impact of any alteration in the services would produce profound effect, for example the percentage of carbon di oxide in air is so negligible.08 per cent as compared to 70 per cent nitrogen, but its increase has led to global warming and its deleterious effects–a small change in atmospheric chemistry leading to a large scale effect. Large change in a small area like change in the composition of a forest by deforestation and afforestation, would affect the floras and fauna in that area leading to impacts on the well-being of the people nearby. For the sake of easy valuation, the ecosystem services are divide into 17 types. There is bound to be overlapping in the services provided by the biomes as all are interlinked and one cannot function in isolation of the other. A quick perusal of the list of ecosystem services listed out by Costanza, will help us understand the magnitude of its importance as all essentials necessary for life finds place in the list.

1. Gas regulation
2. Climate regulation
3. Disturbance regulation
4. Water regulation
5. Water supply
6. Erosion control and sediment retention
7. Soil formation
8. Nutrient cycling
9. Waste treatment
10. Pollination
11. Biological control
12. Refugia
13. Food production
14. Raw materials
15. Genetic resources
16. Recreation
17. Cultural

Valuation of non-market services and goods pose a problem as there is no tangible market to fix a price. In this context willingness to pay for these goods and services has been taken as the methodology to calculate the value. To cite a sample, the value placed to safeguard the tigers vary from individual to individual, one may understand the real value of tigers and wiling to pay $ 100 to protect them, another 20 and another nil as he may consider them useless. The average value of the WTP is taken as the value of the tiger. The unit value of the biome, the value of all the services provided by one hectare of the particular biome, is computed using any one of the following methodology: 1) the sum of the consumer and producer surplus or 2) the net rent or producer surplus or 3) price times quantity as a proxy for the economic value of the service. (1997, Costanza) The unit value was then multiplied by the actual hectares of sea or land area on earth with adequate addition and subtractions to prevent overlapping of services. Thus the total value added up to US$ 33,268 trillion per year. This is in fact 1.8 times higher than the earth's GNP (Gross National Product) which is only US$ 18 trillion per year. In another perspective we can consider replacing all these ecosystem services on earth and it will cost US$33 trillion per year, GNP must be raised to match it. This estimate is the minimal value as the real value would definitely be higher since many intricate services cannot be quantified. Moreover when the resources gets depleted or there is a disruption in the services provided the value too gets increased.

Table 19.1 shows the calculation of the value of ecosystem services for the year 1994, by Costanza *et al.,* and the total value computed is US$ 33,268 trillion per year. Marine area contributes maximum as its area is more compared to terrestrial and its role in nutrient cycling is also maximum. For marine area the value computed is 20,949 trillion US$ per year as against the land area which is 12, 319 trillion US$. Coastal sea area has more contribution towards ecosystem services due to estuaries, algae, coral reef ans shelf as against the open sea whose effect is more in gas regulation. Amongst terrestrial areas forests contribute maximum towards the value of ecosystem services and for tropical forests it is 3818 trillion as that of 894 trillion for temperate ones. Wetlands are another major contributor with 4879 trillion US$ as its value, while lakes and rivers are 1700 trillion. The desert, tundra, ice rocks and urban areas value is so negligible that they find no place in the calculation. We infer from the above table that Coastal sea, forests and wetlands are the maximum contributors of ecosystem services and the biodiversity hot spots are located only in these areas. This exercise would create an awareness about the importance of preservation of biodiversity as its value is immense in terms of monetary value, more than the GNP of all countries of the globe.

Viewing the table from the perspective of services, it is clear that nutrient cycle tops the table with a value of 17, 075 trillion US$, nearly half of the total value, emphasising the importance of the role of every living organism in the food chain. The biological diversity is the basis for nutrient cycling, any disturbance in the form of extinction of species will certainly create a malfunction in the process leading to food scarcity, starvation and death. Cultural value is computed to US$ 3015 trillion and this cannot be the real value since no rate can be fixed for aesthetics, culture, heritage, traditional values and ethics. Another important function is waste assimilation with a

Table 19.1: Valuation of ecosystem services.

Biomass	Area (ha x 10⁶)	1	2	3	4	5	6	7	8	9	10	11	12	13	14	15	16	17	Total Value Per ha ($ ha⁻¹ Yr⁻¹)	Total grand flow Value ($ ha⁻¹ Yr⁻¹)
Marine	36,302																		577	20,949
Open ocean	33,200	38							118			5		15	0			76	252	8381
Coastal	3,102			88					3,677			38	8	93	4		82	62	4,052	12,568
Estuaries	180								21,100			78	131	521	25		381	29	22,832	4,110
Seagrass/ algal	200			567					19,002						2				19,004	3,801
Coral reefs	62			2,750						58		5	7	220	27		3,008	1	6,075	375
Serif	2660								1,431			39		68	2			70	1,610	4,283
Terre- strial	15,323																		804	12,319
Forest	4,855		141		2	3	96	10	361	87		2		43	138	16	66	2	969	4,706
Tropical	1,900		223	5	6	8	245	10	922	87		4		50	315	41	112	22	2,007	3,813
Temperate/ boreal	2,955		88		0										25		36		302	894
Grass/ rangelands	3,898	7	0		3		29			87	25	23		67		0	2		232	906
Wetlands	330	133		4,539	15	3,800				4,177		304		256	106		574	881	14,785	4,879

Contd...

Table 19.1–Contd...

Biomass	Area (ha x 10^6)	1	2	3	4	5	6	7	8	9	10	11	12	13	14	15	16	17	Total Value Per ha ($ ha^{-1} Yr^{-1})	Total grand flow Value ($ ha^{-1} Yr^{-1})
Tidal marsh/ mangroves	165			1,839						6,696			169	466	162		658		9,990	1,648
Swamps/ floodplains	165		265	7,240	30	7,600				1,659			439	47	49		491	1,761	19,580	3,231
Lakes/ rivers	200				5,445	2,117				665				41			230		8,498	1,700
Desert	1,925																			
Tundra	743																			
Ice/rock	1,640																			
Cropland	1,400										14	24		54					92	128
Urban	323																			
Total	51,625	1,341	684	1,779	1,115	1,692	576	53	17,075	2,277	117	417	124	1,386	721	79	815	3,015	33,268	

1: Gas regulation; 2: Climate regulation; 3: Disturbance regulation; 4: Water regulation; 5: Water supply; 6: Erosion control; 7: Soil formation; 8: Nutrient cycle; 9: Wate treatment; 10: Pollination; 11: Biological control; 12: Habitat/refufia; 13: Food production; 14: Raw material; 15: Genetic resources; 16: Recreation; 17: Cultural.

value of US$ 2277 trillion, exemplifies the inherent capability of nature to remove the waste and bring back to original. At this juncture it is proper to remind our selves that we have crossed the limit of nature's resilience, and our planet is fast becoming a waste yard. Excess waste in the atmosphere (Green house gases) over and above its capacity to assimilate is cause of global warming and the threat faced by future generation. Gas regulation, climate, water regulation, food, soil formation, biological control, habitat, pollination are few other valuable services essential for the welfare and existence of life on earth.

Conclusion

Comprehension of ecosystem services and the role of biodiversity in its proper functioning had led to a world-wide awakening among nations in drafting policies to protect biodiversity. In this milieu, there are about six major global conventions towards protection of biodiversity; Convention on Biological Diversity (CBD), Convention on the Conservation of Migratory species (CMS), Convention on International Trade in Endangered Species of flora and fauna a (CITES), International Treaty on Plant Genetic Resources for Food and Agriculture (ITPGR), Convention on Wetlands (Ramsar Convention), and World Heritage Convention (WHC). Unprecedented loss of biodiversity and decline of ecosystem services triggered European nations to take action on a war footing, that in the year 2001 the Heads of EU declared that biodiversity decline should be halted and the time frame of 2010 was set to meet the goal. Achievement of this target is still far–fetched due to difficulties faced by these concerned nations. Lack of capacity building, lacunae in planning, lack of consistent framework for monitoring and assessment, mismatch in fund allocation are few maladies implicated in differential response of nations to biodiversity convention.

A comprehensive, consistent framework that would help achieve the targets set by 2010 and beyond is the need of the hour. One such conceptual framework developed is three pillared, wherein the implementation of biodiversity conventions at the national level with regional co-operation can be realized if the three sectors are given adequate support and integration. 1) *Integration of biodiversity targets in policies and programs of other sectors,* for example, projects like Environment Impact assessment, urban development, forestry, fisheries, trade, biotechnology, tourism, energy, should integrate biodiversity in their work to obtain positive results. 2) *Public awareness and support,* where a learned and informed stakeholders society be created by awareness programs, media, educational training and involvement of society in decision making process and 3) *Adequate funds provision,* nations should adopt measures to protect biodiversity hotspots, incentives and subsidies announced for biodiversity friendly activities, funds raised through private partnerships, so that the estimated US$ 12–45 billion be spent on achieving the targets and not a meager 6.5 billion as of the current global spending. Overall, a coordinated approach by governments, non-governmental organizations (NGOs), academics, conservationists and the public would help in the implementation of biodiversity convention on target. The role of academics in generation of biodiversity information, capacity building, creating framework for assessment and monitoring and valuation of biodiversity and ecosystem services lead to more research work and play a major role in realizing the goal in time.

References

Balaji, S., 2010. Biodiversity challenges ahead, *The Hindu*, May 27.

Costanza, Robert, *et al.*, 1997. The value of the world's ecosystem services and natural capital. *Nature,* 387(May).

Khera, Neeraj, Winkler, Sebastian and Krolopp, Andras, 2010. 'Policy response towards achieving the biodiversity targets–2010 and beyond: Research questions for developing indicators to assess progress'. *Journal of Resources Energy and Development,* 7(1): 11–28.

Valuing Ecosystem Services, 1998. Staff of world resource programme, World Resources, 1998–99.

2013, Biodiversity Conservation for Sustainable Management   Pages *163–166*
Editor: Dr. K. Muthuchelian, *Vice Chancellor, Periyar University, Salem*
Published by: Daya Publishing House, NEW DELHI

Chapter 20

Inventorying Current Status of Biodiversity in Different Ecosystems

P. Suresh, S. Manivannan and K. Vijayakumar
Department of Microbiology,
PKN Arts and Science College, Thirumangalam

Introduction

Biological diversity "bio diversity" can have many interpretations and it is most commonly used to replace the more clearly defined and long established terms, species diversity and species richness. biologists most often define biodiversity as the "totality of genes,species and ecosystem of a region". An advantage of this definition is that it seems to describe most circumstances and present a unified view of the traditional three levels at which biological variety has been identified.

Biodiversity

Biodiversity is the variation of life forms within a given ecosystem, biome, or on the entire Earth. Biodiversity is often used as a measure of the health of biological systems. The biodiversity found on Earth today consists of many millions of distinct biological species. The year 2010 has been declared as the International Year of Biodiversity. Biodiversity is not distributed evenly on Earth, but is consistently rich in the tropics and in specific localized regions such as the Cape Floristic Province; it is less rich in polar regions where fewer species are found.

☆ Species diversity
☆ Ecosystem diversity
☆ Genetic diversity

Rapid environmental modifications typically cause extinctions. Of all species that have existed on earth, 99.9 per cent are now extinct. Since life began on earth five major mass extinctions have led to large and sudden drops in the biodiversity of species. The phanerozoic eon (the last 540 million years) marked a rapid growth in bio diversity of species. In the Cambrian explosion–a period during which nearly every phylum of multi cellular organisms first appeared. The next 400 million years was distinguished by periodic, massive losses of bio diversity classified as mass extinction events. The most recent. The cretaceous–tertiary extinction event. Occurred 65 million years ago, and has attracted more attention than all others because it killed the non avian dinosaurs

Ecosystem

An ecosystem consists of all the organisms living in a particular area, as well as all the nonliving, physical components of the environment with which the organisms ineract, such as air, soil, water, and sunlight. It is all the organisms in a given area, along with the nonliving (a biotic) factors with which they interact; a biological community and its physical environment. The entire array of organisms inhabiting a particular ecosystem is called a community.

Types of Ecosystems

The term was use Ecosystem is a functional unit consisting of living things in a given area, non-living chemical and physical factors of their environment, linked together through nutrient cycle and energy flow.

1. Natural
2. Terrestrial ecosystem
3. Aquatic ecosystem
4. Lentic, the ecosystem of a lake, pond or swamp.
5. Lotic, the ecosystem of a river, stream or spring.
6. Artificial, environments created by humans

First by wildlife scientist and conservationist Raymond F. Dasmann in a lay bio advocating nature conservation. The term was not widely adopted for more than a decade, when in the 1980s it and "bio divdersity" came into common usage in science and environmental policy. Use of the term by Thomas Lovejoy in the Foreword to the book credited with launching the field of conservation biology introduced the term along with "conservation biology" to the scientific community. Until then the term "natural diversity" was used in conservation science circles, including by The Science Division of The Nature Conservancy in an important 1975 study, "The Preservation of Natural Diversity." By the early 1980s TNC's Science program and its head Robert E. Jenkins, Lovejoy, and other leading conservation scientists at the time in America advocated the use of "biological diversity" to embrace the object of biological conservation.

The term's contracted form biodiversity may have been coined by W.G. Rosen in 1985 while planning the National Forum on Biological Diversity organized by the

National Research Council (NRC) which was to be held in1986, and first appeared in a publication in 1988 when entomologist E.O. Wilson used it as the title of the proceedings of that forum.

Since this period both terms and the concept have achieved widespread us among biologists, environmentalists, political leaders, and concerned citizens worldwide. The term is sometimes used to equate to a concern with the expansion of concern over extinction observed in the last decades of the 20th century.

A similar concept in use in the United States, besides natural diversity, is the term "natural heritage." It pre-dates both terms though it is a less scientific term and more easily comprehended in some ways by the wider audience interested in conservation. Furthermore it may be misleading if used to refer only to biodiversity, as natural heritage also includes geology and landforms (geo diversity). The term "Natural Heritage" was used when Jimmy Carter set up the Georgia Heritage Trust while he was governor of Georgia; Carter's trust dealt with both natural and cultural heritage. It would appear that Carter picked the term up from Lyndon Johnson, who used it in a 1966 Message to Congress. "Natural Heritage" was picked up by the Science Division of the US Nature Conservancy when, under Jenkins, it launched in 1974 the network of State Natural Heritage Programs. This network took on a life of its own in the 1990s when it became an independent non-profit organization named Nature Serve. When Nature Serve was extended outside the USA the term "Conservation Data Center" was suggested by Guillermo Mann is now also used by several programs, for example those that operate as part of Nature Serve Canada.

Biological diversity "or "biodiversity" can have many interpretations and it is most commonly used to replace the more clearly defined and long established terms, species diversity and species richness. Biologists most often define bio diversity as the "totality of genes, species, and ecosystems of a region". An advantage of this definition of this definition is that it seems to describe most circumstances and present a unified view of the traditional three levels at which biological variety has been identified.

As one approaches Polar Regions one generally finds fewer species. Flora and fauna diversity depends on climate, altitude, soils and the presence of other species. In the year 2006 large numbers of the Earth's species were formally classified as rare or endangered or threatened species; moreover, many scientists have estimated that there are millions more species actually endangered which have not yet been formally recognized. About 40 percent of the 40,177 species assessed using the IUCN Red List criteria, are now listed as threatened species with extinction –a total of 16,119 species. Biodiversity found on Earth today is the result of 3.5 billion years of evolution. The origin of life has not been definitely established by science, however some evidence suggests that life may already have been well-established a few hundred million years after the formation of the Earth. Until approximately 600 million years ago, all life consisted of archeo, bacteria, protozoans and similar single-celled organism.

The history of biodiversity during the Phanerozoic (the last 540 million years), starts phylum of multicellular organisms first appeared. Over the next 400 million years or so, global diversity showed little overall trend, but was marked by periodic, massive losses of diversity classified as mass extinction events.

Judicial Status

A great of work is occurring to preserve the natural characteristics of Hopetoun Falls, Australia while continuing to allow visitor access.

Biodiversity is beginning to be evaluated and its evolution analysed (through observations, inventories, conservation etc.) as well as being taken into account in political and judicial decisions:

☆ The relationship between law and ecosystems is very ancient and has consequences for biodiversity. It is related to property rights, both private and public. It can define protection for threatened ecosystems, but also some rights and duties (for example, fishing rights, hunting rights).

☆ Law regarding species is a more recent issue. It defines species that must be protected because they may be threatened by extinction. The U.S. Endangered Species Act is an example of an attempt to address the "law and species" issue.

☆ Laws regarding gene pools are only about a century old. While the genetic approach is not new (domestication, plant traditional selection methods), progress made in the genetic field in the past 20 years have led to a tightening of laws in this field. With the new technologies of genetic analysis and genetic engineering, people are going through gene patenting, processes patenting, and a totally new concept of genetic resources. A very hot debate today seeks to define whether the resource is the gene, the organism itself, or in DNA.

Conclusion

The current status of infusion of biodiversity conservation related issues a school college levels that can lead to bio diversity conservation is far from adequate. Ecosystem is having many species and most of them are in state of endangered. As a human and we are part of a ecosystem, conserve nature by joining hands together against the human activity on bio diversity. Biodiversity cannot be learned without the use of good visuals.at the college level the ugc has now developed a common curriculum for all undergraduate courses and has assigned the task of developing a text book for the common paper on environment to the BVIEER.bio diversity and its conservation is substantial input in core module course curriculam of the UGC.

References

Rana, S.V.S., 2005. *Essentials of Ecology and Environment Science,* 2nd edn.

The World Book Encyclopedia (International). Scott Fetzer Company.

2013, Biodiversity Conservation for Sustainable Management   Pages 167–181
Editor: Dr. K. Muthuchelian, Vice Chancellor, Periyar University, Salem
Published by: Daya Publishing House, NEW DELHI

Chapter 21

Threats of Climate Change on Global Water, Food and Economy

P.M. Natarajan[1], N. Ramachandran[2], Shambu Kallolikar[2] and A. Ganesh[3]

[1]Periyar Maniammai University
[2]Project Director and Member Secretary,
Tamil Nadu AIDS Control Society, Chennai
[3]Geography Department, Bharathidasan University

ABSTRACT

The last 50 years of the Earth was likely the hottest in at least the past 1,300 years. Moreover, 11 of the past 12 years rank among the warmest since humans began taking accurate temperature measurements in the 1850s, a record of extremes so pronounced; it is unlikely to be due to chance. Since 1980, the average temperatures have risen well above the 14 degrees Celsius average for the span from 1951 to 1980, which is defined as the norm. During the 1980s, the global temperature averaged 14.26 degrees Celsius. In the 1990s it was 14.38 degrees Celsius. During the first three years of this decade (2000-2002), it has been 14.52 degrees Celsius and now it is 14.76 degrees Celsius.

Many natural calamities like droughts, floods, cyclones, heat waves, generation of unknown diseases, glacial retreat, forest fires, coral bleaching etc., have been intensified and they are the standing examples which suggest that they are primarily due to the impact of manmade warming and climate change.

The causes for Global Warming are split up into two groups, man-made or anthropogenic and natural.

The natural causes are causes created by nature. One natural cause is a release of methane, a greenhouse gas from arctic tundra and wetlands as well as from agriculture. Another natural cause is that the Earth goes through a cycle of climate change. This climate change usually lasts about 40,000 years.

There are many man-made causes by which a huge quantity of greenhouse gages emission. Emission of greenhouse gases is one of the biggest man-made Pollution. Pollution comes in many shapes and sizes. Burning fossil fuels such as coal, or oil is one thing that causes pollution. When fossil fuels are burned they give off a green house gas called CO_2. When coal or oil is mined methane escapes and cause pollution.

Another major man-made cause of Global Warming is population. More people mean more food, and more methods of transportation etc. That means more methane due to agriculture and more carbon dioxide due to the burning of fossil fuels for the transportation of agricultural operations. Agriculture is going to be damaged by Global Warming, but now agriculture is going to help cause Global Warming. The farm animals which help farm operations also produce methane gas which is a greenhouse gas. Another source of methane is manure. Because more food is needed we have to raise food by using manure. Another problem with the increasing population is transportation. More people means more cars and more cars means more pollution.

Since CO_2 contributes to global warming, the increase in population makes the problem worse because we breathe out CO_2. Also, the trees that convert our CO_2 to oxygen are being demolished because we are converting the land occupied by forest for agriculture and for our homes and buildings. We are not replacing the trees (an important part of our eco system), so we are constantly taking advantage of our natural resources and giving nothing back in return. In this way due to anthropogenic causes the climate is threatened by the huge generation of greenhouse gases.

As a result of the greenhouse gases emission the temperature is rising in the last 50 years and threatening the water and food resources in a bigger way. The economy is also under threat.

The volume of fresh water available in the form of snow and glaciers in the snow covered land areas in the Earth is 63.57 lakh TMC (thousand million cubic feet). This water is under threat due to climate change. The value of water in terms of food grain production is about Rs. 174 million crores. As per the World Health Organization standard of supplying domestic ware supply of 135 lpcd (litre per capita per day), it is possible to supply to the present world population for 554 years, or to India for 3,045 years, or to Tamil Nadu for 58,321 years, or to Chennai city for 7.31 lakh years.

Due to the rise of temperature it has been assessed that the world food grain production is going to reduce to about 125 million tons to 208 million tons per annum and as a result of this the hungry people' population is going to increase to 1.57 billion during 2025 from the present 1.02 billion.

According to Nicholas Stern, Chief Economist and Senior Vice-President of the World Bank, if we take no action to control emissions, each metric ton of CO_2 that we emit now is causing damage worth at least $85 and he further indicates that an investment of one percent of global GDP is required to mitigate the effects of climate change, with failure to do so risking a recession worth up to twenty percent of global GDP. According to ET, John Llewellyn Lehman Brothers global economist, for every 2 degree Celsius rise in temperature would result in a 3 per cent dip in global GDP. If the emission is not controlled some researchers indicate that there will be an economy loss of about $140 billion per annum globally.

Therefore, this paper explains about the threats of global warming and climate change on the global water, food and economy resources and the mitigation strategies to arrest the climate change.

Keywords: *Global warming and climate change, Threats of climate change on global water, food and economy, Sea level rise, Water refugees, Mitigation of climate change.*

Introduction

There will always be uncertainty in understanding a system as complex as the world's climate. However there is now strong evidence that significant global warming is occurring. Global warming is now being found almost everywhere in the world, from the tops of mountains, where glaciers are in retreat, to ocean deeps, where the average water temperatures have increased all the way down to depths of 3,000 metres because of the warming effect of a hotter atmosphere. The evidence comes from direct measurements of rising surface air temperatures and subsurface ocean temperatures and from phenomena such as increases in average global sea levels, retreating glaciers, and changes to many physical and biological systems. It is likely that most of the warming in recent decades can be attributed to human activities. This warming has already led to changes in the Earth's climate. The existence of greenhouse gases in the atmosphere is vital to life on Earth–in their absence average temperatures would be about 30 degrees Celsius lower than they are today. But human activities are now causing atmospheric concentrations of greenhouse gases–including carbon dioxide, methane, troposphere ozone, and nitrous oxide–to rise well above pre-industrial levels.

Carbon dioxide levels have increased from 280 ppm in 1750 to over 375 ppm today–higher than any previous levels that can be reliably measured (*i.e.* in the last 420,000 years) and possibly the past 20 million years. The projected future increase in temperature as a result of increases in GHGs (greenhouse gases) is much more rapid and larger than any change during the past 10,000 years (IPCC 2001), and the increase would occur relative to an already warm, non-ice-age climate. Without the naturally occurring "greenhouse effect," the average temperature of the Earth would be about 35°C (63°F) colder than at present, or an inhospitable –20°C (–4°F) (McElroy 2002). Carbon dioxide is a greenhouse gas that has been increasing remarkably, mostly as a result of the combustion of fossil fuels such as coal, petroleum and natural gas. The atmospheric concentration of carbon dioxide is increasing at the rate of about 1.3 times per annum now. If no other measures than those taken at present are introduced, the Carbon dioxide is expected to reach 500 ppmv (parts per million by volume) in 2050 and 700 ppmv in 2100, and to continue to increase for many centuries to come. Since pre-industrial times, atmospheric concentrations of carbon dioxide (CO_2), methane (CH_4), and nitrous oxide (N_2O) have climbed by over 36 percent, 148 percent, and 18 percent, respectively due to human influence. Humans have already caused so much damage to the atmosphere that the effects of global warming will last for more than 1,000 years.

Increasing greenhouse gases are causing temperatures to rise; the Earth's surface warmed by approximately 0.6 to 0.76 centigrade degrees over the twentieth century. The Intergovernmental Panel on Climate Change (IPCC) projected that the average global surface temperatures will continue to increase to between 1.4 degrees Celsius and 5.8 degrees Celsius (2.5–10.4°F) above 1990 levels, by 2100.

It is undisputed that humans are entirely responsible for the increase in atmospheric CO_2 over the past few centuries. In pre-industrial times, large natural sources of CO_2 were balanced by equally large natural removal processes, such as photosynthesis in plants, maintaining a stable level of CO_2 in the atmosphere for thousands of years. Human-produced emissions upset this balance. Because human-produced emissions are not completely absorbed by natural processes, they accumulate in the atmosphere, increasing the concentration of CO_2.

CO_2 has a long lifetime in the atmosphere. A quarter of present-day emissions will remain in the atmosphere after several centuries. Some will still be there in a millennium. Some other green house gages (such as sulfur hexafluoride) have even longer lifetimes. Delaying action will make it more difficult to stabilize GHG concentrations at levels that would prevent severe consequences (O'Neill and Oppenheimer 2002). We cannot just sit back and wait for advanced technologies that reduce GHGs– they may never be developed without mandatory emissions limits and incentives for developing GHG-reducing technologies. These require public acceptance and political will, both of which take time to develop.

According to IPCC estimate, the current annual fossil-fuel emissions amount to about 7 billion tons of carbon. Under the IPCC's business-as-usual estimate, they would rise to about 16 billion by 2050 and 29 billion by 2100, partly because of a greater use of coal. The corresponding atmospheric concentrations of carbon dioxide would reach 735 parts per million (ppm) by 2085, in contrast to the pre–industrial level of 280 ppm and today's level of 380 ppm.

Carbon dioxide can remain in the atmosphere for many decades. Even with possible lowered emission rates we will be experiencing the impacts of climate change throughout the 21st century and beyond. Failure to implement significant reductions in net greenhouse gas emissions now will make the job much harder in the future.

Therefore every one should understand that global warming and climate change are as real as the Sun and protect the Mother Earth from man made warming.

If the greenhouse emission is not stopped it is going to threaten the global water, food and economy resources as well as to intensify almost all the natural calamities.

Global Warming Consequences as per the IPCC

The IPCC (Intergovernmental Panel on Climate Change) released its report in April 2007. This report does confirm the cause and effects of global warming, which have already been known for years. The report urges mankind to start acting quickly to reduce the emissions of greenhouse gases so as to mitigate severe effects on our environment.

The General global warming consequences are 1. Increasing number of deaths as a consequence of heat waves, floods, droughts, tornadoes and other extreme weather conditions 2. More and larger fires in woods 3. Within a couple of decades, hundreds of millions of people will not have enough water 4. Reduction of the biological diversity on Earth: 20 to 30 percent of all species are expected to be extinguished. This will have severe consequences on the respective food chains 5. The increase of the sea level is expected to force tens of millions of people per year to move away from coastal areas within the next decades 6. Melting of glaciers: Small glaciers will disappear entirely, larger ones will shrink to about 30 per cent of their current size 7. Change in agricultural yields will force many people (in particular for warmer countries) to migrate into other areas of the world. Hundreds of millions of people are facing starvation by the year 2080 as an effect of global warming 8. Come back of diseases like malaria into areas, where they have previously been extinguished.

Effects of Global Warming in Geographical Regions

The IPCC has predicted and summarized the general effects of global warming on geographical regions and they are given below.

☆ *Plains:* The plains will have a lot more fires than expected because of global warming. Global Warming will add extra heat to that region, in addition to the heat they already have there.

☆ *Coastal:* The coastal regions will flood because of all the extra rain. This will cause many inhabitants to leave their homes and flee.

☆ *Polar:* A large portion of the Polar Ice Cap might melt making sea level rise 16-20 feet. A rise of 20 feet in the sea level could cover most of the state of Florida.

☆ *Forests:* Some forests will dry up and become deserts by the middle of this new century. Also because of the increase in thunderstorms there could be more forest fires because of the lightning and the dry wood from the trees.

☆ *Mountains:* The snow covered mountain peaks will probably melt causing floods and very dangerous mudslides.

☆ *Deserts:* Most deserts will become much hotter than they usually are making it much harder to cross than it is today.

☆ *Fringe:* A fringe land, which is the fertile area in between a desert, such as an oasis, will probably dry up and become part of the desert and no longer be fertile.

Threats of Climate Change on Water and Food Resources

Of all the threats of global warming, its threat on water and food resources is considered the biggest one.

On Water Resources

Due to rapid population growth, the potential water availability of Earth's population decreased from 12,900 m3 per capita per year in 1970 to 9 000 m3 in 1990, and to less than 7,000 m3 in 2000. The global availability of freshwater is projected to drop to 5,100 m3 per capita per year by 2025. This amount would be enough to meet individual human needs if it were distributed equally among the world's population.

In densely populated parts of Asia, Africa and Central and Southern Europe, current per capita water availability is between 1,200 m^3 and 5,000 m^3 per year.

According to Population Action International, based upon the UN Medium Population Projections of 1998, more than 2.8 billion people in 48 countries will face water stress or scarcity conditions by 2025. Of these countries, 40 are in West Asia, North Africa or Sub-Saharan Africa. Over the next two decades, population increases and growing demands are projected to push all the West Asian countries into water scarcity conditions. By 2050, the number of countries facing water stress or scarcity could rise to 54, with their combined population being 4.5 billion people—about 48 per cent of the projected global population of 9.4 billion(UNEP, 2002).

Global warming is expected to account for about 20 percent of the global increase in water scarcity in this 21st century. It is predicted that global warming will alter precipitation patterns around the world, melt mountain glaciers, and worsen the extremes of droughts and floods. Alteration in precipitation is likely to decrease the precipitation where already less and in such a situation the scarcity region may further experience severe drought.

Global water consumption increased six fold in the last century—more than twice the rate of population growth—and will continue growing rapidly in coming decades. Yet readily available freshwater is a finite resource, equivalent to less than one percent of the water on Earth. But the economically available water for human use is only 0.007 per cent.

The peculiar status of the world is that, water and populations are unevenly distributed across the globe; arid and semi-arid regions receive only two percent of all surface runoff yet account for 40 per cent of the global land area and house half of the world's poor. Finally, our existing freshwater resources are under heavy threat from overexploitation, pollution, and global warming. Given these trends, equitably providing adequate water resources for agriculture, industry and human consumption poses one of the greatest challenges of the 21st century.

Rising global temperatures is speeding the melting of glaciers and ice caps and results in early ice thaw on rivers and lakes. Global glacial studies clearly suggest that a quarter of the world's glaciers could disappear by 2050 and half by 2100 as a result of the projected temperature rise of 1.4 to 5.8 degree Celsius at that point of time. In that case the freshwater resources status of the entire globe will be very grim during that period.

The land snow and glaciers in the ice caps are the sources of water for domestic, irrigation, industry, fish culture, navigation, hydropower and environment in many countries. One billion Chinese, Indians and Bangladeshis, 250 million people in Africa, and the entire population of California, United States, are among the 3 billion people who rely on the continuous flow of fresh and clean snow melt mountain water.

Now glacier retreat is occurring everywhere in the Earth. Between 1980 and 2001, the thickness of 30 major mountain glaciers decreased by an average of 6 metres. The Kutiah Glacier in Pakistan holds the record for the fastest glacial surge. In 1953, it receded more than 12 kilometres in three months, averaging about 112 metres per day. Between 1962 and 2000 the Kilimanjaro has lost approximately 55 per cent of its glaciers. Between 1980 and 2001 by an average about 6 metres thickness of glaciers decreased in 30 major mountains.

In humid parts of the world, mountains provide 30 per cent to 60 per cent of downstream freshwater. In semi-arid and arid environments, they provide 70 per cent to 95 per cent. Such a freshwater supply ice caps are retreating and hence many people who are dependent on the above source of water for their water, food and livelihood securities are going to suffer heavily in future.

Half the world's population (about three billion people) depends on rivers starting from mountain glaciers as their freshwater source. Himalayan glaciers feed 7 major Asian rivers—the Ganges, Indus, Brahmaputra, Salween, Mekong, Yangtze and Huang He—ensuring a year-round water supply for two billion people. Shrinking glaciers, usually attributed to global warming, threaten the uniformity of the flows of these major rivers, and hence threaten the water supplies of billions of people. The shrinking glaciers of the Andes Mountains in Latin America, the Rocky Mountains in the western US and the Alps of Europe probably account for the remaining one billion people whose water supplies are put at risk as a result of shrinking glaciers.

In the whole of the Himalayan Range, there are 18,065 glaciers with a total area of 34,659.62 km2 and a total ice volume of 3,734.4796 km3–1.32 lakh TMC (Qin Dahe 1999). This includes 6,475 glaciers with a total area of 8,412 km2, and a total ice volume of 709 km3 in China. The major clusters of glaciers occur in and around the following ten Himalayan peaks and massifs: Nanga Parbat, the Nanda Devi group, the Dhaulagiri massif, the Everest-Makalu group, the Kanchenjunga, the Kula Kangri area, and Namche Bazaar.

About 67 per cent of the nearly 12,000 square miles of Himalayan glaciers are receding in the last 10 years in this mountain and, as the ice diminishes, glacial runoffs and river flows in summer will also go down, leading to severe water shortages.

According to estimates by China's Academy of Sciences, that country's highland glaciers are shrinking by an amount equivalent to all the water in the Yellow River each year. Hence, the countries surrounding the Himalaya are going to be heavily affected.

The mountains and ice caps cover about 5. 55 lakh sq km and contain about 1, 80,000 cubic km (6.36 million TMC) frozen water. This may be the source for the future generation in the present context of dwindling water resources. At the present rate of glacier retreat it is estimated that within 30 to 50 years period, all the ice caps in the globe are going to disappear and as a result of this many countries are going to loose a substantial quantity of fresh water. Since this source of water may not be available to the future generation due to the global warming induced huge glacier retreat, many countries are going to face shortage of water for domestic, irrigation, industrial, hydropower generation and fish culture. Therefore, absence of glacier melts is going to affect the livelihood security of farmers as well as the agricultural economy of many countries very severely since many developing countries depend on its use for their growth.

Salinisation of Fresh Water Due to Intrusion and Inundation of Seawater

Studies indicate that warming is projected to cause the global mean sea level to rise by between 0.1 metre and 0.9 metre from 1990 to 2100. Sea-level rise will affect coastal regions throughout the world, causing flooding, erosion, and saltwater intrusion into aquifers and freshwater habitats.

As a result of the anticipated sea level rise due to global warming during 2100 there will not only migration of people, but also the coastal area aquifers not only in India but in the entire global coastal sedimentary aquifers where there is appreciable quantity of fresh groundwater are likely to be further salinised to a greater extent and depth.

Already the high tides in the Indian and in the other coastal regions of the globe are salinising the groundwater slowly. In addition to that the 26th December 2004 tsunami seawater inundation has salinised about 15 to 20 TMC of fresh groundwater in the South Indian coastal aquifers and about 5 TMC of groundwater in Tamil Nadu alone. Similarly there is salinisation of groundwater in the areas where it occurs frequently.

Glacial Retreat will Affect the Sharing of Water

Already about 20,000 TMC of water is being shared globally and there are programmes to share about another 30,000 TMC of water in the immediate future. Therefore, the glacial retreat is not only going to affect the ongoing diversion of water from one basin to other in the globe but also it is going to affect the future programmes of inter basin transfer of water in many countries. This may likely to affect the proposed diversion of the surplus Indian rivers with the water scarcity basins of this country also. As per the Indian interbrain water transfer programme, it is proposed to transfer about 303 TMC of water to Tamil Nadu State. The Himalayan as well as the other snow clad Indian mountains' snow and glacier retreat in India is likely to affect the transfer of water from the surplus Indian rivers not only to Tamil Nadu but also to the water scarcity regions and basins of this country.

Value of Snow and Glacier Water in Terms of Food and Domestic Water Supply

The volume of fresh water in the form of snow and glaciers in the land area of the Earth is 63.57 lakh TMC (thousand million cubic feet). This water is under threat due to climate change. The value of water in terms of food grain production is about Rs.174 million crores. As per the World Health Organization standard of supplying domestic ware supply of 135 lpcd (litre per capita per day), it is possible to supply to the present world population for 554 years, or to India for 3,045 years, or to Tamil Nadu for 58,321 years, or to Chennai city for 7.31 lakh years.

Hence, the Climate Change is likely to be a big threat not only to the snow water but also to the coastal aquifers. Therefore, not only India but the globe is likely to suffer a lot in future as a result of water scarcity due to the melting of snow water as well as due to the salinisation of the fresh water coastal aquifers. Hence, the impact of the uncontrolled anthropogenic Climate Change is not only going to generate a huge *"global warming refugees"* due to the abnormal sea level rise in the future but also *"water refugees"* as a result of severe water scarcity problem everywhere in the globe.

On Food Grain Production

If one degree Celsius increases in the atmosphere, wheat production loss will be of 4-5 million tons and there will be a loss of 10-40 per cent in crop production by 2100. It is estimated, for example, that the production of winter wheat will decrease by 55 per cent in India and 15 per cent in China by the year 2100 in the most populous countries. This will in turn have serious impact on Japan, which depends heavily on other countries for its food supply. Warming of 4 degree Celsius or more is likely to seriously affect global food production.

A study published in Science suggest that, due to climate change, "southern Africa could lose more than 30 per cent of its main crop, maize, by 2030. In South Asia losses of many regional staples, such as rice, millet and maize could drop 10 per cent".

In Japan, the impact of global warming is already being seen in the production of rice, the country's stable food, and wheat. While rice harvests are expected to increase in the Hokkaido and Tohoku regions, wheat production is likely to decrease in all regions. It is also possible that global warming will trigger frequent natural disasters, including accelerated activity of weeds and harmful insects, allowing harmful insects from the tropical and subtropical zones to spread to the temperate zone and damage harvests.

Crop ecologists at the International Rice Research Institute in the Philippines have recently reported that rice fertilization falls from 100 per cent at 34 degrees Celsius (93 degrees F) to essentially zero at 40 degrees (104 degrees F). Scientists in the U.S. Department of Agriculture are seeing a similar effect of high temperature on other grains. The scientific rule of thumb is that a 1 degree Celsius rise in temperature above the optimum reduces grain yields by ten per cent.

According to the recent estimate, the global warming is going to decrease the food production of the developing countries at an average of 10 to 25 per cent by the 2080s. The decline in India could be about 30 to 40 per cent, Sudan 56 per cent, Senegal 52 per cent and USA about 25 to 35 per cent.

The corn plant, a highly productive crop that accounts for 70 per cent of the U.S. grain harvest, is particularly vulnerable to heat. In a heat-stressed field, leaves curl in order to reduce moisture loss through evaporation. Under these conditions, photosynthesis declines and the plant switches from a growth path to a survival mode, reducing yields.

In the long run, the climatic change could affect agriculture in several ways: 1. Productivity-in terms of quantity and quality of crops, 2. Agricultural practices–through changes of water use (irrigation) and agricultural inputs such as herbicides, insecticides and fertilizers, 3. Environmental effects–in particular in relation of frequency and intensity of soil drainage (leading to nitrogen leaching), soil erosion, reduction of crop diversity, 4. Rural space-through the loss and gain of cultivated lands, land speculation, land renunciation, and hydraulic amenities, 5. Adaptation-organisms may become more or less competitive, as well as humans may develop urgency to develop more competitive organisms, such as flood resistant or salt resistant varieties of rice.

The food demand to the projected population of 8 billion during 2025 is 3 billion tons. The area irrigated in the world is about 276.79 million hectares (World Recourses Institute 2004, UN, FAO). At the rate of 0.45 ton and 0.75 ton per hectare of grain loss due to the temperature of rise of about 1 degree Celsius the anticipated production loss per annum will be about 125 million tons to 208 million tons respectively. Therefore, the expected food grain production will be not 3 billion tons but minus 125 million tons to 208 million tons. At the rate of 350 gram per capita per day about 10 million to 16.3 million people will be under hunger at that point of time due to so much loss in grain production. But according to Nicolas Stern there will be additionally about 550 million hungry people in addition to the present 1.02 billion increasing to 1.57 billion. About 4 tons of food grain per ha of land has to be produced to achieve the production of 3 billion tons during 2025. If this much production is not achieved, about 20 per cent to 30 per cent of the people of the globe at that point of time will be under hunger, if the average world production rate of 2 tons per ha is maintained. As a result of temperature rise India will loose about 6 million tons and in Tamil Nadu the grain production loss will be varying between one lakh tons to two lakh tons per annum. If there is no check on global warming, its effect on water and food grain production will be the biggest threat and this will affect the very existence of the people.

Impact of Warming on the Global Economy

Many estimates of aggregate net economic costs of projected damages and benefits from climate change across the globe are now available. These are often expressed in terms of the social cost of carbon (SCC), the aggregate of future net benefits and costs, due to global warming from carbon dioxide emissions, which are discounted to the present. Peer-reviewed estimates of the SCC for 2005 have an average value of US$43 per ton of carbon (tC) (*i.e.*, US$12 per ton of carbon dioxide, tCO_2) but the range around this mean is large. For example, in a survey of 100 estimates, the values ran from US$-10 per ton of carbon up to US$350/tC.

One of the most widely noted projections on this issue is the Stern Review, a 2006 report by the former Chief Economist and Senior Vice-President of the World

Bank Nicholas Stern, predicts that climate change will have a serious impact on economic growth without mitigation. The report suggests that an investment of one percent of global GDP is required to mitigate the effects of climate change, with failure to do so risking a recession worth up to twenty percent of global GDP.

Many economists have supported Stern's approach, or argued that Stern's estimates are reasonable, even if the method by which he reached them is open to criticism.

The United Nations IPCC concludes with 33 to 67 per cent confidence that the aggregate market sector effect of a small increase in global temperatures could be "plus or minus a few percent of world GDP". Developed countries are more likely to experience positive effects and developing countries are more likely to experience negative effects. Larger temperature rises would be more adverse across the board.

The IPCC says GDP growth will slow by up to 0.12 per cent a year by 2030 if greenhouse gases are stabilized.

Nicolas Stern suggests that climate change threatens to be the greatest and widest-ranging market failure ever seen, and he provides prescriptions including environmental taxes to minimize the economic and social disruptions. According to him there are schemes that allow people to reduce emission in CO_2 by spending less than $25 a ton and the benefits over time of actions to shift the world onto a low-carbon path could be in the order of $2.5 trillion each year. If we take no action to control emissions, each metric ton of CO_2 that we emit now is causing damage worth at least $85.

According to ET, John Llewellyn Lehman Brothers global economist, for every 2 degree rise in temperature would result in a 3 per cent dip in global GDP. The next 2 degrees would do even more damage to the economy. However for India the effects are likely to be much more harmful. For every 2 degree rise in temperature the effect on GDP is 5 per cent and for the next 6 degrees it would be 15-16 per cent.

According to a report by University of California, Berkeley, agricultural and resource economics professors David Roland-Holst and Fredrich Kahrl, the losses due to climate change in Calibornia State USA are as follows.

An estimated $5 billion in assets are at risk, and costs could reach $600 million a year in what the researcher calls the "high-warming scenario." Energy: $21 billion in assets at risk, with annual damage ranging from $2.7 billion to $6.3 billion. Potential impacts could include less hydropower due to less rainfall; more hot days requiring greater use of air conditioning; and more winter storms causing more power shortages. Transportation: $500 billion at risk to ports, airports, roads and bridges. Tourism and recreation: $98 billion in assets are at risk, with annual damage ranging from $200 million to $7.5 billion. Real estate and insurance: $2.5 trillion in assets are at risk, and water damage could cost $1.4 billion a year, while fire damage could result in $2.5 billion in damages. Agriculture, forests and fisheries: $113 billion in assets are at risk, with annual damage ranging from $300 million to $4.3 billion. Public health: Annual costs due to atmospheric changes range from $3.8 billion to $24 billion a year

According to Climate scientists, economists and policy researchers the limiting of the long-term global warming is achievable at a "negligible" cost. Now, the responsibility for action lies in the hands of politicians, they say.

According to the Intergovernmental Panel on Climate Change report 2007, the stabilization of the greenhouse gases in the atmosphere will cost between 0.2 per cent and 3.0 per cent of global GDP by 2030.

In the next decade, the annual cost of global warming will hit $150bn a year–that's five times the annual earnings of the entire population of Nigeria.

Approaches to Arrest the Warming

As per the IPCC report, CO_2 concentrations since the 1800's have increased from 280 parts per million (ppm) to 380 ppm, causing global temperatures to increase by about 0.76 degree Celsius. To undo that warming, we would have to return the CO_2 concentration to its pre-industrial level. This would require removing 800,000 million metric tons of CO_2 from the atmosphere.

Model calculations indicate that to avoid a temperature increase of 3.6°F, we must stabilize CO_2 concentrations at about 450 parts per million (ppm) or less. To achieve this target by 2050, emissions must be about 50 percent less than today, and by the end of the century 75 per cent less. To avoid the tipping point, global CO_2 emissions should peak no later than 15 years from now, and then begin to decrease. Though it will be a tall order to reduce to this extent, for the betterment of the human race we have to act immediately to curtail emission and bring it to this level.

Reducing CO_2 emissions by 75 per cent will require a profound change in the way we produce and use energy, but there is no need for panic or despair. If we get started now, we can make this transition slowly, a per cent or two each year. It's a job that our children and grandchildren will continue to work on through the end of the century, but we can start today.

If we don't stabilize CO_2 concentrations at about 450 parts per million (ppm) or less then the entire Greenland will disappear. Rising temperatures could lead to melting permafrost in the arctic tundra, releasing large deposits of carbon dioxide and methane, an even more powerful greenhouse gas. To avoid the worst consequences of global warming, we will have to limit the increase in average global temperature to roughly 2°C above pre-industrial levels. This requires preventing greenhouse gas concentrations in the atmosphere from exceeding 450 parts per million (ppm) of carbon-dioxide equivalent (CO_2e).

Key to achieving this emissions goal will renewed investment in public transport, renewable energy technologies and more efficient vehicles. Further by using more carbon-intensive fossil fuels as well as mature renewable energy technologies such as large hydro, biomass combustion and geothermal and also using other renewable sources include solar assisted air conditioning, wave power and nanotechnology solar cells, although they all still require more technological or commercial development it would be possible to reduce emission.

Yet another option could be carbon capture and storage technology. This involves capturing carbon dioxide before it can be emitted into the atmosphere, transporting it to a secure location, and isolating it from the atmosphere, for example by storing it in a geological formation. Irrespective of climate change, over $20 trillion is expected to be invested in upgrading global energy infrastructure from now until 2030. The additional cost for altering these investments in order to reduce greenhouse gas emissions would range from negligible to an increase of five to 10 per cent.

If we spend about 1 per cent to 3 per cent of GDP globally or a 12 per cent annually as per Pachauri it would be possible to arrest the emission now, and in this way can expect a return of more than 400 per cent. Imagine having a savings account by these returns.

Even if we stabilized greenhouse gas concentrations today–a virtual impossibility– the rate of warming would slow, but not stop for another 30 years. This delayed warming caused by (among other factors) how long it takes for the ocean to heat and cool, is called "warming in the pipeline". The IPCC estimates that warming in the pipeline will increase global temperatures by an additional 1.0°F, no matter what action we take. But we can–and better–stop it there. So *we can't reverse the warming we have already caused.*

Naturally, the more CO_2 we produce, the higher CO_2 concentration in the atmosphere. Four billion tons of carbon emitted into the atmosphere will raise CO_2 concentrations by 1 ppm (see CO_2 Arithmetic, *Science* Magazine). So to bring down the carbon content of the atmosphere to the level of of the pre industrial years a huge quantity of carbon dioxide emission has to be reduced.

To avoid the tipping point, global CO_2 emissions should peak no later than 15 years from now, and then begin to decrease. By 2050, emissions must be about 50 percent less than today, and by the end of the century 75 percent less.

"Is capping greenhouse gas concentrations achievable? Absolutely yes," says Saleem Huq, director of the climate change programme at the UK-based International Institute for Environment and Development. "But there are two aspects to 'achievable'– the technological and the political. The political is a tougher question."

Reduction by all concerned developed countries of the total emission volume of CO_2, CH_4, N_2O, HFCs, PFCs, SF_6, in carbon dioxide equivalent, by at least 5 per cent of the 1990 emission levels (or the 1995 level for CFCs) by 2000–2012. Japan has to reduce 6 per cent, USA 7 per cent and European Union 8 per cent.

Conclusion

The very first United Nation's Millennium Development Goal addresses poverty and hunger. Of the 1.2 billion living below the poverty line, the largest number–900 million–live in rural areas. Not all of these are farmers–many are landless, and rely on wage by employment or products obtained from the local environment. The remaining 300 million of the world's poor currently living in urban areas also have a stake in agriculture, depending as they do on access to plentiful and affordable food.

As on today about 1.1 billion people are living without access to clean drinking water and it is projected to increase to 4.5 billion during 2050. Already due to the drinking of contaminated water about 12 million people are dying every year as a result of diseases like malaria, cholera, diarrhea etc. Already about 1.02 billion hungry people are living in the Earth and it is going to increase to 1.57 billion due to the food grain reduction and water scarcity as a result of climate change. Now due to warming many natural calamities have intensified.

As a result of the findings of the above scenarios in the Earth, there seems to be a consensus among the leading developed countries that the temperature increase caused by global warming must not exceed 2° C (3.6° F). For example the European Union (EU), the G8 industrial and the G 5 developing countries have committed itself to this threshold even earlier than the commencement of the 15th international conference on climate change in Copenhagen during December 2009. Though this conference ended without making any agenda to cut the emission target, but this conference has agreed to generate about 100 billion US dollars every year during 2020 for the global warming mitigation purposes of the under developed and developing countries. So the world leaders are aware that the warming is already affecting and going to affect the people.

To reach this target the annual global CO_2 emissions have to be reduced from about 28 billion in 2006 to 20 billion of CO_2 by the year 2050 and to 10 billion of CO_2 by the year 2100 according to IPCC. At the first glance, this does not look like a major reduction. However one should keep in mind that the world population will grow from 6.4 billion people in 2007 to about 9.4 billion people in 2050. At the same time more and more developing countries will progress their industrialization and as a result they will want to copy our western life style causing high CO_2 emissions. At that point of time only green energies should be used to reduce the emission further. Carbon capture is also an approach to reduce emission.

The present world-wide average CO_2 emission per capita is about 4 tons per year. For North America it is about 26 tons and for Europe about 10 tons per year per capita. By 2050, the world-wide average CO_2 emission per capita needs to be reduced to 2 tons per year. In the following years, the emissions will need again to be cut by half.

The Ocean is vast and hence has huge potential as a carbon sink. According to the IPCC Third Assessment Report (2001), an estimated 6.3 billion tons of carbon is released into the atmosphere every year through human activities, such as combustion of fossil fuels, approximately half of which (3.1 billion tons) is absorbed by Earth. The ocean ecosystem is thought to absorb 1.7 billion tons and the terrestrial ecosystem particularly forest is thought to absorb 1.4 billion tons.

The Empire State Building is in New York, USA. The owners of this building are now spending $13 million to improve its energy efficiency with the aim of providing a model that could spread across America and around the world. It is expected to cut the energy use by almost 40 per cent, reducing bills more than $4 million and paying back the cost of expenditure in three years. That is a figure that is relevant not just to the Empire State but to the whole of New York City and other large metropolises like it. Almost 80 per cent of New York's energy consumption is through its buildings, mainly in the larger of the leaky older structures. Hence by this energy auditing, it is possible to cut the carbon footprint of the Empire State Building by more than 1, 00,000 metric tons over the next 15 years, the equivalent of taking 20,000 cars off the road. If this approach is adapted just a fifth of the largest buildings in America, it would save 2.3 billion metric tons of carbon emissions, equivalent to the amount of greenhouse gas pollution produced by the whole of Russia each year. In case of such modification in all the largest buildings in the entire world there will be a tremendous cut in carbon emission. Since buildings account about 8 per cent of the total emission, it is suggested that all the big buildings in the globe should be suitably modified into green buildings (Pilkington, Ed, 2010, Can the tallest be the greenest too? The Hindu dated 28.7.2010).

So to arrest the water scarcity, food grain reduction, sea level rise and the intensification of natural calamities due to the manmade emissions of green house gases as well as to improve the global economy the public, governments, politicians and non governmental organizations should work together.

According to scientific studies, the sun is going to burn brighter for another 5 billion years. Then it will expand in size to a red giant, encompass the orbit of Mercury and at that point of time the Life on Earth will have been extinguished due to rise in temperature of the sun. Within 1 billion years, all water will have evaporated, migrated to stratosphere, and then to space. At that time even Mother Earth will be also burnt due to the enormous heat of the sun. Let us not hasten the natural event to take place earlier than the natural process and kill our Mother Earth and end all the lives by increasing the warming by man made emission through green house gases (The Life and Death Planet Earth—by Peter Ward and Donald Brownlee and Doomday Argument—By Brandom Karcar 1983).

References

IPCC, 2007. *Intergovernmental Panel on Climate Change* (IPCC) report.

Malhotra, Reenita, 1999. 'The high stakes of melting Himalayan glaciers. *CNN The New Scientist,* June 5.

Man Mohan. Himalayan glaciers may disappear by 2035, Tribune Special Going, Going, Gone, Climate Change and Glacier Decline by World Wildlife Fund.

Natarajan, P.M. and Shambu Kallolikar, 2008. India likely to run short of water before 2050. *The Hindu,* October, 13.

Stern, Nicolas. STERN REVIEW: The Economics of Climate Change Risks of water scarcity, UNEP WATER AND HUMAN SECURITY, Aaron T. Wolf, Department of Geosciences Oregon State University, USA.

2013, Biodiversity Conservation for Sustainable Management   Pages *182–195*
Editor: Dr. K. Muthuchelian, *Vice Chancellor, Periyar University, Salem*
Published by: Daya Publishing House, NEW DELHI

Chapter 22

Influence of Climate Change on Microbial Diversity

H.C. Lakshman

*P.G. Department of Studies in Botany, Microbiology Laboratory,
Karnatak University, Dharwad – 580 003*

ABSTRACT

The impact of biodiversity loss on the functioning of ecosystem and on their ability to provide ecological services to humans has become a central issue in ecology today. The diversity of primary produces (algae) and decomposers (bacteria) in aquatic microcosms and found complex interactive effects of algal and bacterial biomass production. Both algal and bacterial diversity had significant effects on the number of carbon source used by bacteria, suggesting nutrient cycling associated with microbial exploitation of organic carbon sources as the link between bacterial diversity and algal production. More concretely, niche breadth might be interpreted as the diversity of enzymes that a microbial species has and that allows it to break down a diversity of organic compounds with different C:N ratios and niche height might be interpreted as a species potential enzymatic activity. The effect of microbial diversity on the nutrient recycling efficient from organic compounds to decomposers. It is also important to realize that this conclusions apply to constant environments such as those typically found in laboratory experiments. One additional dimension of the role of microbial diversity in natural systems is related to microbial adaptations spatial and temporal variation in the environment. For many process that occurs under anaerobic conditions in micro-spaces, bacteria live in a dormant state until conditions become favorable. Thus, microbial diversity might be important in maintaining ecosystem process in the face of spatial and temporal environmental variability.

Introduction

Since the dawn of life, some 3-5 billion years ago our planet has giant asteroids, erupted with explosive volcanoes at the rate of Mount saint Helen –sized blasts per month, accumated in its atmosphere one the lethal chemicals in the history of life endured three ice ages some of most magnificent species originated, *i.e*, Microbial life existed more than 3.5 billions years ago, 1 billions years after the formation earth, 3 billions years before the plants and animals appeared on earth.

International Convention on Biological Diversity (CBD) Microbial diversity is defined as the variability among microorganisms from all sources including terrestrial, marine and aquatic ecosystem. According Wilson, 1988. Biodiversity is defined as the variability among living organisms. This includes diversity within species, between species and of ecosystem.

Microorganisms are essential part of the environment contributing for the maintenance of stable ecosystem. They are distributed everywhere *i.e.* soil, water and air. They are present deep inside the earth as well as in deep sea vents. They play important role in recycling biological elements such as oxygen, carbon, nitrogen, sulphur and phosphorus known as biogeocycles. For example 70 per cent of cells weight and 24 per cent of oxygen is present in atmosphere which is available to all the microbes. About 78 per cent of nitrogen (N_2) is present in atmosphere while 9-15 per cent of cells dry weight contains essential cellular element nitrogen which contains amino acids, nucleic acids and some co-enzymes.

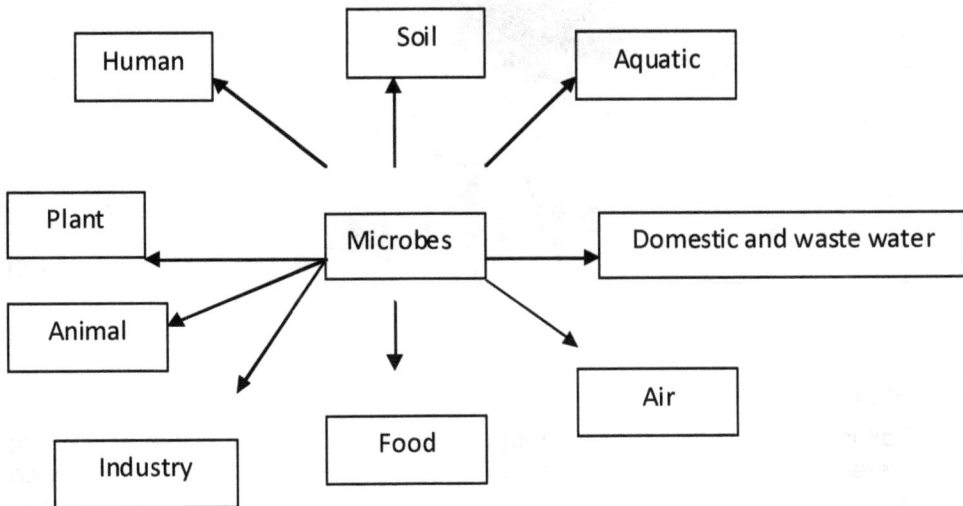

Figure 22.1

Importance of Microbial Diversity

Microorganisms provide a resource reservoir from which individual species can be selected to serve biotechnological purposes other than microbial communities. Therefore, it is important to make genus, species and community composition.

Biological microbial diversity that is sensitive to environmental change consists of three major concepts; (i) scale, (ii) components and (iii) view point.

Biodiversity is continually transformed by a changing climate. Conditions change across the face of the planet, sometimes in large increment, sometime in small increment, resulting in the rearrangement of biological association. But now a new type of climate change, brought about by human activities, is being added to this natural variability, threatening to accelerate the loss of biodiversity already under way due to other human stressors.

Climate involves variation in which the atmosphere is influenced by and interacts with other parts of the climate system, and "external" forcing (Figure 22.2). The internal interactive components in the climate system includes the atmosphere, the oceans, sea ice, the land and its features (including the vegetation, albedo, biomass and ecosystem), snow cover, land ice and hydrology (including rivers, lakes and surface and subsurface water). The components normally regarded as external to the system include the sun and its output.

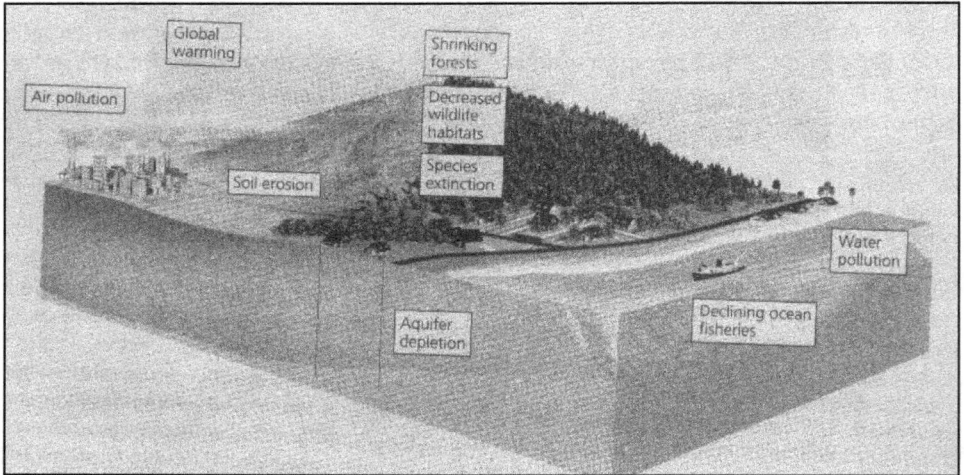

Figure 22.2: Microbial populations in fertile soil.

Microbial Diversity in Soil

Soil is a organic living matrix that is essential part of terrestrial ecosystem considered to be a store house of microbial activity. The soil is not a dead inert material. Actually it is full of life. 1 gram of soil has about 200-500 billions of microorganisms. Man depends upon the soil for his food. The soil depends upon microorganisms for its fertility. Agriculture would not be possible without microorganisms in the soil. The soil is a tremendous *growth medium*. It has organic matter, soil solution and soil air. All these components are affected by the activities of microorganisms. So the soil is a *constantly changing medium*. The soil solution in a good agricultural soil has ions like K^+, Na^+, Mg^{++}, Ca^{++}, Fe^{++}, S^-, NO_3^-, SO_4^+, CO_3^-,

PO_4^- and others. These ions are very essential in culture media. Thus the soil is an excellent natural medium for microorganisms.

Type	Number per gram
Bacteria: direct count, dilution plate	$25 \times 10^8 15 \times 10^6$
Actinomycetes	7×10^5
Fungi	4×10^5
Algae	5×10^4
Protozoa	3×10^4

Practically all kinds of bacteria may be found in soil. Some of the physiological types are:

Mesophiles	*autotrophs*	*protein digesters*	*nitrogen fixers*
Thermophiles	*heterotrophs*	*polysaccharide digesters*	*hydrogen bacteria*
Psychrophiles	*aerobes*	*sulphur transformers*	*iron bacteria*
Pathogenicones	*anaerobes*	*carbon transformers*	*phosphorus transformer*

Many of the soil bacteria perform useful function. Some important functions are

1. Decomposition of organic matter.
2. Conversion of soil constituents into materials useful for plant life.
3. Biogeochemical cycling of elements like carbon, nitrogen, phosphorus, iron, sulphur, manganese etc.
4. Production of antibiotics.

Soil H (Lithosphere)

Soil constitutes the major habitat of terrestrial microorganisms. Soil is a favorable habitat for the microbes. Higher number occurs in the organically rich surface layers than in the underlying mineral soils. Particularly high number of microbes occurs in association with plant roots as shown below.

Distance from Plant Root (mm)	Microbes 1/g Dry Wt Soil		
	Non-Filamentous Bacteria (x 10^7)	*Streptomycetes (x10^7)*	*Fungi (x10^6)*
0	16.0	4.7	3.6
0-3	5.0	1.6	1.8
3-6	3.8	1.1	1.7
9-12	3.7	1.1	1.3
15-18	3.4	1.0	1.3

Genus	Percentage of Total no. of Microbes
Arthrobacter	5-60
Bacillus	7-67
Actinomycetes	10-33
Pseudomonas	3-15
Agrobacterium	1-20
Alcaligenes	2-12
Flavobacterium	2-10
Corynebacterium	<5
Micrococcus	<5
Staphylococcus	<5
Xanthomonas	<5
Mycobacterium	<5

Fungi constitute the major proportion of the microbial biomass in soils. Fungi occur as free-living or associated with plant roots. Most common soil fungi are Fungi Immperfecti, but numerous Ascomycetes and Basidiomycetes also occur.

In refrigerators at 5°C foods remain unspoiled. In a freezer at –5°C the crystals formed tear and shred microorganism. It may kill many of the microbes. However, some are able to survive. *Salmonella* spp and streptococci survive freezing. Deep freezing at –60°C, it reduces biochemical activities of microbes.

Fungi

These include a group of organisms from molds to fleshy fungi. Majority soil fungi grow in acidic soil with aerobic condition. Actinomycetes fungi grow in saline soil. They exist in both mycelial and spore stage. They are heterotrophic in nutrition. They may be parasitic, saprophytic and symbiotic.

☆ The mycelium penetrates through the soil and forms a net-work. This gives the soil the crumb structure. Agriculturally it is very important. It improves the physical condition of the soil. It also improves the physical condition of the soil. It also improves the water holding capacity.

☆ They break down sugar, organic acids, disaccharides, starch, pectin, cellulose, fats and the lignin molecule. These carbon sources are particularly resistant to bacterial degradation. Fungi can utilize and degrade these major plant constituents.

☆ By the degradation of plant and animal remains, the fungi participate in the formation of humus from fresh organic residues.

Algae

Algae are generally found on the surface of moist soils. They are photosynthetic organisms. So they found near upper layer or just below the surface layer of soil. The major types of algae are as follows.

Green Algae	Yellow Green Algae	Blue Green Algae	Diatoms
Chlamydomonnas	Heetrococcus	Anabaena	Cymella
Chlorella	Heterothrix	Nostoc	Navicula
Chlorococcum	Botrydiopsis	Oscillatoria	Pinnularia
Cladophora	Bumilleria	Tolypothrix	Synedra
Coccomyxa	Bumileriopsis	Nodularia	Fragilaria

Protozoa

Many types of protozoa are found in the soil. Flagellates and amoeba are usually larger in number. The number per gram of soil ranges from a few hundred to several hundred thousand in moist soils rich in organic matter. Depending upon the conditions of the soil the protozoa may exist as vegetative or cyst forms. They use decaying organic matter for food.

Some of the typical soil protozoa are:

Mastigophora	Sarcodina	Cilliata
(Flagellates)	(Amoebae)	Balantiophorus
Bodo	Amoeba	Colpidium
Allanton	Biomyxa	Colpoda
Cerocobodo	Difflugia	Gastrostyla
Cercomonos	Euglypha	Halteria
Monos, euglena	Harmanella	Oxytricha
Oikomonas	Lecythium	Pleutotricha
Spiromonas	Nucbaria	Voticella
Spongomonas	Trinema	Uroleptus
Chlamydomonas	Naegleria	Enchelys

Protozoa do not serve any major in the soil. They serve to regulate the size of bacterial population. Since protozoa do engulf bacteria they maintain some equilibrium of the bacterial flora of the soil. Thus they have a role in biological equilibrium.

Diversity of AM Fungi in Soil

Arbuscular mycorrhizal fungi form the main component of soil microbiota in most agro-systems. Hence AM Fungi have been shown to have strong influence on plant diversity. Since these fungi are obligate symbioants, their population and diversity are determined by plant species, human activities also affect these fungi. These modify the structure and functioning of plant communities in a complex and unpredictable way. AM Fungi, by forming an extended, intricate hyphal network, can absorb mineral nutrients from the soil and deliver them to their host plants in exchange for carbohydrates. Facilitated nutrient uptake, particularly with respect to immobile nutrients such as phosphorous, is believed to be the main benefit of this symbiosis for plants. Apart from this, they can also enhance drought resistance, resistance against root pathogens and tolerance to heavy metal toxicity in plants. AM fungi also play a role in the formation of stable aggregates building up of macrporous structures of soil that allow penetration of water and air, and prevent erosion. In a given ecosystem these fungi play an important role in carbon allocation, nutrient cycling and maintenance of diversified ecosystem. The presence of these fungi and their genetic and functional diversities are important for both plant communities and ecosystem productivity.

Human activities like application of fertilizer, crop rotation and soil management may also alter the population and diversity of arbuscular mycorrhizal fungi. Modern intensive farming practices are evidently a threat to AM fungi, as indicated by studies on these fungi. However, little is known about the effect of management practices on the species diversity of fungi. In recent days, AM fungal population and diversity are declining because of agricultural/land use intensification.

In tropical soils, application of organic matter either in the form of FYM, compost or organic amendments stimulates proliferation of AM Fungi. This is probably because of the low organic matter content in tropical soils. Addition of organic amendments such as paddy straw, maize straw and pongamia leaf increased the mycorrhizal activity. Of the three amendments studied, the addition of pongamia leaf encouraged AM Fungi to the maxium level followed by maize straw.

As the land use intensification increased, the AM fungal species diversity also decreased. Apart from this they also observed more spore abundance in low input grasslands than in low to medium input farming system and high input forming system. Even the AMF species diversity index was more in grasslands. Among two different farming system, AMF species diversity index was more in sites with crop rotation rather than sites with monocropping. It has been studied that the populations than soils under conventional management. Spore load and colonization of maize roots by AM fungi were higher in non-tilled than in moldboard ploughed and chisel disked soil.

Glomalin, n Arbuscular mycorrhizal fungal soil protein plying an important role in soil aggregation was significantly affected by land use pattern. Glomalin concentration were highest in native forest soils, moderate in afforested soils and lowest in agricultural lands. Soils C and N were highly correlated with glomalin across all soils and within each land use type, indicating that some glomalin may be under similar control as soil C. These results also show that glomalin may be useful as an indicator of land change.

Microbial Diversity in Aquatic System

Water is the elixir of life. It is an essential part of protoplasm and create a state for metabolic activates to occur smoothly. Therefore, no life can exist without water. In addition, there are thousands of microorganisms which live in water and transported through it. Major area (about 3/4) of earth surface is covered by water manly by oceans to some extent by lakes, rivers, streams, ponds etc. however, water is constantly in continuous circulation.

Heavy bloom of cynobacterium, *Oscillotoria erythraea*. This has phycoerythrin and phycocyanin pigments, produces a condition called Red Sea. Red tides, brown, amber of Greenish yellow colouration are also due to abundant microorganisms.

Estuaries: Estuary is a semi-enclosed coastal water body having connection with open sea. It receives fresh water with all particulate suspensions through rivers. Temperature, salinity, turbidity and nutrient load fluctuate. So fluctuation in the occurance of microorganisms is also seen. In areas receiving domestic waste with organic nutrients contain the following organisms: *Bacteria, Coliforms, Faecal Streptococci, Bacillus, Proteus, Clostridium, Sphaerotilus, Thiothrix* and *Thiobacillus*.

Soil Bacteria : Azotobactor, Nitrosomonas and Nitrobactor.

Fungi: ascomycetes, Phycomycetes and fungi imperfecti.

Aquatic Microorganisms in Sea

Marine environment covering 70 per cent of the earth surface and 11.2 km deep in same zones. The ocean is the largest and most stable of all biomes. Sea is the largest natural environment inhabited by microbes. Marine microorganisms comprise a comparatively untapped reservoir of commercially valuable compounds. Many bacteria are able to produce and secrete polymers and enzymes. Some marine bacteria are potent producers of DNAases, Lipases, alginases and protease.

Lakes and ponds of temperate region show thermal stratification which influences the microbial population in different seasons. In spring and autumn mixing occurs resulting in massive growth of algae called "Bloom". Lakes and ponds enriched with nutrients show "eutrophication". Common fresh water microorganisms are *Pseudomonas, Flavobacterium, aeromaonas* and *alcagines.*

The microbial flora includes bacteria, algae, protozoa, molds and yeasts. The microorganisms may show horizontal and vertical distribution

1. *Horizontal distribution*: the number of microorganisms is more in coastal waters and it gradually decreases towards the open sea. The bacterial count is less than 10 per ml. in open sea away from the coast. In coastal waters the count rises to about 100,000 per ml. the horizontal distribution is influenced by wind, tide, currents, seasons, solar radiations, concentrations of salt, distance from the coast and discharge of river into the sea.

2. *Vertical distribution*: depth of the sea varies and so the distribution of microorganisms also varies. It depends on pressure, temperature, biotic factors and sedimentation of organic matter.

Marine Plankton

Plankton are of two types, namely phytoplankton and zooplankton. Phytoplanktons form group of microorganisms which convert radiant energy to chemical energy and which support the entire population of fishes. Ex-*Diatoms, cynobacteria, dinoflagellates, siicoflagellates, chrysomonads, cryptomonads, chlamydomonos*, etc.

The following are the advantages over the gram positive bacteria:

1. Better suited to live in nutritionally dilute aquatic medium.
2. Hydrolytic enzymes are retained in the protoplasm.

3. Lipopolysaccharide outer membrane gives protection from fatty acids and antibiotics. Common marine forma are *vibrio, acinetobacter, pasudomonas, flavobacterium, alteromonas* and *staphylococcus.*

Zooplankton

Protozoans form the major group of zooplankton. Foraminifera, radiolaria and cilliata occur in large numbers in photosynthetic zone during day time. Zooplankton feed on phytoplankton. Mold spores and mycelial fragment are present throughout the photosynthetic zone. Several species Deuteromycetes, phycomycetes, and myxomycetes have been recorded in the sea.

In the ocean itself, the bacteria adhere to the surface of particles. The most important bacteria in the sea are the pleomorphic, Gram-negative, usually motile psychrophiles resembling species of genera *Vibrio* or *Mycoplasma. V. marinus* is a common example. The luminous bacteria or photogenic bacteria have been isolated from oceans using sea-water-agar medium with peptone. *Photobacterium phoaphoreum* and *Vibrio pierantonii* are isolated from luminous marine fish. In waters with salinity equivalent to that in sea water, these organisms luminesce (emit light).

Actinomycetes

Actinomycetes are generally considered as the intermediate group of organisms between eubacteria and fungi. They are like fungi in producing hyphae and conidia or sporangia. The mycelium is about the same as that of bacterial cell. Actinomycetes whose hyphae undergo segmentation, resemble bacterial cells. Actinomycetes make up approximately 20-60 percent of the total microbial population of the soil. The so called "earth smell" of soil is due to the production of perpenoids and extracelluar enzymes by actinomycetes. The commonest genera of actinomycetes are sreptomycetes, Nocardia, Micromonospora, actinomyces, actinoplanes, microbispopra and streptosporangium. Among these species of streptomyces are known for antibiotic production.

Microbial Diversity in Air

The atmosphere as a habitat is characterized by high light intensities, extreme temperature, variations, low amount of organic matter and a scarcity of available water making it is a non-hospitable environment for microorganisms and generally unsuitable habitat for their growth. Nevertheless, substantial numbers of microbes are found in the lower region of the atmosphere.

Microbes of air within 300-1,000 or more feet of the earth's surface are the organisms of soil that have become attached to fragments of dried leaves, straw or dust particles, being blown away by the wind. Species vary greatly in their sensitivity to a given value of relative humidity, temperature and radiation exposures. More microbes are found in air over land masses than far at sea. Spores of fungi, especially *Alternaria, Clodosporium, pencillium* and *Aspergillus* are moe numerous than other forms over sea with in about 400 miles of land in both polar and tropical air masses at all akltitudes upto 10,000 feet.

Microbes found in the atmosphere.

Type of Microbes	Percentage
Bacteria	
Gram–positive pleomorphic rods, such as *Corynebacterium*	20
Gram-negative rods, such as *Achromobacter, Flavobacterium*	5
Endospore-forming genera, like *Bacillus*	35
Gram-positive cocci, like *micrococcus*	40
Fungi	
Cladosporium	80
Alternaria	5
Penicillium	2
Others *(Aspergillus, Chaetomium, Fumago, Fusarium, Helminthosporium, Drechslera, Sclerotinia, Stachybotrys, Memnoniello, Trichoderma, Verticillium)*	13

Microbes found in air over populated land areas, below altitude of 500 feet in clear weather include spores of *Bacillus* and *Clostridium*, ascospores of yeasts, fragments of mycelium and molds and streptomycetaceae, pollens, protozoan cysts, algae, *Micrococcus, Corynebacterium*, in the dust and air of schools and hospital wards or the rooms of the persons suffering from infectious diseases, microbes such as tubercle bacilli, streptococci, pneumococci have been demonstrated.

Diversity of Endophytic Fungi

Endophtes are defined as microorganisms that colonize internal part of tissues. Endophytes were defined as asymptomic microorganisms that inhibit at least for one period of their life cycle. An additional definition is endophytes are considered as either bacteria or fungi that invade the tissues of living plants and that cause unapparent and asymptomatic infections to plant.

Endophytes were initial considered as neutral not causing benefits nor sowing detrimental effects to their hosts. Further studies indicate that, in many cases, they have important role in host plant protection, hosts acting against predators and pathogen. Landmark reports on the subject started in 1981. From 1981 to 1985, which may be considered to be a historical period, it was shown that the protection against herbivorous, can be provided by endophytic microorganisms.

Although endophytic microorganisms may be isolated from seeds in some plant species, usually they are not vertically transmitted trough seeds. Colonization of different plants tissues and organs begins by endophytic penetration through artificial and natural openings, such as root emission zone, injuries caused by agricultural practices and root growth. After penetration, these endophytes colonize the host, being found in all tissues and plant organs, inhibiting the apopalst, plant vessels and in cases interior of the cells. Inside the host plant or during colonization, endophytes interact with epiphytic, pathogenic and/or other endophytic microorganisms, resulting in an equilibrium is broken, endophytic microorganisms may be pathogenic.

Endophytic fungi can be detected by direct microscopic examination of plant tissues, but most of the work with endophytes is carried out by isolation of fungi from hosts. Usually the first is the sterilization of the outer surface of plant followed by the transfer of the plant fragments to appropriate culture media. After incubation, fungi can be isolated, purified and characterized. According to the objectives of the research and plant host species, several different variations of the basic procedure may be applied.

There are reports that endophytes isolated from lichens, moss, ferns, gymnosperms. Their occurrence may be affected by organ or tissue age, by the season of the year, by plant genotype, anthropogenic and several other factors. It has been proposed that each plant species has endophytic microorganisms still not classified and with potential biotechnological interest. Since, it is known that the plant diversity in tropical and subtropical area is higher than in temperate climates, the biodiversity of endophytic fungi associated inside these plants should also be higher than that observed in plants from temperate regions.

Future Prospects of Microbial Diversity

☆ There are important implications for planning in order for a great deal of microbial diversity. Planning has to include much longer time as well as current short ones 50 and 100 years as well 5 and 10. The planning has to include scales relevant to process, continental in some cases, down to local.

☆ The establishment of global Microbial resource centers (MCRCEN). Similarly the world Health Organization (WHO) declared in 1980 that smallpox had been eradicated from the earth. The news was greeted with much jubilation around the world. About 20 year later, WHO announced plans to destroy the last remaining stock of smallpox virus particles. The news met with controversy from various sectors (Ogunseitan, 2002) the debates were framed by two very different schools of thought. At one end were those who believed that stockpiles of smallpox virus remain undeclared by purpose of developing biological weapons. Therefore, as a precaution, the known stock of the virus should not be destroyed in case vaccines would need to be produced quickly in the event of biological warfareor other emergencies at the end were those individuals who opposed any deliberate actions that humans might take to cause the extinction of any other organism, including viruses. Several countries also maintain microbial conservation centers. The database on ribosomal RNA sequences diversity in different ecosystems all over the world. Microbial diversity research to levels that approach those dedicated to botanical and zoological species. However, deeper insights into the nature of speciation and process that generate molecular and physiological diversity among microorganisms are needed to facilitate the establishment of more comprehensive, evidence-based global inventories of microbial diversity.

☆ Responses by individual species to climate change may disrupt their interactions with others at the same or adjacent trophic levels. When closely interacting organisms display divergent responses or susceptibilities to change, the outcome of their interactions may be altered, as long term on data on both terrestrial and marine microorganisms. Both temperature and humidity affect their reproductive physiology and population dynamics.

☆ Estimates, say that 1 per cent of the earth's microbial population could have yet been cultured. Development of new isolation techniques may lead to the discovery of new microorganisms that may open new doors. We all know that the microorganisms are essential partners with higher organisms in symbiotic associationship. More knowledge in the field of this associationship will lead to improvement in health of plants, livestock and humans.

☆ The impact of biodiversity loss on the functioning of ecosystem and on their ability to provide ecological services to humans has become a central issue in ecology today. The diversity of primary produces (algae) and decomposers (bacteria) in aquatic microcosms and found complex interactive effects of algal and bacterial biomass production. Both algal and bacterial diversity had significant effects on the number of carbon source used by bacteria, suggesting nutrient cycling associated with microbial exploitation of organic carbon sources as the link between bacterial diversity and algal production. More concretely, niche breadth might be interpreted as the diversity of enzymes that a microbial species has and that allows it to break down a diversity of organic compounds with different G: N ratios and niche height might be interpreted as a species potential enzymatic activity.

☆ The effect of microbial diversity on the nutrient recycling efficient from organic compounds to decomposers. The rate of nutrients loss from organic compounds approaches zero, microbial diversity always has a positive effect on nutrients recycling efficiency and, hence, on primary production, secondary production, producer biomass and decomposer biomass. The role microbial diversity in natural systems is related to microbial adaptations to spatial and temporal variations in the environment. For many processes that occur under anaerobic conditions in micro-spaces, bacteria live in a dormant state until conditions become favourable.

References

Abbot, L.K. and Robson, A.D., 1991. Field management of mycorrhizal fungi. In: *Rhizosphere and Plant Growth*, (Eds.) D.L. Klester and P.B. Cregan. Kluwer Academic Publishers, Dordrecht, The Netherlands, p. 355–362.

Andrew, W.B.J., Jonathan, D., Todd, Lei Sun, M., Nefeli, N.K., Andrew, R.J.C. and Rachel, R., 2008. Molecular diversity of bacterial production of the climate changing gas, dimethyl sulphide, a molecule that impinges on local and global symbioses. *J. Exp. Bot.,* 59(5): 1059–1067.

Azevedo, J.L., 1997. Endophytic fungi and their roles in tropical plants. In: *Progress in Microbial Ecology*, (Eds.) T. Martins, A. Sato, J.M. Tiedje, L.C.M. Hagler, J. Dobereiner and I. Sanchez. Sociedade Brasileira de Microbilogia, Sao Paulo, pp. 279–287.

Azevedo, J.L., 1998. Biodiverdade microbiana e potential biotechnologico. In: E*cologia Microbiana*, (Eds.) I.S. Melo and J.L. Azevedo. Embrapa–CNPMA, Jaguariuna, pp. 117–137.

Bethlenfalvay, G.J., 1993. Mycorrhizae in the agricultural plant-soil system. *Symbiosis*, 14: 413–425.

Clay, K., Hardey, T.N. and Hammold, Jr., A.M., 1985. Fungal endophytes cyperus and their effect on the insect herbivore. *American Journal of Botany*, 72: 1284–1289.

Dubey, R.C. and Meheshwari, D.K. (Eds.), 2002. *A Textbook of Microbiology*. S. Chand and Company Ltd., New Delhi, India, 684 pp.

Gamboa, M.A. and Bayman, P., 2001. Communities of endophytic fungi in leaves of a tropical timber tree (*Guarea guidonia*) Meliaceae. *Biotropica*, 33: 352–360.

Hallmann, J., Quadt-Halmann, A., Mahaffee, W.F. and Kloepper, J.W., 1997. Bacterial endophytes in agricultural crops. *Canadian Journal of Microbiology*, 43: 895–914.

Lakshman, H.C., 2009. Selection of suitable AM fungus to *Artocarpus heterophyllus* Lam.: A fruit/timber for an ecofriendly nursery. M.D. Publisher, New Delhi, pp. 62–73.

Lakshman, H.C., Hosamani, P.A. and Kadam, L.B., 2005. Microorganisms and their role in phosphate solubalization. In: *Proceedings of Biotechnological Applications in Environmental Management*, (Ed.) P.C. Trivedi, pp. 71–79.

Lakshman, H.C., Sandeepkumar, K. and Hosamani, P.A., 2009. The use of biofertilizers in improving some fiber yielding plants. In: *Proceeding of ICAR Conference*, April 14–16, NAL Bangalore, pp. 127–129.

Maria, G.L. and Shridhar, K.R., 2003. Endophytic fungal assemblage of two halophytes from west coast mangrove habitats, India. *Czech Mycology*, 55: 241–251.

Michel Loreau, 2010. Microbial diversity, producer-decomposer interactions and ecosystem processes: A theoretical model. *Proc. Roy. Soc., London, B.* 268: 303–309.

Romero, A., Carrion, G. and Rico-Gray, V., 2001. Fungal latent pathogens and endophytes from leaves of *Parthenium hysterophous* (Asteracea). *Fungal Diversity*, 7: 81–87.

Singh, R.P., (Eds.), 2007. *Microbiology and Control of Microorganisms*. Kalyani Publication, New Delhi, pp. 204.

Thamas, E. Lovejoy and Hannah, Lee (Eds.), 2005. *Climate Change and Biodiversity.* TERI Press, New Delhi, India, pp. 420.

Tilman, D., 1999. Global environmental Impacts of agricultural expansion. The need for sustainalble and efficient practices. In: *Proceedings of the National Academy of Sciences, USA*, 96: 5995–6333.

Vijaya Ramesh, K. (Ed.), 2004. *Environmental Microbiology.* MJP Publishers, Chennai, India, pp. 390.

Wofle, A.L., 2002. Species diversity and community composition of arbuscular mycorrhizal fungi in tropical forest fragments and adjacent pastures. *In. 87th Annual Meeting of the Ecological Society of America and Annual International Conference of the Society for Restoration*, Tuscon, A.R, U.S.A, August 4–9, pp. 62.

2013, Biodiversity Conservation for Sustainable Management   *Pages 196–200*
Editor: **Dr. K. Muthuchelian,** *Vice Chancellor, Periyar University, Salem*
Published by: **Daya Publishing House, NEW DELHI**

Chapter 23

Old Building are Green Buildings

M. Pavaraj, Ga. Bakavathiappan and S. Baskaran

Post-graduate and Research Department of Zoology,
Ayya Nadar Janaki Ammal College (Autonomous), Sivakasi – 626 124

ABSTRACT

A study was made on the temperature inside old buildings and new buildings with and without trees. The study revealed that in old buildings the temperature was low when compared with new buildings. Among the new buildings, the buildings without trees showed high temperature when compared the new buildings with trees. The study established that old buildings are green buildings and in the new buildings trees help to reduce the temperature inside house.

Keywords: *Green buildings, New buildings, Climate change, Global warming, Trees.*

Introduction

Climate change is one of the most important global environmental challenges facing humanity with implications for food production, natural ecosystems, freshwater supply, health, etc. The impact would be particularly served in the tropical areas, which mainly consists of developing countries, including India. National Mission on Sustainable Habitat aims to promoting energy efficiency as a core component of urban planning; the plan calls for; extending the existing Energy conservation Building code (Fulekar and Kale, 2010). Nowadays the energy crisis is one of the most important

* Corresponding Author: E-mail: pavarajphd@gmail.com

problems of the world. Because of this, design of green buildings is the most important challenge to reduce of energy consumption in buildings. Now there is a movement for Green Buildings which conserve electricity. It is interesting to note that in earlier period people constructed house which are very cool when compared to ambient temperature. This results in conserving electricity (Ahmad *et al.,* 2009). Global climate change has become more apparent over the last few decades. Most climate experts agree that the humans, atleast in part, become the root cause for this development. The experts are calling for immediate and far-reaching action to fight global warming and its remedial measures. One of the most important tasks is to reduce greenhouse gas emissions. An increasing concentration level in the atmosphere is said to be the main reason for the raising temperatures. For instance, the CO_2 concentration in most industrialized countries has increased by more than 20 per cent in the last 60 years. (Nelson and Rakau, 2010).

Buildings over their life cycle account for a large share of global greenhouse gas emissions. The European Commission reports that buildings are responsible for the largest share of the EU's final energy consumption (42 per cent) and for about 35 per cent of all greenhouse gas emissions. Most European countries have also tightened environmental regulation for new buildings and refurbishments of old buildings. In recent report by the United Nations Environment Programme (UNEP) finds faults with the property industry for being too slow in addressing its increasing environmental foot-print. The Intergovernmental Panel on Climate Change (IPCC) reports estimates that by 2020 the primary energy use for the buildings sector will double from 103 EJ (1990) to 208 EJ (1 EJ or exajoule is equivalent 1,018 joules). The corresponding rise in carbon dioxide emissions from the building sector will go up from 1,900 million tonnes of carbon dioxide (MtC) to 2,700 MtC. The population increase, rapid urbanization, and the extensive building that both engender are going to continue. As a result, energy use for lighting, domestic appliances, and air-conditioning will rise. Containing and reducing the environmental foot-print of the building sector is by no means an impossible task. Environmentally driven emerging technologies can improve energy efficiency and reduce energy consumption in residential and commercial sectors. More than one-third of energy is consumed in buildings worldwide, accounting for about 15 percent of global greenhouse gas emissions. In cities, buildings can account for up to 80 per cent of CO_2 emissions. The built environment is therefore a critical part of the climate change problem—and solution. Most existing buildings were not designed for energy efficiency, but by retrofitting with up-to-date products, technologies and systems, a typical building can realize significant energy savings. Improving the energy efficiency of buildings is a priority for reducing both greenhouse gas emissions and energy costs. Hence, in the present study, the temperature inside old buildings and new buildings with and without trees was undertaken.

Materials and Methods

In the present study three different houses in Sivakasi residential areas like, old house, new house without tree and new house with tree were selected and measured the temperature and relative humidity using Digital Hygro thermometer (J4111H Mextech). The temperature was recorded from inside and outside of the houses during 12.00 pm–1.00 pm for 10 days.

Results and Discussion

In the present study the temperature within the old and new buildings with and without trees was studied (Table 23.1). The temperature within the old house was very low when compared to the new houses with and without trees. So the old buildings in Sivakasi town are comparatively cooler, than reducing the electricity consumption for fans and lights. They have wooden ceiling with greater height than the new houses. The results in low temperature. The mean difference in temperature between inside and outside the old house is 4.45° C, which is comparatively higher than the new houses with tree (2.37) and without trees (2.56), and the old buildings are cooler. Buildings complying with high energy-efficiency and other environmental standards decrease CO_2 emissions and are often referred to as "green buildings" (Auer *et al.*, 2008 and European Commission 2007a). Chan *et al.* (2003) stated that homes are built to suit their intended use, the local climate and the natural environment. The healthy indoor environments of a green building can increase well-being of its occupants. Energy efficiency is a key component of green buildings.

Among the new houses, temperature was found to be low in the houses with trees whereas in the houses without trees, high temperature was observed. A well-placed tree, shrub, or vine can deliver effective shade, act as a windbreak, and reduce your energy bills. Carefully positioned trees can save up to 25 per cent of the energy a typical household uses for energy. Research shows that summer daytime air temperatures can be 3° to 6° cooler in tree-shaded neighborhoods than in treeless areas (Wilson *et al.*, 2003). Cooling of air temperature due to the effect of trees has been well documented in the past through various studies. A tree can be regarded as a natural "evaporative cooler" using up to 100 gallons of water a day (Kramer and Kozlowski, 1960). This rate of evapotranspiration translates into a cooling potential of 230,000 kcal/day. This cooling effect, observed in a study by Geiger, is the primary cause of 5°C differences in net peak noontime temperatures observed between forests and open terrain, and a 3°C difference found in noontime air temperatures over irrigated millet fields as compared to bare ground (Geiger, 1957). As the percentage of land taken up by structures and paving an increase in dense urban areas, so too does the ambient temperature. This "urban heat island effect" can be reduced by the presence of trees in several ways. The canopy of a grove of trees shades the ground and structures while the natural release of water vapor from trees can help cool the surrounding air. By blocking the wind, tree foliage reduces the infiltration of air into homes. Most of the heat gain inside a house comes from sunlight (or solar energy) hitting the roof and streaming through the windows. Energy conservation measures that block the sun before it strikes the roof or windows are the most effective ones to implement. Trees and other plants that provide shade are the most effective long-term measures for reducing your home's energy consumption for heating and cooling (Christopher, 2006).

Mature trees and shrubs can have a dramatic effect on utility bills, according to the U.S. Department of Energy Office of Energy Efficiency and Renewable Energy. For example, an energy-saving landscaping design can cut heating bills by about one-third during cold-weather months. The potential savings during warm-weather months are equally dramatic: A well-planned landscape can reduce an unshaded

Table 23.1

Date		Old House			New House (With Tree)			New House (Without Tree)		
		Temperature °C	RH per cent	Difference	Temperature °C	RH per cent	Difference	Temperature °C	RH per cent	Difference
13.07.2010	outside	43.3	27	9.9° C	38.8	36	2.9° C	39.4	37	2.4° C
	Inside	33.4	48	21 per cent	35.9	39	3 per cent	37	39	2 per cent
14.07.2010	outside	40.6	40	7.4° C	36.6	41	3.4° C	39.4	36	2.1° C
	Inside	33.2	50	10 per cent	33.2	48	7 per cent	37.3	40	4 per cent
15.07.2010	outside	38.7	34	4.8° C	37.1	39	2.4° C	47.4	22	5.1° C
	Inside	33.9	47	13 per cent	34.7	46	7 per cent	42.3	27	5 per cent
17.07.2010	outside	35.1	42	2° C	34.6	46	1.4° C	36.1	40	1.6° C
	Inside	33.1	48	6 per cent	33.2	49	3 per cent	34.5	44	4 per cent
18.07.2010	outside	36.4	48	2.7° C	34	48	1.2° C	34.7	49	1.7° C
	Inside	33.7	60	12 per cent	32.8	54	6 per cent	33	53	4 per cent
19.07.2010	outside	32.4	60	1.5° C	31.9	66	0.2° C	32.8	62	0.7° C
	Inside	30.9	63	3 per cent	31.7	66		32.1	62	
21.07.2010	outside	36.4	43	5° C	35.8	43	3° C	38.6	37	2.4° C
	Inside	31.4	56	13 per cent	32.8	56	13 per cent	35.0	46	9 per cent
22.07.2010	outside	35.8	41	3.5° C	35.8	41	3.5° C	38.8	36	3.2° C
	Inside	32.3	49	8 per cent	32.3	49	8 per cent	35.6	46	10 per cent
23.07.2010	outside	36.7	33	2.4° C	36.7	33	2.4° C	41.2	24	3.0° C
	Inside	34.3	38	5 per cent	34.3	38	5 per cent	38.2	32	8 per cent
26.07.2010	outside	38.5	29	3.3° C	38.5	29	3.3° C	40.6	26	3.4° C
	Inside	35.2	39	10 per cent	35.2	39	10 per cent	37.2	33	7 per cent
X		33.04	41.42	4.45	33.95	47.7	2.37	36.22	42.2	2.56
SD		1.34	10.28	2.65	1.64	9.06	1.10	2.90	10.41	1.21

home's summer air conditioning costs by 15 to 50 percent, depending on how tight the structure is and how well it's insulated. The study revealed that in old buildings the temperature was low when compared with new buildings. Among the new buildings, the buildings without trees show high temperature when compared the new buildings with trees. The study established that old buildings are considered as green buildings whereas in the new buildings trees play a prominent role to reduce the temperature within the house. Trees are a good investment. Studies by real estate agents and professional foresters estimate that trees raise a home's resale value seven to 20 percent. Thus, by planting trees, homeowners can help fight against global warming.

Acknowledgements

The authors express the profound thanks to the Management and Head of the Department of Zoology. Ayya Nadar Janaki Ammal College (Autonomous) Sivakasi for providing facilities to carry out this work.

References

Ahamed, T., Mahyar, G., Mojtaba, K. and Jamshid, M., 2009. Effect of radiator positions on heat distribution in the building using numerical model. *World Academy of Science, Engineering and Technology,* 58 : 1006–1009.

Auer, J., Heymann, E. and Tobias, J., 2008. Building a cleaner planet: The construction industry will benefit from climate change. DB Research. *Current Issues,* Frankfurt.

Chan, W.R., Price, P.N., Sohn, M.D. and Gadgil, A.J., 2003. Analysis of U.S. Residential Houses Air Leakage Database. LBNL report 53367.

Christopher, S.J., 2006. Tree Placement on Home Grounds. G6900. Columbia, Missouri: University of Missouri. http://extension.missouri.edu/explore/agguides/hort/g06900.html.

European Commission, 2007a. A lead market initiative for Europe. Brussels, Belgium.

Fulekar, M.H. and Kale, R.K., 2010. Impact of climate change: Indian scenario. *University News,* 48(24): 15–23.

Geiger, R., 1957. *The Climate Near the Ground,* 4th edn. Harvard University Press, Cambridge.

Kramer, P.J. and Kozlowski, T., 1960. *Physiology of Trees.* McGraw-Hill Publishers, 35 pp.

Nelson, A. and Rakau, O., 2010. "Green buildings" A niche becomes mainstream. *Energy and Climate Change,* p. 1–24.

Wilson, A., Thorne, J. and John, M., 2003. *Consumer Guide to Home Energy Savings,* 8th edn. ACEEE, Washington, D.C.

2013, Biodiversity Conservation for Sustainable Management   Pages *201–205*
Editor: **Dr. K. Muthuchelian,** *Vice Chancellor, Periyar University, Salem*
Published by: **Daya Publishing House, NEW DELHI**

Chapter 24

Conservation and Management of Biodiversity Resources: Problems and Prospects

B.S. Krishna Moorthy[1] and P. Ravichandran[2]
[1]Department of Biology, [2]Department of Economics,
GTN Arts College, Dindigul – 624 005

Introduction

Biodiversity is defined as the variety and variability among living organisms from all sources, including inter *alia*, terrestrial, marine and other aquatic ecosystems and the ecological complexes of which they are a part; this includes diversity within species, between species and of ecosystem (NBA, 2004).

Biodiversity both affects and is affected by climate change; On the one hand climate change is a major cause of biodiversity loss, on the other, the conservation and sustainable use of biodiversity offers flexibility to climate variability and natural disasters. The climate change, food production and economic crisis are wakeup calls to the need for factoring in sustainability in development choices.

Impacts of Climate Changes on Biodiversity

The warmer climate encourages the growth of pests which destroy forests in precedented numbers. A good example of that is a pine beetle infestation of forests in British Columbia, Canada, which would have killed 50 per cent of the pines by 2008. Increased numbers and intensity of forest fires and reduced diversity of wild life, another significant effect that climate change will force upon the mountainous ecosystem is the melting of their snow cover and retreat and disappearance of glaciers.

Ocean acidification has been implicated for damaging some of the most beautiful creature on the planet-coral reefs, as well as other shell-forming organisms. Threats to species are principally due to a decline in the areas of their habitats, fragmentation of habitats and declines in habitat quality, and in the case of some mammals, hunting (Kumar *et al.,* 2000).

In some areas, invasion by exotic species of plants also results in habitat degradation. Prominent examples are the spread of Peruvian thorny tree *Prosopis juliflora* in the dry parts of North India where it replaces native species such as *Acacia nilotica* (babool), and the spread of the South American flowering bush *Lantana camara* in the sub–Himalayan belt.

The loss of the habitats leads to extinction of the amphibians (Tropical frogs) in Central America dependent on these forests for their survival. Polar bears have become "Poster children" for the melting of Artic ice due to climate change. Melting ice reduce the ability of polar bears to find enough food as they prefer to use ice as a platform to hunt for prey (Garshelis, D.L. 2008).

Another example is Orangutans may be seriously affected by the spread of viruses and bacteria which normally thrive in warmer conditions. This among many other things, may push these animals even closer to the brink to extinction (Kumar, A. 2008). The list of animals at risk of climate change will, of course, be longer and longer as the planet gets hotter and hotter.

Economic Prospects of Biodiversity Resources

Percentage of pharmaceutical sector's turnover ($650 billion annually) derived from genetic resources: 20 to 50 per cent. Namibia's protected areas contribute 6 per cent of GDP in tourism alone with a significant potential for growth (CBD, 2008). Income from Namibia's conservancies (and conservancy-related activities): US$ 4.1 million (TEEB, 2009) Percentage of total export from foreign tourist spending: estimated 24 per cent (GEF/UNDP, 2008). Contribution of the Great Barrier Reef to the Australian economy (value of tourism, other recreational activities and commercial fishing): AU$ 6 billion (World Resources, 2008). Sixty percent of ecosystem services have been degraded in fifty years and the cost of failure to halt biodiversity loss on land alone in last 10 years is estimated to be $500 billion (Turpie *et al.,* 2004).

Livelihoods and Employment

Nearly a sixth of the world's population depends on protected areas for significant percent of their livelihoods (GBRMPA, 2007). Over a billion people in developing countries rely on fish as a major source of food and 80 per cent of the world fisheries are fully or over exploited (TEEB, 2009). Total economic output: 145 million $–2.6 per cent of Botswana GNP. Number of people in the world whorely on timber and non-timber forest products: 1.6 billion and annual rate of deforestation: 13 million hectares roughly the area of Bangladesh (Balmford, A. *et al.,* 2004).

Health, Nutrition and Vulnerability

Percentage of people in Africa estimated by WHO to rely on traditional medicines (plants and animals) as the main source of their health care needs: 80 per cent.

Number of people worldwide who depend on drugs derived from forest plants for their medicinal needs: 1 billion (Mayers, J. and Vermeulen, S. 2002). About 8 per cent of the 52,000 medicinal plants used today are threatened with extinction (U N, 2009).

Role of Tribes in Conservation and Management of Biodiversity

The tribal population in India is 84.51 million, which constitutes 8.14 per cent of tribal population. There are about 449 tribes and sub tribes in different parts of India. Half of India's tribal people live in the forests and forest fringes and their economy is linked with the forests. TamilNadu has 6,51,321 tribal population as per 2001 census which constitutes 1.02 per cent of the total population. There are 36 tribes and sub tribes in Tamilnadu. Literacy rate of the population is 27.9 per cent. Most of the tribals in Tamilnadu are cultivators, agriculture labourers or dependent on forests for their livelihood. There are six primitive tribes in Tamilnadu. The tribal groups in Tamilnadu are distributed in almost all the districts and they have contributed significantly in the management of the forests (Annamalai, R. 2004).

Current debates over the Tribal Rights Bill suggest that neither conservationists nor tribal rights activists have seriously considered a solution involving the transfer of tourists revenues to local communities although some (Saberwal *et al.*, 2001) have called for much more meaningful local involvement in conservation management.

Table 24.1: SWOT analysis for conservation and management of biodiversity.

Strength	Weakness
Biodiversity Hotspots	Over exploitation of resources
Genetic resources used for medicines, Food, fuels, etc.	Hunting, Deforestation, NHAI
Carbon storages to reduces warming	Unknown knowledge of indigenous product
Maintain the ecosystem level balancing.	Threatened species
Biodiversity richness, endemism	Climate changes, Global warming
Biodiversity studies	Sea acidification, Acid rain
	Urbanization, Industrialization
	Forest fires, natural disaster
Opportunity	*Threat*
National parks, zoos, sanctuaries.	Habitat destruction - forest, wetlands.
Biosphere Reserve	Destruction of Coastal areas.
Wildlife Tourism, Reef Barrier Tourism.	Uncontrolled commercial exploitation.
Livelihood source of Tribes.	Species eradication.
Economical values of pharma.	Ecosystem level disturbed (eg: Phytoplanktons affected by UV radiations give a result on insufficient food for small fish and zooplanktons give ultimate results on mitigation of birds and other predators).
Foreign exchange earning source.	Algal Bloom (*Noctiluca* sp.)
Employment of indigenous peoples.	

Even today many local and indigenous communities in the Asian countries meet their basic needs from the products they manufacture and sell based on their traditional knowledge. Herbal drugs obtained from plants are believed to be much safer; this has been proved in the treatment of various ailments (Mitalaya KD, 2003). Rural communities, in particular paliyar tribes, depend on plant resources mainly for herbal medicines, food, forage, construction of dwellings, making household implements, sleeping mats, and for fire and shade. Rural people not only depend on wild plants as sources of food, medicine, fodder and fuel, but have also developed methods of resource management, which may be fundamental to the conservation of some of the world's important habitats (Gemedo-Dalle T, 2005).

Conclusion

We come to conclude that, the protected areas (*e.g.* parks and nature reserves) have been the cornerstone of efforts to conserve the world's species and ecosystems. They also play a key role in sustaining local livelihoods and contributing to economic and social well-being. Protected areas also have an important role in reducing risks from natural disasters and in helping counteract climate change impacts with avoided deforestation and support to maintaining ecosystem services within and beyond their boundaries. Protected areas need to be carefully planned, and properly managed in order to ensure biodiversity and people benefits. It is important to address pollution, climate change, irresponsible tourism, poorly located infrastructure and increased demand for land and water resources, all of which exert continued pressure on protected areas and the ecosystem services that they provide. When well planned and implemented carefully, the benefits of protected areas greatly outweigh the costs.

References

Annamalai, R., 2004. *Tamil Nadu Biodiversity Strategy and Action Plan*. Tamil Nadu Forest Department.

Balmford, A.E. Houghdele and Stanley, R., 2004. The worldwide costs of marine protected areas. *Proceedings of the National Academy of Sciences*, 101(26).

CBD, 2008. *Annual Review of National Reports*. Fourth Assessment, 12 pp.

Garshelis, D.L., Ratnayeke S. and Chauhan, N.P.S. (2008). In: IUCN 2008. IUCN Red List of Threatened Species. Downloaded on 26 January 2009.Listed as Vulnerable (VU A2cd+4cd, C1 v3.1)

GEF/UNDP, 2008. *Biodiversity: Delivering Results*, available at http://www.undp.org/gef/documents/publications/bd_web.pdf

Gemedo-Dalle, T., Maass, B.L. and Isselstein, J., 2005. Plant biodiversity and Ethnobotany of *Borana pastoralists* in Southern Oromia, Ethiopia. *Economic Botany*, 59: 43–65.

Great Barrier Reef Marine Park Authority, 2007. *Protecting the Great Barrier Reef Marine Park: A Precious Resource*. Corporate Brochure.

Kumar, A., Singh, M. and Molur, S., 2008. In: IUCN 2008. IUCN Red List of Threatened Species. Downloaded on 4th January 2009.

Kumar, A., Walker, S. and Molur, S., 2000. Prioritization of endangered species. In: *Setting Biodiversity Conservation Priorities for India*, (Eds.) S. Singh, A.R.K. Sastry, R. Metha and V. Uppal. New Delhi, WWF–India, 2: 341–425.

Mayers, J. and Vermeulen, S., 2002. International Institute for Environment and Development IIED. Power from the Trees: How good forest governance can help reduce poverty, IIED11:27pp.

Mitalaya, K.D., Bhatt, D.C., Patel, N.K. and Didia, S.K., 2003. Herbal remedies used for hair disorders by tribals and rural folk in Gujarat. *Indian Journal of Traditional Knowledge*, 2: 389–392.

National Biodiversity Authority, 2004. *The Biological Diversity Act 2002 and Biological Rules 2004*, 57 pp.

Saberwal, V., Rangarajan, M. and Kothari, A., 2001. *People, Parks and Wildlife: Towards Coexistence*. Orient Longman, New Delhi.

TEEB, 2009. The Economics of Ecosystems and Biodiversity for National and International Policy Makers–Summary: Responding to the Value of Nature, available at http://www.teebweb.org/LinkClick.aspx?_leticket=I4Y2nqqliCg per cent 3D and tabid=924 and language=en–US.

Turpie, R.S., Topfezer, W. and Thomas, R.N., 2004. Economic Analysis and Feasibility Study for Financing Namibia's Protected Areas. in UNDEP–UNEP Poverty–Environment Initiative, Mainstreaming Poverty–Environment Linkages into Development Planning: A handbook for practitioners, 55–57p.

U.N., 2009. *The Millennium Development Goals Report 2009*, p. 41–13.

World Resources Institute (WRI), 2008. In collaboration with United Nations Development Programme, United Nations Environment Programme, and World Bank.. World Resources 2008: Roots of Resilience–Growing the Wealth of the Poor, 121–123p.

2013, Biodiversity Conservation for Sustainable Management   Pages *206–213*
Editor: Dr. K. Muthuchelian, *Vice Chancellor, Periyar University, Salem*
Published by: Daya Publishing House, NEW DELHI

Chapter 25

Biodiversity and Bioresource Values of Tree Species in Tropical Forests of Southern Eastern Ghats, India

L. Arul Pragasan and N. Parthasarathy

Department of Ecology and Environmental Sciences,
Pondicherry University, Puducherry – 605 014

ABSTRACT

This study aims to assess biodiversity of tree species in six major tropical hill forests of southern Eastern Ghats, India, namely, Bodamalai, Chitteri, Kalrayan, Kolli hills, Pachaimalai and Shervarayan hills, which vary in their composition of predominant species and in the degree of human disturbance, and also to determine the resource values of tree species of southern Eastern Ghats, along with phytogeographic distribution and conservation significance. The entire stretch of southern Eastern Ghats was divided into smaller grids of 6.25 km × 6.25 km, and the six sites totaled to 120 grids. Within each grid, a belt transect of 0.5 ha (5 m × 1000 m) area was laid and all trees e" 30 cm girth at breast height were enumerated. A total of 272 tree species (e" 30 cm gbh) that belonged to 181 genera and 62 families were recorded in the total 60 ha area inventoried. The diversity indices such as Shannon, Simpson and Fisher's alpha indices were 2.44, 0.03 and 42.1, respectively for the whole 60 ha area. Similarity indices such as Jaccard and Sorenson showed that sites CH and KA are more similar in terms of species composition. The total stand density and basal area for the total 60 ha area were 27,412 stems (457 stems ha^{-1}) and 1,012.12 m^2 (16.9 m^2 ha^{-1}), respectively. Both the stand density and basal area of tree species varied significantly across the six hill complexes. The spearman correlation revealed that the species richness and basal area were negatively correlated with the site disturbance scores but not density. Seventy-three per cent (198 species) and 64 per cent (175 species) of total tree species were considered respectively as valuable resources of ecological and economic importance. Of the total 272 species, 188 species with 15,860 individuals (58 per cent), and 64 per cent of economically important species and 95 per cent of ecologically importance species depend on faunal community for their seed dispersal in southern Eastern Ghats, revealing high dependence of tree species on fauna. Phytogeographical analysis showed that nearly 60 per cent of the tree species inventoried from southern Eastern Ghats are common to Sri Lanka revealing their close geographic affinity. This large-scale tree diversity inventory provides a baseline data for a variety of investigations and is expected to be useful for effective forest management and biodiversity conservation of southern Eastern Ghats region. Species recovery program through plant propogation by tissue culture is necessary for conservation of RET species of southern Eastern Ghats.

Keywords: *Biodiversity, Bioresource values, Conservation, Eastern Ghats, Tree community, Tropical hill forests.*

Introduction

The conservation of biodiversity is essential for the proper functioning of ecosystems and for the maintenance of the environmental services they provide (Hooper *et al.,* 2005; Lopez-del-Toro *et al.,* 2009). Populations of naturally growing woody species valued for their contribution to human livelihoods are threatened with extinction (Tabuti 2007). While biodiversity loss is a global phenomenon, its impact may be greatest in the tropics where the majority of species are distributed (Collen *et al.,* 2008). Rapid loss of biodiversity in tropical forests is recognized as one of the serious environmental and economic problems all over the world (Hare *et al.,* 1997). Historical and contemporary losses in forest cover associated with human activities occur in many regions of the world, particularly in tropical regions (Rudel and Roper, 1997; Lamb *et al.,* 2005).

Primary forests, particularly those of the Western Ghats and the Eastern Ghats of peninsular India are disappearing at an alarming rate due to anthropogenic activities and are replaced by forests composing inferior species or their land use pattern changed (Parthasarathy, 1999; Chittibabu and Parthasarathy, 2000). Floristic inventory is critical for conservation planning and management of forest ecosystems. Hence, this study was undertaken to assess diversity and resource values of tree community in six major tropical hill forests of southern Eastern Ghats, namely, Bodamalai, Chitteri, Kalrayan, Kolli hills, Pachaimalai and Shervarayan hills, and also to determine their seed dispersal modes, along with phytogeographic distribution, which are expected to provide a baseline data that can be of tremendous advantage for conservation planning and effective management of forest ecosystems of southern Eastern Ghats.

Methodology

Study Area

The present ecological research was carried out in southern Eastern Ghats (11 08.5'–12 06.0' N; 78 07.5'–78 48.5' E), which covers six major hill complexes *viz.* Bodamalai (BM), Chitteri (CH), Kalrayan (KA), Kolli hills (KO), Pachaimalai (PM) and Shervarayan hills (SH). The hills of southern Eastern Ghats are composed of masses of Charnockite associated with gneisses and varied metamorphic rocks. Soil of southern Eastern Ghats is red loamy and lateritic. Tribal settlements are common in all the hills. The climate data of Salem, the nearest station to the study sites, available for 20 years (1988-2007), reveal that the mean annual temperature is 28.3°C and the mean annual rainfall is 1058 mm. The bulk of the rainfall is received from August to October. The mean annual rainy days for the same period are 61 days.

The southern Eastern Ghats harbour five major forest types–tropical evergreen, semi-evergreen, mixed deciduous, dry deciduous and thorn forests. The vegetation of southern Eastern Ghats is being affected by a wide range of human disturbances such as forest land encroachment, hill agriculture, plantation, over-exploitation of non-timber forest produce (NTFPs), herding of cattle/goats inside the forests, quarrying, location on factories in the proximity of forests, and further, by the invasion of weeds, construction of buildings, roads, dams, etc. and most of them are escalating in intensity.

Methods

The entire stretch of southern Eastern Ghats was divided into 6.25 km × 6.25 km grids. A total of 120 grids were obtained from the six hill complexes of southern Eastern Ghats. In each grid, all live trees e" 30 cm girth at breast height (gbh) were enumerated from a belt transect of 0.5 ha (5 m × 1 km) area. To facilitate inventory, each transect was sub-divided into fifty 5 m × 20 m quadrats. Floras such as Hooker (1879), Gamble and Fischer (1915-1935), Trimen (1974), Nair and Henry (1983), Henry *et al.* (1987; 1989) Dassanayake and Fosberg (1987) and Matthew (1991) were used for identification of trees. Voucher specimens were collected and confirmed with the herbarium of our department and French Institute (IFP), Puducherry, and also online herbarium catalogue of Royal Botanic Gardens, Kew. They are deposited in the herbarium of Department of Ecology and Environmental Sciences, Pondicherry University.

Resource value of each species was analysised as ecological and economic importance. Ecological importance of a species was valued based on rewards such as fruit, nectar, pollen, etc. as food source for faunal community. For example, *Artocarpus heterophyllus*, *Ficus* spp. were valued for their high fruit production. Economic importance of a species was valued as timber, medicinal, edible, fuelwood or other economic values. Site disturbance scores were obtained by assessing various disturbances (on a 1-5 scale) which include resource extraction, forest land use change, tribal settlements and their dependence on forests and degraded forest area. The site with high disturbance score reflects the extent of high level of forest disturbance.

Results and Discussion

Species Diversity

A total of 272 tree species ($\geq$ 30 cm gbh) that belonged to 181 genera and 62 families were recorded in the total 60 ha area inventoried (Table 25.1). Species richness ranged from 64 species in BM to 169 species in KA. One way ANOVA revealed that the species richness varied significantly across the six sites ($F_{(5, 823)}$ = 4.854, P = 0.0002). The diversity indices such as Shannon, Simpson and Fisher's alpha indices were 2.44, 0.03 and 42.1, respectively for the whole 60 ha area. Similarity indices such as Jaccard (Cj) and Sorenson (Cs) showed that sites CH and KA (Cj–56.8 and Cs–72.4, respectively) are more similar in terms of species composition among the six hill complexes, whereas sites BM and KO (Cj–30 and Cs–46.2) showed high dissimilarity in species composition. *Hildegardia populifolia* (Sterculiaceae) an endemic species to Eastern Ghats, and *Grewia laevigata* (Tiliaceae) are the only two RET species recorded in this study.

Table 25.1: Summary of tree diversity inventory ($\geq$ 30 cm gbh) in the six hill complexes of southern Eastern Ghats, India–Bodamalai (BM), Chitteri (CH), Kalrayan (KA), Kolli hills (KO), Pachaimalai (PM) and Shervarayan hills (SH).

Variable	BM (5 ha)	CH (10 ha)	KA (17 ha)	KO (9 ha)	PM (12 ha)	SH (7 ha)	Total for 60 ha
Species richness	64	143	169	157	131	165	272
Total abundance	1449	5022	8951	3824	5388	2778	27412
Density (stems ha⁻¹)	290	502	527	425	449	397	457
Total basal area (m²)	27.75	216.74	257.87	220.05	150.18	139.53	1012.12
Basal area (m² ha⁻¹)	5.6	21.7	15.2	24.4	12.5	19.9	16.9

Family Richness

A total of 62 plant families were represented by the 272 species enumerated in the 60 ha inventoried (Table 25.1). Taxonomically, Euphorbiaceae constituted the most diverse family with 25 species, followed by Rubiaceae and Moraceae (17 species each), Rutaceae (14) and Lauraceae (7). In terms of tree abundance, Mimosaceae with 4,126 (15.1 per cent) stems dominated the tropical forests of southern Eastern Ghats, followed by Euphorbiaceae (12.6 per cent), Rubiaceae (7.8 per cent), Rutaceae (6.5 per cent) and Melastomataceae (4.9 per cent).

Stand Density and Basal Area

The total stand density of trees for the six hills of southern Eastern Ghats was 27,412 individuals in the 60 ha area (Table 25.1). The mean tree density was 457 stems ha⁻¹. Stand density was as low as 290 stems ha⁻¹ in BM to as high as 527 stems ha⁻¹ in KA, and it varied significantly across the six hill complexes ($F_{(5, 823)}$ = 4.85, P < 0.0002). The total basal area for the 60 ha area inventoried from the six hill complexes was 1,012.12 m². The mean stand basal area was 16.9 m² ha⁻¹ and it ranged from a low value of 5.6 m² ha⁻¹ in BM to a high of 24.4 m² ha⁻¹ in KO. Basal area of tree species varied significantly across the six hill complexes ($F_{(5, 823)}$ = 2.71, P < 0.02).

Tree Diversity and Forest Site Disturbance

Site disturbance scores obtained by assessing various disturbances (on a 1-5 scale) reveal that site BM scored a maximum score of 53, followed by sites PM, CH, KA, KO and SH. The site with high disturbance score reflects the extent of high level of forest disturbance. The spearman correlation between species richness, species density and basal area with site disturbance score revealed that the species richness and basal area were negatively correlated with the site disturbance scores, but not density.

Resource Values

Seventy-three per cent (198 species) of the total diversity of southern Eastern Ghats are ecologically very important (Table 25.2). Site KO had high diversity of ecologically important species followed by sites SH, KA, CH, PM and BM. In case of population abundance the largest site KA dominated the other sites. Species such as *Artocarpus heterophyllus* and *Ficus* spp. are valued for their fruit production supporting a number of vertebrate fauna including macaque and other small mammals and a variety of birds, *Canthium dicoccum* var. *dicoccum* and *Memecylon edule* Roxb. occur in high abundance are ecologically important for producing profuse flowers which provide nectar for variety of insects.

A total of 175 species (64 per cent of total species) with 20,869 individuals (76 per cent) were considered as economically important tree species in southern Eastern Ghats (Table 25.2). Among them, 139 species with 16,235 individuals (78 per cent) have medicinal importance, 61 species with 6825 individuals (33 per cent) have timber value, 24 species with 728 individuals (5 per cent) yield edible fruits, 22 species with 8112 individuals (39 per cent) are used for fuelwood and 22 species with 2813 individuals (13 per cent) are used for other economic values as spice, coffee, match stick, basket, dye, oil, etc. The seeds of *Terminalia chebula* and *Strychnos nux-vomica* are heavily harvested in Kalrayan hills for their high medicinal property. Population of one of the highly valued timber trees, *Tectona grandis is* threatened by illegal extraction in southern Eastern Ghats.

Dispersal Mode

Three types of seed dispersal modes were recognized in southern Eastern Ghats, namely anemochory (seeds dispersed by wind), autochory (seeds dispersed by the exploding fruit) and zoochory (seeds dispersed by animals). The predominant dispersal mode is zoochory with 15,860 individuals (58 per cent) in 188 species (69 per cent), followed by autochory with 8163 individuals (30 per cent) in 51 species (19 per cent) and anemochory with 3389 individuals (12 per cent) in 33 species (12 per cent) (Table 2). The three dispersal modes varied significantly across the six sites for tree species richness ($F_{(2,15)}$ = 39.829, p>0.0001) and abundance ($F_{(2,15)}$ = 6.867, p>0.01). The site KO recorded the maximum of 121 species of zoochory, revealing high dependence of faunal community on tree species. Sixty-four per cent of economically important

Table 25.2: Diversity and abundance of tree species classified by dispersal mode, ecological and economic importance from the 60 ha inventoried in the six major hill complexes of southern Eastern Ghats.

Site	Diversity							Abundance						
	BM	CH	KA	KO	PM	SH	Total	BM	CH	KA	KO	PM	SH	Total
Dispersal mode														
Anemochory	6	19	25	16	14	18	33	166	481	1507	281	551	403	3389
Autochory	14	29	30	20	23	34	51	592	1070	3181	744	1930	646	8163
Zoochory	44	95	114	121	94	113	188	691	3471	4263	2799	2907	1729	15860
Ecological importance	45	99	121	125	97	123	198	620	3563	4690	2883	3031	1910	16697
Economic importance														
Medicinal	42	90	101	81	71	96	139	1131	3222	5373	2092	2745	1672	16235
Timber	16	34	44	37	28	45	61	248	1655	2370	699	866	987	6825
Edible	7	12	14	11	9	16	24	58	174	232	95	40	129	728
Fuelwood	11	13	17	13	15	14	22	465	1533	2655	1121	1755	583	8112
Others	4	13	13	12	10	13	22	179	513	1143	248	550	180	2813

species and 95 per cent of ecologically importance species depend on faunal community for their seed dispersal in southern Eastern Ghats.

Phytogeographic Analysis

Out of the 272 tree species, just 3 species have Pantropical distribution (*i.e.* geographic distribution in Africa, Asia and Americas), 12 species shared Neotropics (species restricted to the Americas (New World)) and 27 species are restricted to Palaeotropics (species distributed in Africa and Asia (Old World)). A total of 161 species (59 per cent) of the tree species inventoried from southern Eastern Ghats are common to Sri Lanka (including 37 endemic species) revealing the close geographic affinity of tree species of southern Eastern Ghats with that of Sri Lanka.

The present study provides valuable data on diversity, density, resource values, seed dispersal modes and phytogeographic distribution of tree species of southern Eastern Ghats, which will be useful for the sustainable utilization of plant resources and conservation planning and management of tropical forests of southern Eastern Ghats. Futher, it also forms a base-line data for several scientific researches. Species recovery program through plant propogation by tissue culture are of immense need for conservation of RET species of Indian Eastern Ghats.

Acknowledgements

Authors thank the Department of Biotechnology, New Delhi for financial support through a project (No.BT/PR6603/NDB/51/089/2005), Tamil Nadu Forest Department for site permission to conduct field research, French Institute, Puducherry for herbarium consultation, and Royal Botanic Gardens, Kew for providing online access of their herbarium catalogue.

References

Chittibabu, C.V. and Parthasarathy, N., 2000. Attenuated tree species diversity in human-impacted tropical evergreen forest sites at Kolli hills, Eastern Ghats, India. *Biodivers. Conserv.*, 9: 1493–1519.

Collen, B., Ram, M., Zamin, T. and McRae, L., 2008. The tropical biodiversity data gap: addressing disparity in global monitoring. *Trop. Conserv. Sci.*, 1: 75–88.

Dassanayake, M.D. and Fosberg, F.R., 1987. *A Revised Handbook to the Flora of Ceylon.* Oxford and IBH Publ. Co. Pvt. Ltd., New Delhi.

Gamble, J.S. and Fischer, C.E.C., 1915–1935. *Flora of Presidency of Madras*, Vols.1–3. Adlard and Son Ltd., London.

Hare, M.A., Lantagne, D.O., Murphy, P.G. and Chero, H., 1997. Structure and tree species composition in a subtropical dry forest in Dominican Republic: Comparison with a dry forest in Puerto Rico. *Trop. Ecol.*, 38: 1–17.

Henry, A.N., Chithra, V. and Balakrishnan, N.P., 1989. *Flora of Tamil Nadu, India,* Vol. 2. Botanical Survey of India, Coimbatore.

Henry, A.N., Kumari, G.R. and Chithra, V. 1987. *Flora of Tamil Nadu, India*, Vol. 3. Botanical Survey of India, Coimbatore.

Hooker, J.D., 1879. *Flora of British India.* L. Reeve and Co., London.

Hooper, D.U., Chapin, F.S., Ewel, J.J., Hector, A., Inchausti, P., Salvorel, S., Lawton, J.H., Lodge, D.M., Loreau, M., Naeem, S., Schmid, B., Setala, H., Symstad, A.J., Vandermeer, J. and Wardle, D.A., 2005. Effects of biodiversity on ecosystem functioning: a consensus of current knowledge. *Ecological Monograph,* 75: 3–35.

Lamb, D., Erskine, P.D. and Parrotta, J.A., 2005. Restoration of degraded tropical forest landscapes. *Science,* 310: 1628–1632.

Lopez-del-Toro, P., Andresen, E., Barraza, L. and Estrada, A. 2009. Attitudes and knowledge of shade-coffee farmers towards vertebrates and their ecological functions. *Trop. Conserv. Sci.,* 2: 299–318.

Matthew, K.M., 1991. *An Excursion Flora of Central Tamil Nadu, India.* Oxford and IBH Publ. Co. Pvt. Ltd., New Delhi.

Nair, N.C. and Henry, A.N., 1983. *Flora of Tamil Nadu, India,* Vol. 1. Botanical Survey of India, Coimbatore.

Parthasarathy, N., 1999. Tree diversity and distribution in undisturbed and human-impacted sites of tropical wet evergreen forest in southern Western Ghats, India. *Biodivers. Conserv.,* 8: 1365–1381.

Rudel, T. and Roper, J., 1997. The paths to rain forest destruction: Cross-national patterns of tropical deforestation, 1975–90. *World Develop.,* 25: 53–65.

Tabuti, J.R.S., 2007. The uses, local perceptions and ecological status of 16 woody species of Gadumire Sub-country, Uganda. *Biodivers. Conserv.,* 16: 1901–1915.

Trimen, H., 1974. *A Handbook to the Flora of Ceylon.* Bishen Singh Mahendra Pal Singh, Dehradun.

2013, Biodiversity Conservation for Sustainable Management   Pages 214–218
Editor: Dr. K. Muthuchelian, Vice Chancellor, Periyar University, Salem
Published by: Daya Publishing House, NEW DELHI

Chapter 26

People's Biodiversity Register: A Programme of Empowering People for Managing their Bioresources

K.N. Deviprasad
Vivekananda College,
Putter, D.K, Karnataka

Introduction

The survival of world's human population depends on the biological resources. The human life has been influenced by biodiversity from time immemorial. There is considerable knowledge about biodiversity in small section of the village community who are directly depending on it. All knowledge and wisdom ultimately flow from practices but their organization differs among different streams of knowledge. Folk knowledge is maintained and transmitted and increased almost entirely in the course of applying it in practice. Folk ecological knowledge and wisdom are therefore highly sensitive to changing relationship between people and their ecological resources base. Today both are eroding at a fast rate for two reasons, firstly people now have to newer resources, such as modern medicine etc., and secondly people are increasingly losing control over the local resource base. Loss of traditional knowledge of practices of sustainable utilization of biodiversity is also an equally serious concern. However folk knowledge and wisdom with their detailed locality and time specific content are of value in many contexts. They must be supported in two ways by creating more formal institutions for their maintenance and most importantly by creating new contexts for their continued practice. The programme of People's Biodiversity Register (PBR) is such an attempt.

What is PBR?

People's Biodiversity Register (PBR) is an indeed record of traditional knowledge and practices of sustainable use of local bioresources and conservation of biological resources. The process of preparations of PBR as well as the resultant documents could serve a significant role in promoting more sustainable flexible, participatory systems of management and in ensuring a better flow of benefits from economic use of the living resources to the local communities.

People's Biodiversity Register (PBR) is a programme in tune with the objectives of the Convention on Biological Diversity (CBD) urging all parties including India and some 200 other countries to

1. Respect preserve and maintain the knowledge innovations and practices of indigenous and local communities embodying traditional lifestyles relevant for the conservation and sustainable use of biological diversity.
2. Promote the wider application of such knowledge, innovations and practices with approval and involvement of the holders.
3. Encourage the equitable sharing of the benefits arising from the use of such knowledge innovations and practices.

PBR is a record of knowledge, perceptions and priorities of local people about biodiversity, its utilization and conservation. The traditional knowledge of the local people regarding biodiversity play very important role in biodiversity conservation. But there is no documentation for this. Hence we have to document the traditional knowledge of local people on biodiversity utilization and conservation and forward to future generation. The tradition knowledge documented in the PBR should be incorporated while framing the rules and regulations and also incorporated during implementation of new developmental projects both in government and private sectors. We lose the rights on important plant species and animal species because multinational companies claim the patent on these important species. The proper documentation of utilization of these species in the PBR can help in to prevent the patent claim of multinational companies. In India Biodiversity ACT–2002 was implemented in 2004. Each Gram panchayath in India should have the People's Biodiversity Register according to the biodiversity act.

Modules PBR

1. Peoplescape
2. Lifescape
3. Ecological history
4. Management options
5. Developmental aspirations
6. Conflicts and consensus
7. Strategy and Action plan

Peoplescape

It includes different segments of the concerned human communities and their relationships to the local base of natural in particular living resources. People scape module explains the village profile, civic amenities, livelihood activities of the village,user groups, knowledgeable individuals jn user groups, gender aspects in user groups, seasonal activities, influences of outsiders on the village and sociocultural aspects of the local people.

Lifescape

It includes levels of abundance of different elements of biodiversity, in particular those groups with which local communities are familiar in different elements of the landscape and waterscape. Landscape means the local mosaic of different kinds of land and water habitats from which the concerned people bring in most living resources such as fuel wood, dung, honey, freshwater fishes,forest products, medicinal herbs, small timber, cane, bamboo etc. Lifescape module explains land ownership, use and composition, species diversity distribution and social value of landscapes and animal and plant species of the villages.

Ecological History

This module explains the past and present processes of biodiversity erosion and the driving forces behind the changes. It includes bench marks, earlier land composition and its utilization, changes and driving forces, conservation efforts and conservation efficacy. This module is mainly related to the ongoing changes in the local landscape waterscape and lifescape and the forces driving these changes.

Management Options

It includes local people's perceptions of options for development and management of the natural resource in a biodiversity friendly fashion. It includes biological measures, social measures, restrictions, institutional mechanism of decision making, assessment of implementation system of finances, reforms in the panchayath raj set up, sustainable harvest of bioresources.

Developmental Aspirations

It includes personal aspiration of local people and how these would affect their relationship with the natural, especially the living world. It includes development and its impact on biodiversity and biodiversity friendly development.

Conflicts and Consensus

Opinions of different groups of people do not often match on many issues and this leads to social conflicts on biodiversity conservation. This module also includes consensus of different groups of people for managing biodiversity.

Strategies and Action Plan

It includes landscape elements prioritized for conservation, species prioritized for conservation, landscape elements and species not prioritized, future scenario, feasible conservation strategy and policy changes required for biodiversity conservation.

Methodology of PBR Preparation

The People's Biodiversity Register exercise consists the following techniques.

1. Construction of the study team.
2. Field exercise
3. Report writing and publication.

Construction of the Study Team

According to Biodiversity Act 2002, in each gram panchayat level there should be Biodiversity Management committee. The study team is constructed by involving the members of Biodiversity Management Committee. The study team members belonging different categories like local knowledgeable persons, local school college teachers, environmentalists, youth community of the village, government agencies involved with development and management of the natural resources.

Field Exercise

In PBR, the information on the following components is documented

1. Village habitation map prepared with the participation of local people relating locations of temples, schools, bus stop, post office, punchayat building etc.
2. Village landscape map prepared incorporating various landscape elements by involving local people.
3. Identification of user groups [livelihood communities] in the village.
4. Identification of knowledgeable individuals in each user group.
5. Holding the interviews with knowledgeable persons group wise and individual wise.
6. Recording species, their uses and values across user groups.
7. Recording harvest levels, relevance for trade, prices effects.
8. Recording ecological history to understand the past and present processes of its erosion and the driving forces behind the changes.
9. Recording the management options of user groups for protecting biodiversity.
10. Recording development aspirations of the people and user groups for relating development and it's impact on biodiversity.
11. Recording conflicts and consensus of different groups of people for managing biodiversity.
12. Prioritizing species of plants and animals by different user groups for conservation and also for creating social awareness for sustainable utilization of bioresources.
13. Recording the feasible conservation strategy and policy changes required for biodiversity conservation.

Report Writing and Publication

People's Biodiversity Registers in local languages should be deposited with concerned village panchayats, schools and colleges and be open to scrutiny by all interest parties. Commercial enterprises wishing to access this information may do so on payment of a fee and on signing an appropriate agreement worth reference to benefit sharing.

References

Gadgil, M., 1996. Recording India's wealth: People's biodiversity register. *Amruth,* Vols. 1 and 5.

Gadgil, M., Thomas, Winfred and Ghate, Utkarsh, 1996. *Srustigyaan: A Methodology Manual for Documenting People's Priorities for Biodiversity Utilisation and Conservation.*

2013, Biodiversity Conservation for Sustainable Management   Pages *219–225*
Editor: **Dr. K. Muthuchelian,** *Vice Chancellor, Periyar University, Salem*
Published by: **Daya Publishing House, NEW DELHI**

Chapter 27

Ecotourism: A Potential Way for Biodiversity Conservation and Sustainable Development

A.A. Kazi and P.N. Mehta

ASPEE College of Horticulture and Forestry,
Navsari Agricultural University, Navsari – 396 450, Gujarat

ABSTRACT

A wide range of natural resources *viz.* wildlife, water, vegetation, flora, fauna and landscape are potential areas in various ecotourism activities. 82 per cent of Indian forests have the possibility of development as ecotourism sites, the major national parks and sanctuaries fetch the maximum visitors. In India, the travel and tourism industry accounts for nearly 6 per cent jobs and 4.8 per cent of GDP. The World Tourism Organization predicts that in the next decade India will attain 7.9 per cent growth per annum in tourism-related GDP and 5.1 per cent annual growth in government expenditure to tourism. We are to host nearly 5.08 million tourists in 2010 and this figure is to rise to 8.90 million by 2020. The positive values of nature based ecotourism needs to be incorporated in each wild destination which will lead to affirmative outcome that can combat with key issues of our planet, *i.e.* climate change and biodiversity conservation. Wildlife protection is also a vital issue in almost all the Protected Areas. Ecotourism advocates check against indiscriminate poaching, unwanted interference in jungle and can save flagship species of wild flora and fauna by continuous vigilance and monitoring. The employment opportunity generation to the tribes living in and around the forested regions is highly prolific by sustainable promotion of ecotourism sites which also reduce the pressure on the forest activities. Livelihood upliftment of such inhabitants is one of the key objectives of ecotourism development. The governments of many states have constituted Ecotourism policies which in countless case studies imply for dramatic change in the socio-economic standards of the forest dwellers. The impact assessment study in the *Gir* Forest was carried in 2008 and has given a clear understanding that tourism helps spread information, knowledge and awareness to sensitize the society to comprehend the significance of the *Gir* and the need to protect such assets. The number of incidence of poaching has also been reduced by promotion of ecotourism in the *Gir*. The findings include lacuna in the system adopted by forest department and suggestions for improvement and adoption of scientific techniques in ecotourism.

Keywords: *Ecotourism, Wildlife, Conservation, Gir, Protected Areas (PAs), Biodiversity, Development.*

Introduction

Indian subcontinent has very important place on the wildlife map of the world. There are about eight hundred species of mammals, over two thousand one hundred species of avians, many species of fish, reptiles, amphibians and more than 30,000 forms of insects found in India. The Asiatic Lion, the armoured one horned rhinoceros, the magnificent Royal Bengal tiger, the massive elephant and the large Indian gaur are big five of Indian forests. India is the only country in the world where lion and tiger both are found. The white tiger is also unique to India. In 20th century many species, which were abundant, came to the verge of extinction and so the concept of transforming many categories of forests into National Park, Sanctuary, protected areas, reserves, etc. came into reality. 4.70 per cent of total geographical area of India is occupied by these Protected Areas (PAs). Development of such PAs and dramatic reduction of populations of wild creatures has bred intense interest in people for roving and sighting such wildlife in wilderness. By the wide acceptance of masses this science of nature-based-tourism has been industrialized very swiftly and if we go through last few years' data, it has been playing crucial role in scaling the GDP of our country. Ecotourism is well-defined by IUCN as *environmentally responsible travel and visitation to relatively undisturbed natural areas that promotes conservation, has lower visitor impacts and provides beneficially active socio-economic involvement of local populations.*

India is one of the 12 mega biodiversity countries of the world, and a country with a rich and renewed cultural heritage. Yet, the potential of ecotourism to help conserve these resources and assist in economic development remains largely untapped both in terms of domestic and international tourism market. The governments of many states have taken policy initiatives to promote ecotourism. The government of Himachal Pradesh announced 'Policy on Development of Ecotourism' in 2001, government of Madhya Pradesh announced 'Eco and adventure Tourism Policy', Gujarat Government declared 'Ecotourism policy' in 2007, etc. Forest Departments of almost all the states have been positively endeavouring to appreciate ecotourism undertakings since the advantages of ecotourism are directly connected with the well-being and sustainable upliftment of dwellers who are living in and around the fine-forested areas. It also claims preservation of cultural heritage, protection of wild and endangered flora-fauna, avoids damage to diversity of the region and furthermore very importantly, builds the robust intimacy with Mother Nature.

The World Tourism Organization predicts that India will be hosting 5.08 million tourists in 2010 and the figure will rise to 8.90 by 2020. If ecotourism is properly implemented, it can integrate conservation with development and can make due difference in the ecological assets of the country.

Strategies

Non-consumptive nature based tourism facilitates the experience to appreciate beauty, tranquillity and dynamic balance of nature leading to an aware, informed and responsible citizen. The basic strategies can be summed up as:

Identification of Site

Well-diversified forested land can be taken as ecotourism sites for conservation purpose. The long coast line and many beautiful beaches are untouched and potential to be developed as green beaches, the plantation of mangrove species and salt tolerant species like *Casuarina equisetifolia* not only improve biodiversity of the bank but also add scenic beauty and help in livelihood activities to locals. Sanctuaries and National Parks are of course good sites, but the tourism capacity of each PA is to be calculated by thorough studies. Haphazard tourism has negative impact on ecology, and it should be taken care of with the help of due legislation. Till date the uniqueness of India has not been properly utilised for tourism, certain secluded places require to be explored and tourism is to be introduced.

Visitor Education

Education is essential at each and every eco-site. Knowledgeable Education Officer ought to be appointed and before entry all the visitors should be given information regarding site's significance about wildlife, biodiversity, unique landscape, ecological significance and cultural exceptionality of the area. Target oriented programmes, visit in various circuits and science based and thematic information should be incorporated through various means. Visitor education is the only tool that eventually satisfies the key objectives of ecotourism.

Facilitation

Accommodation, basic needs, medication and essential amenities are lacking in many ecotourism sites. Somewhere the sites are well-equipped but ill-managed. Proper maintenance of site attracts decent tourists and has less waste as well. It is general notion in India that one doesn't get good facilities if managed by government, this notion should be rectified by keen interest by authorities and up to the mark facilities at various sites. Transparency in the booking system, proper transportation, availability of guides and medical care are rudimentary facilitation points.

Adventure

Land and water based adventure activities like trekking, mountaineering, rock climbing, white water rafting, Para sailing, surfing, etc. are increasingly becoming the very part of the Eco-experience. Fortunately India has lofty mountain ranges, long sea-coast and flowing rivers where this tourism can flourish.

Carrying Capacity

The over or messy use of resources leads to affect biodiversity adversely. Physical, social and economic carrying capacity of each area is to be estimated by using scientific methods. Indicators of visitor's impact on site health and positive outcome of tourism promotion are to be evaluated periodically to have sustainable tourism. Quality and quantity of economic flows, impact on local society and culture, impact on wildlife habit and amendment in the local plant diversity should be incorporated discreetly while assessing carrying capacity.

Monitoring

Besides informed participation of relevant stakeholders, accountable political leadership is required to ensure positive participation and consensus build up for this industry. It is a continuous process and continuous monitoring, introduction of necessary preventive and corrective measures are must. Detailed protocol for monitoring shall be the part of management plan of each ecotourism hub.

Super-Diversity and Sustainable Development

There is lot of difference between aspiration and effort and preaching and practice. Ecotourism is 'concept' played by theorists or politicians. The level of upshot out of ecotourism sites in Indian has to be elevated up to the mark. Although World Bank has supported JFM and Indian eco development projects, the core issues of community have not been addressed judiciously. The communities consider ecotourism as a source of livelihood supplement and not looking to compete markets and that is why the community-held ecotourism is playing negligible role. We have been inept in using our diverse resources and see to development (And that too in sustainable manner). If the methodical mechanism is developed and put into practice, ecotourism has a great potential to contribute in Indian economy. The ecotourism learnings of past and methodology of today are needed to be fused not just only for securing sustainable development for our lag-behind communities but also using it as an influential tool for every man of society.

Legislative Involvements in Ecotourism Activities

Legal and policy frameworks at national and state levels which are regulating ecotourism activities can be listed as:

☆ Wildlife (Protection) Act, 1972, Amendment 1993 and 2002

☆ Forest (Conservation) Act, 1980

☆ Environment (Protection) Act, 1986

☆ Coastal Regulation Zone Notification, 1991

☆ Environment Impact Assessment Notification, 1994 and 2006

☆ National Forest Policy, 1998

☆ Ecotourism policy and guidelines, 1998

☆ National Tourism Policy, 2002

☆ The Biological Diversity Act, 2002

☆ National Environment Policy, 2006

☆ Ecotourism Policies of different States

Aftermath of Impact Assessment Study of Gir

A study on 'Impact of tourism on *Gir* forest' was conducted in the year 2007-08. The detailed study had many findings pertaining to conservation and tourism aspect. Few of crucial aftermaths are:

☆ Tourists come to *Gir* with more appropriate and focused purposes. Hence, the seriousness, dedication and concern of the tourists towards *Gir* are clearly more conservation oriented. While more pilgrims come with varied and disguised purposes other than worship. Thus the chances, of pilgrims being more harmful, are greater.

☆ 80 per cent of the visitors are aware about the existence of code of conducts. Also, there are guides who keep check on tourist's behaviour. Hence there are few chances where tourist can break the rules and cause harm to the Park.

☆ 68 per cent of the pilgrims were totally unaware about the rules regarding entry timings. Even after re-visiting (56 per cent of the pilgrims are not first-time visitors) the same place most of them are unaware about such rules.

☆ Lot of information was gathered by the tourists regarding lions, other animals, flora and fauna, the wild habitat, etc. The major source from where the tourists can acquire information about *Gir* is guide. Thus by visiting the place with guides also adds to the knowledge of the tourists to an extent.

☆ 16 per cent of tourists visited the park without guides. It suggests that there are people who enter the forest area without permission.

☆ By a visit to the park, tourists and pilgrims became aware about the problems faced by the PA of *Gir*. Hence this shows that apart from gaining knowledge the tourists also became aware and sensitized about the issues which according to them are disturbances. Such kind of understanding of a place and its problems has effectively come through a visit by the visitors.

☆ No signboards are installed regarding code of conducts, free booklets and brochures meant for distribution among tourists are not distributed, code of conduct poster only in two languages *i.e.* Gujarati and English, no provision for the national language; Hindi, communication for film show is not passable and local products encouragements were major gap in this PA.

Threats to Gir

1. Formalities like the checking of luggage, inflammable items, car, and food was not performed with almost all tourists. This shows that there is lack of proper security checking system.

2. Almost no formalities of any kind were carried out to perform security check on pilgrims. The number of pilgrims and the vehicle brought in enters the park without any checking. Such situation can prove very harmful to the

forest and the wildlife. The incidences of poaching and other illegal activities can rise up. In absence of proper checking the pilgrims tend to behave more carelessly and irresponsibly as they are assured that there is no one to watch and punish them.

3. In case of tourists it can be said that there are less chances of problems to happen because there is one guide in every group. There is a continuous watch of the guide during the visit in the park and so tourists would not get any such chances to create problems. But in case of pilgrims there is no watch over their activities. Thus the chances of pilgrims causing damage to the PA are far more than that of the tourists.

4. Almost all pilgrims who have the habit of chewing tobacco or smoking throw their waste in open. Inflammable items like burning bidis can lead to man-made forest fires. Thus such careless and ignorant attitude of pilgrims is a big threat to the PA of *Gir.* This also suggests that there is either lack of awareness and lack of proper monitoring of pilgrim's activities.

5. The tourists mainly travel in gas based vehicles. The Forest Department has made it compulsory to either have a gas based cars or latest models of petrol cars and diesel EURO-III. According to the Forest Department there is not much difference between the pollution caused by a petrol and LPG cars. Thus this really helps to bring down the level of pollution in the park and keeps the environment cleaner. The number of cars that can enter the park for safari is also restricted as per the carrying capacity decided by the Forest Department.

6. The maximum vehicles that go to Tulsishyam are diesel based. The pollution caused by such cars is huge. And considering the total numbers of cars (which is also not regulated) that enters Tulsishyam the impact is threatening. And as can be seen in almost all pilgrims came across animals on the road between the check post and temple. So it suggests that, the area is home to many wild animals and as they were spotted en route, hence, their chances of being affected due to pollution caused by these vehicles are obvious.

7. There is no proper system for waste disposal—plastic waste is used as a fuel by the local people to burn their *chullas* in the food stalls at Tulsishyam.

Conclusion

The 73rd and 74th amendment to the Constitution of India accords right to local self-government institutions, bringing into jurisdiction matters related to land, water, socio-economic development, infrastructure development, social welfare, social and urban forestry waste management and maintenance of community assets. Ecotourism falls under the purview of these subjects and therefore decision making is by local self-governance is important. If the science of ecotourism is followed appropriately, the Biodiversity loss can be shrunk rapidly. Ecotourism provides employment opportunities at the door steps of locals and the sustainability of their socio-economic upliftment can be greatly achieved. The study of impact assessment on *Gir* forest gives a clear understanding about the impact; direct as well as indirect impact of both the forms of tourism on Forest. The study says that both forms of tourism have positive as well as negative impact on Forests. Wildlife tourism a very limited direct negative impact on the park that is limited to solid waste, pollution and long term impact on the wildlife behaviour. The direct positive impacts of wildlife tourism on the *Gir* Forest include the economic benefits and the spreading of conservation message. Considering the indirect impact of wildlife tourism, the study says that the tourists get attached to *Gir* and they start playing the role as ambassadors of *Gir*. Thus tourism helps to spread information, knowledge and awareness to create a group of people who are aware and sensitized about forests and their problems.

References

Bhatt, Seema and Syed, Liyakat, 2008. *Ecotourism Developments in India.* Centre for Environment Education, Ahmedabad.

Chandra, R., 2007. *Wildlife and Ecotourism.*

D'Essence, 2004. *Ecotourism: An Opportunity in India.*

GEER Foundation, 2007. *Gujarat Ecotourism Policy* GR No. WLP–2005–1764–G1 (1818).

Haripriya, G.S. *A Note on Ecotourism.*

Vijay Kumar and Chouhan, J.S., 2007. *Eco-tourism policies and practices in India: A critical review. Jour. Trop. Forestry,* 23 (I and II).

Sinha, G.N. *Eco-cultural Tourism as a Means for Sustainable Development.*

2013, Biodiversity Conservation for Sustainable Management   Pages *226–230*
Editor: Dr. K. Muthuchelian, *Vice Chancellor, Periyar University, Salem*
Published by: Daya Publishing House, NEW DELHI

Chapter 28

Women's Wisdom on Indigenous Plants and her Role in Biodiversity Conservation

J. Pushpa, V. Sekar and R. Selvin
Department of Agricultural Extension,
AC and RI, Madurai

ABSTRACT

This paper describe the role of women and her wisdom on uncultivated medicinal plants on human health care practices as well as diet. This study was conducted at ten villages of foot hills of Alagarkovil by using participatory approach, group approach, group discussions, participatory learninig and personal interview methods. Results indicate that women are having their ancestral wisdom to cure many diseases by using local wild plants and also used as a diet. These indigenous practices of health were found to be appropriate on account of low cost, good efficacy and easy local availability. T he plants *viz., Syzygium cumini, Jammun, Tribulus terrestris, Gymnema* Sylvester,Lucas aspera,*Erythrina variegate,Jatropa Curcas, Ocimum basilicum, Solanaum surattense, Datu*ra sp, *Mirabilis jalapa, Euphorbia hirta*, *Cynodon dactylon, Indigofera tinctoria, Cassia auriculata, Calotropis giganta, Solanum surattense, Eclipta prostrate, Erythrina variegate, Acalypha indica, Abutilon indicum, Lucas aspera* and *Amaranthus gangeticus* were used by the women for medicinal value in the study areas. The major factors responsible for appropriateness of indigenous practices on health of human are economically viable (87.25 per cent),easy to handle (62.5 per cent), widely used 66.25 per cent locally available reported by 77.5. It could be concluded that it is important to explore the folk and gender based knowledge about indigenous flora as this play an important role in the diversity maintenance, cultural preservation and development.

Keywords: *Women's wisdom, Indigenous knowledge, Ethnomedicine, Medicinal plants, Traditional knowledge.*

Introduction

The renaissance of medicinal plants and indigenous practices for healthcare is becoming more pronounced in the recent years, not only in the developing countries but also in the under developed countries. It is evident that making of almost 90 per cent of medicaments have basically natural origin. In India, majority of the women depend on the medicines based on locally available indigenous plants and practices for curing different diseases and health related problems. These people have unique ancestral wisdom gained through trail and error by words of mouth and different kind of folk media.

Women have a profound knowledge of plants, medicines and their environment. Traditionally, they have been using a variety of indigenous plants and animals and have a direct stake in their preservation. Women's knowledge is an integral part of the culture and history of a local community for their particular identity. Studies have revealed that the women have greater interest in using ethno medicine, preserving and conserving the local forest plants and other natural resources for perpetual use. In addition, women are traditional caretakers of genetic and species diversity in biodiversity. The knowledge of women about necessary growing conditions and nutritional characteristics of various species have given them a crucial fund a experiences in selection of use of natural resources with special reference to botanical herbs and plants at village level. Focus group discussion (FGD) was adopted to explore the qualitative data regarding cultural, social, social capital and historical aspects of ethno medicine. The study was conducted with the following objective

1. To document the indigenous medicinal plants and their utility parts
2. To assess the factors responsible for appropriateness of indigenous practices

Methodology

This study was conducted at ten villages of foot hills of Alagarkovil by using participatory approach, group approach, group discussions, participatory learning and personal interview methods. Focus group discussion (FGD) was adopted to explore the qualitative data regarding cultural, social, social capital and historical aspects of ethno medicine. A transect walk was made to develop the resource flow map of village and explore the local plants used in various ethno medicine The resource flow map has helped in learning the current use of natural resource with special reference to botanical herbs and plants at village level. Interview was conducted with elderly women in the villages with the sample size of sixty respondents were interviewed to know their perception regarding the attributes responsible for appropriateness.

Findings and Discussion

By focused group interview method, women's knowledge on plants and their medicinal value was obtained and recorded their medicinal value which is unique in nature

Table 28.1: Medicinal value of uncultivated plants.

Sl.No.	Plants Name	Local Name	Uses
1.	*Tribulus terrestris*	Seerru nerunjal	Diabetic diseases control
2.	*Gymnema Sylvester*	Chiru kuruchan	Fever, cough
3.	*Lucas aspera*	Thumbai	Inner body heat controlled
4.	*Erythrina variegate*	Kalyani murungai	To increase Secretion of milk during pregency
5.	*Jatropha curcas*	Kattamanaku	Cure Wounds, skin problem
6.	*Ocimum basilicum*	Thirunitrupacsali	Cough, head ache
7.	*Solanaum surattense*	Kandam Kathiri	TB, Vengundram problem
8.	*Datura* sp	Umathai	Bone buldge breathing problem
9.	*Mirabilis jalapa*	anthimantharaii	Concepationswounds in the body
10.	*Euphorbia hirta*	Amman pacharesei	Sexual Problem
11.	*Cynodon dactylon*	Arugam pull grass	For blood circulation
12.	*Indigofera tinctoria*	Avuri	Nervous problem solved
13.	*Cassia auriculata*	Avaram poo	Diabetics control
14.	*Calotropis giganta*	Aeruku	Wounds, Heavy buldge
15.	*Eclipta prostrate*	Karichilan kani	For brain sharpness
16.	*Erythrina variegate*	Kaliana Murigangai	Seethapathi problem,heat problem
17.	*Acalypha indica*	Kubaimeni	Chori, cheranku, stomac worms control
18.	*Abutilon indicum*	Thuthi	Body inner heat reduced
19.	*Lucas aspera*	Thmbai	Inner Heat Reduced
20.	*Amaranths gangetic*	*Chiru keerai*	Control stone problem
21.	*Cardiospermum halicacabum*	Muttagattan	Concepation

It could be observed from the table that women had belived that *Tribulus terrestris* and *Cassia auriculata control the diabetics. Cardiospermum halicabum and* Mirabilis jalapa *have controlled the concepation problem of human being. Ocimum basilicum* and Gymnema Sylvester used to l control the fever and cough.Certain plant parts controlled the stone problem, blood circulation, skin problem etc.

Factors Responsible for Appropriateness of Indigenous Practices

The majority (77.50 per cent) of the people reported that economic viability of indigenous health management practices is the most important indicator for its appropriateness (Table 28.1). Each practice has logical rationality to cope with the diseases/problems reported by 75 per cent of the people, while, no adverse effect, good efficacy and local availability are third and fourth important factors responsible, respectively, for the appropriation of practices as perceived by people. Compatibility with socio-economic conditions, past experiences, efficacy nature and compatible with cultural conditions were found to be the most important indicators for appropriateness of indigenous practices as perceived by poor people were noted as wide potential in the applicability, easy to handle and reproducibility in nature. All these attributes contribute significantly for the existence and use of natural resources by following the indigenous approach.

Perception of Resource-Poor People regarding Factors Responsible for Appropriateness of Indigenous Practices for Human Health

Table 28.2: Factors responsible for appropriateness of indigenous practices for human health.

Indicators of Appropriateness	Percentage of Response	Ranks
Economically viable	77.50	I
No adverse effect	71.25	III
Good efficacy	71.25	III
Easy to handle	52.25	IX
Widely used	56.25	VIII
Locally available	67.50	IV
Rationality	75.00	II
Faith in quality	51.25	VIII
Non-failuring nature	63.75	VI
Compatible with socio-economic conditions	65.00	V
Compatible with past experiences	65.00	V
Compatible with cultural experiences	60.00	VII
Reproducibility nature	50.00	X

Conclusion

☆ Therefore, it is of paramount need that locally available indigenous health practices should be incorporated in the research system to test their scientific rationality and assure the validity.

A participatory research project can be designed with the help of people to carryout the research, treatment and health maintenance. The social capital and informal rural social institutions, as studied can be taken into account while preparing such plans, to make the efforts participatory and successful. Many elements need to come together for positive changes to preserve the treasure of ethomedicine. On account of above discussions, the following policies could be framed, which may offer better growing and sustaining the indigenous practices dealing with human and animal medicines.

Various necessary educational learning elements are required to be blended with existing system of school education. For this, training and capacity building of primary school teachers may be done to given them an opportunity for transferring the knowledge among younger generations. To guide and promote substantial dialogue between school teachers and community elders for transferring the indigenous knowledge systems of local plants, foods and medicines to the children for making the knowledge dam and avoid erosion.

☆ Invite the women traditional healers and knowledge holders in the school for giving the lecture as well as interacting with teachers and school children so that knowledge network could be developed.

☆ Various educative medias, video games and cartoons may be developed pertaining to traditional knowledge about ethnomedicine for transferring the knowledge from one generation to another through the young children.

☆ Effort of the community base conservation of those plant.A knowledge network of like minded women can be developed through various print and electronic media for preserving and promoting the indigenous practices of human health.

☆ The value addition through scientific research and development can make the local medicines more popular. It will also help in assuring the intellectual property rights after knowing scientific active ingredients in a particular medicinal plant. The equitable benefit share in terms of both tangible and non-tangible benefit should be assured with the help of Prior Informed Consent(PIC) before putting the women's knowledge into the public domain.

References

Ganesan, S., Suresh, N. and Kesaven, L., 2004. Ethnomedicinal survey of lower Palani hills of Tamil Nadu. *Indian Journal Traditional Knowledge*, p. 299–304.

Ramphele, M., 2004. Women's indigenous knowledge: Building bridges between traditional and modern. In: *Indigenous Knowledge Local Pathways to Global Development Knowledge and Learning Group*. Africa Region, The World Bank, Washington DC, pp. 13–17.

Singh, R.K., Dwivedi, B.S. and Singh, R., 2002. Traditional wisdom of farmers: An experience towards the sustainable development of livestock. *Indian Journal of Traditional Knowledge*, 1(1): 70–74.

2013, Biodiversity Conservation for Sustainable Management   Pages *231–247*

Editor: **Dr. K. Muthuchelian,** *Vice Chancellor, Periyar University, Salem*

Published by: **Daya Publishing House, NEW DELHI**

Chapter 29

Marine Biodiversity and Bioresources for Coastal Socio-Economical Development in India

P. Nammalwar*

Project Leader (INCOIS),
Institute for Ocean Management, Anna University,
Chennai – 600 025, Tamil Nadu

ABSTRACT

The coastal marine ecosystems play a vital role in India's economy by virtue of their natural resources, potential habitats and wide biodiversity. India has a long coastline of 8129 kms with Exclusive Economic Zone (EEZ) of 2.5 million sq.km which is an important area both for exploration and exploitation of natural resources. Marine biodiversity affords enormous economical, environmental and aesthetic value to human kind. Humans have long depended on marine aquatic resources for food, medicine and materials as well as for recreational and commercial purposes such as fishing and tourism. Marine organisms also rely upon the great biodiversity of habitats and resources for food, materials breeding and larval disposal environment. This interdependence is essential and maintaining a balance between them is cardinal. But the marine ecosystems are deteriorating at an alarming rate. The factors responsible for it are over exploitation of species, introduction of exotic species, pollution from urban, industrial, and agricultural areas as well as habitat loss and alteration of

* Corresponding Author: E-mail: drnrajan@gmail.com

water diversion, excessive use of water resources etc. As a result, valuable marine aquatic resources are becoming increasingly susceptible to both natural and man made environmental changes. The present paper deals with the strategies to protect and conserve marine biodiversity which are necessary to maintain the balance of nature and support the availability of natural resources for future generations in India.

Introduction

The coastal marine areas contain some of the world's most diverse and productive biological systems. They include areas of complex and sophisticated ecosystems, such as enclosed sea and tidal systems, estuaries, salt marshes, coral reefs, sea grass beds and mangroves that are sensitive to human activities, impact and interventions. Pressure on these systems is growing more intense. The coastal marine environment plays a vital role in India's economy by virtue of their resources, productive habitats and wide biodiversity.

The world oceans and seas are linked to many bodies of freshwater through coastal areas and the two forms an independent ecosystem that spawns much of the world's marine life. The Rio De Janera Earth Summit, (1992), the World Bank and other development practioners are emphasizing an Integrated Coastal Zone Management (ICZM) approach. Integrated Coastal Zone Management (ICZM) provides unifying framework for protecting and managing the world oceans and coastal areas consistent with environmentally sustainable management. Since, 1993, the World Bank has promoted the establishment of Integrated coastal Zone planning and management in client countries through (a) awareness creation and capacity building (b) investment and (c) partnerships. These efforts have paralleled support for marine environmental protection, including pollution control and conservation of marine biodiversity.

The marine and coastal areas play an even more important role today, since they provide protein from fish and other seafoods. The current problems of environment destruction in tropical coastal seas, and the effects on the productivity of fish and other seafood from these areas are therefore are of primary importance. Continued destruction of estuaries and lagoons, mangrove forests, sea grass beds and coral reefs in the tropical Third world countries will mean the difference between life and death for millions of poor people and for many others, the difference between a life in reasonable health and one of malnutrition, disease and starvation. There is now considerable evidence that for marine areas in the tropics, there is a clear correlation between the productivity of coastal ecosystems (particularly in mangrove forests, sea grass beds and coral reefs) and the productivity of fisheries. The waters has become increasingly clear is the relative importance of near shore areas in the tropical areas. The use of Integrated Coastal Zone Management (ICZM) as a toolbox to develop coastal resources in a sustainable manner and to mitigate conflicts between users has proven to be a possible solution in many countries. The ICZM is a method that can lead to sustainable development because it has the advantage of securing government participation as well as stakeholder involvement.

The sustainability of coastal zone is a growing concern worldwide. There is rapid ongoing destruction of many of the marine and coastal resources essential to human beings throughout the third world countries. Siltation and nutrition rich emission from agriculture, waste discharges from industries and from urban conurbations are among the most important causes of coastal resources degradation. The major underlying factor is the rapid population growth that is taking place in most tropical countries. The courses are particularly venerable and often experience the highest growth rate of more than 5 per cent per year. Coastal degradation cannot be solved within the traditional sectors like fisheries and shipping. What is required are integrated coastal zone management and projects, to address all the factors that have impacts on coastal zones. Major steps have been taken in several countries to halt negative trends and we will probable see more well implemented ICZM programmes that will address the coastal resource user conflicts. In India, as rapid development and population continues in coastal areas, increasing demands are expected on natural resources and on the remaining natural habitats along the coast. Unless corrective measures are undertaken, environmental degradation and over exploitation will erode marine and coastal biodiversity, undermine productivity and intensify socio-economic conflicts over the increasingly scarce resources of the coastal areas. Current sectoral approaches to the management of coastal and marine resources have generally not proven, capable of conserving the marine and coastal biological diversity. This problem is more serious in Indian context that has a long coastline of 8129 kms with Exclusive Economic Zone (EEZ) of 2.5 million sq.kms. This zone suffers from the absence of an integrated attention to conservation and development. Since these regions form a vital link between the terrestrial and aquatic ecosystems their conservation is essential to maintain the ecological balance and biodiversity. A well-defined biodiversity lessons learnt in other regions of the world is proposed for implementation in several different types of regional scale coastal marine ecosystems. Various conservation and management strategies for sustainable use of coastal marine biodiversity are suggested for socio-economic development in India.

Maintenance of Species Diversity

A major need for biodiversity maintenance is protection of special or critical, littoral habitats including mangrove forest, coral reef, sea grass meadows, shallow water bodies like shallow water lagoons and beaches. While it is useful and practical to focus on individual habitat types or species, one must not forget that they exist only as components of wider coastal systems. The complexity of biotic systems and interrelatedness of their components require that each coastal water ecosystems be managed as a system. The need to preserve the biological systems and the method for doing so were terrestrially derived. Therefore, they required modification to fit to coastal habitats. Few oceanic species are in danger of extinction because of habitat damage. But along the coast and beaches many species (Turtles) jeopardized by habitat degradation and loss. Five aspects of marine biological diversity are paramount for consideration. 1.The diversity of marine fauna is much greater than for terrestrial fauna at higher taxonomic levels. 2. The marine fauna is also much well known 3. Most marine species are widely dispersed. 4. Most marine communities are highly patchy and variable in species composition. 5. The response type to environmental perturbations is relatively small. Major strategic objectives for ICZM is to preserve the habitats of the species that have been designated as especially valuable are in danger of extinction. Therefore, an important motivation for designating "Ecologically Critical Areas" (ECAS) for special conservation is species protection; other purposes might be protection of specially productive or scenic natural resources. Threats to the productivity of unique biological systems of the coastal zone, species and their habitats arise from development activities and their side effects, including reef and beach mining, shoreline filling, lagoon pollution, sedimentation and marine construction activities that are quite distinct from those on land. The strategy plan must recognize that species and the habitats of coastal zones are so different from their terrestrial counterparts as to require different and special forms of conservation. For example, coral reefs, beaches, coastal lagoons, submerged seagrass meadows and intertidal mangrove forests have no counterparts in terrestrial systems. In addition to habitat management, there may be other appropriate actions to be taken under the ICZM programmes, for example, banning exploitation of endangered species.

Man made changes have the effect of reducing the number of species inhabiting the area. For example, eutrophication leading to lowered oxygen concentration in the water and the sediments limits the number species to those few to tolerate those conditions. In areas where large-scale aquaculture is practiced, the cultured species often occupy a disproportionately large amount of the habitat, forcing out many of the natural species. A case in point is in the mangrove forest of Southeast Asia where there has been great development of ponds for shrimp culture. Many of these operations rely on the natural spawning of shrimps to provide juveniles for introduction to the ponds, yet the mangrove ecosystems are so damaged in many areas that the natural shrimp populations are declining. In general, most adverse impacts on coastal ecosystems are characterized by decreasing species diversity. Monitoring of species diversity is therefore a useful technique for assessing damage to the system and maintenance of good species diversity is a positive management objective.

Of several plants and animals inhabiting the coastal ecosystems including the coral reefs, mangroves and estuaries only some species are exploited for human use. However, such species are often irrationally exploited with powerful harvesting techniques, sometimes leading to collapse of their respective habitats. It is, therefore, appropriate to deal with the question of biodiversity management from the point of view of "species-habitat" units in the case of the more sensitive and vulnerable ecosystems in the coastal zone.

Protected Areas in India

Protection of species requires protection of their special habitats as well as preventing the hunting and harassing them. World concern regarding loss of biodiversity is neither felt equally in all nations nor all members of society equally share it. It stems from realization that humans have been transforming the oceans as dumping grounds for their wastes and modifying the natural composition of its environment. Preserving biodiversity is an important reason for protecting natural area. Endangered species are major beneficiary of coastal habitat protection, for example: Coastal birds, turtle and even marine mammals. Other protections for species are mostly regulatory that is providing legal protection against killing and disturbing endangered species whether inside or outside a designated area.

Some marine ecosystems–coral reefs for example have high species diversity but despite increasing marine pollution and degradation of coastal habitats, there is little evidence for an imminent major loss of marine biodiversity at the species level. This may be due to partly to lack of knowledge but an important factor is the typical marine life history. Marine species live in wide-open systems, the sea being continuous around the earth and have greater ranges and fecundity than the terrestrial species. This confers on them a greater resilience to exploitation and environment change. Similarly endemism is rare. Many marine species particularly fish and invertebrates are so called strategists, producing large number of seeds, but having short lives. Recent species extinction is almost unknown among marine organisms with a planktonic larval stage and among the many species of migratory, highly mobile and wide spread fishes. Species that are large bodied, long lived, and slow breeding, producing few offspring with much parental investment. A few known cases of marine extinction in historic times include marine mammals.

The first Marine National Park in India came into existence in the Gulf of Kutch (Pirotan area) in 1980 followed by Gulf of Mannar and Wandoor Marine National Park in the South Andamans. A marine park is a reserve and should be managed along several ecological principles and should serve many relevant purposes such as habitat and species preservation, scientific research, recreation and financial gains. Though these three marine habitats have been declared as protected areas, delineation of the core areas and the park limits and regulations on various human activities in the protected areas remain to be implemented. Proposals have been initiated to establish marine parks and preserves in Malvan-Vengrula (Coast of Maharashtra), Mincoy, Kavaratti, Chetlat, Kadamat and Kalpeni (in Lakshadweep).

Areas Rich in Biodiversity

The priority towards the conservation of marine biodiversity is to identify/locate the areas, which are highly critical and rich in species distribution and their favourable habitats. Species diversity can be made as the criteria for ranking the areas for conservation. An area that harbors smaller assemblage of species is to be given more importance, evolutionary significance, ecological importance and endangerment. An area with lower diversity might be better at providing ecological services important to people. Species diversity differs markedly on both ecological and biogeographic special scales. On an ecological scale, it is important to conserve coral reefs and mangrove forests, which are unusually species rich for its ecosystem type. On a biogeographic scale, using species diversity/richness as the sole criterion for priority conservation. Areas of high biodiversity may not be most critical to the sea as a whole, for various reasons. Coral reefs generally have high species diversity but tend to be low in other biological attribute such as endemism. The salt marshes, mangrove forests and seagrass beds can have special importance than coral reefs because they serve as significant nursery areas and their productivity supports important food webs a. Moreover, some areas are especially important seasonally because they are critical to key elements of marine biological diversity, even if their diversity is low.

Category-1: Marine protected areas (National parks and sanctuaries).
(MAPs having entire areas in intertidal/subtidal or sea water-mangroves, coral reefs, lagoons, estuaries, beaches etc.,) (Singh, 2002)

Sl.No.	Name of the MPA (District) State/UT	Declaration	Area (sq km)	Ecosystem
1.	Mahatma Gandhi Marine NP Wandoor (South Andaman) Andaman	1983	281.50	Tropical evergreen forest, mangroves, Coral reefs, creeks and seawater.
2.	Rani Jhansi Marine NP (Ritchies Archipelago) Andaman	1996	256.14	Evergreen forest, mangroves and Coral reefs.
3.	Lahabarrack (Salt water crocodile) Sanctuary (South Andaman) Andaman	1987	100.00	Dense mangroves (tidal forest), littoral forest, creeks, marine water and tropical evergreen forest.
4.	Gulf of Kachchh Marine NP (Jamnagar) Gujarat	1982	162.89	Mangroves, Coral reefs, mudflats ,Creeks, beaches and scrub forest.
5.	Marine Sanctuary Gulf of Kachchh (Jamnagar) Gujarat	1980	295.03	Mangroves, Intertidal area, marine water, coral patches and sandy beach.
6.	Bhitar Kanika NP (Cuttak) Orissa	1988	145.00	Estuary, delta and mangroves.
7.	Bhitar Kanika Sanctuary (Kendrapara) Orissa	1975	672.00	Estuary, mangroves, terrestrial forest and ecotone with marine environment.
8.	Gahirmatha Marine sanctuary (Kendrapara) Orissa	1997	1,435.00	Sea water, sandy beach, estuary mangroves and ecotone with marine environment.
9.	Chilka (Nalabund) WLS (Khundra, Puri, Ganjam) Orissa	1987	15.50	Island, Lagoon and Brackishwater.
10.	Gulf of Mannar NP (Ramanathapuram/Tuticorin) Tamil Nadu	1980	6.23	21 islands, coral reefs, mangroves, sea grass beds and beaches.
11.	Pulicat Lake (Bird) Sanctuary Tiruvellore, Tamil Nadu	1967	17.26	Lake of Brackishwater of rain and seawater, mangrove and estuarine environment.
12.	Point Calimere Sanctuary (Nagapattinam) Tamil Nadu	1967	17.26	Tidal swamp, mangroves, creek and evergreen forests.
13.	Coringa Wildlife sanctuary (East Godavary) Andhra Pradesh	1978	235.70	Mangroves, estuary, back water, creek and mud flats.
14.	Krishna Wildlife Sanctuary (Krishna/Guntur) Andhra Pradesh	1999	194.81	Mangroves, back water, creeks and mud flats.
15.	Pulicat Lake Bird Sanctuary (Nellore) Andhra Pradesh	1976	500.00	Brackishwater of rain and seawater, mangroves, estuarine and algal beds.

Contd...

Category-1–*Contd...*

Sl.No.	Name of the MPA (District) State/UT	Declaration	Area (sq km)	Ecosystem
16.	Sunderbans Nantional Park-Tiger Reserve (North and South 24 pargana) West Bengal	1973/1984	1,330.10	Mangroves, estuarine, creeks, swampy islands and mudflats
17.	Halliday Sanctuary (South 24 Pargana) West Bengal	1976	5.95	Mangroves, estuaries, swampy islands and mudflats
18.	Lothian Island Sanctuary (South 24–Pargana) West Bengal	1998	38.00	Mangroves, Estuaries, creeks, swampy islands and mudflats
19.	Sajnakhali Sanctuary (South 24 Pargana) West Bengal	1976	362.4	Mangroves, estuaries, creeks, swampy islands and mudflats.

Category II: Marine protected areas.
(Islands MPAs in Andaman and Nicobar and Lakshadweep Islands, which have major parts in marine ecosystem and some part in terrestrial ecosystem)

Sl.No.	Name of the MPA (District) State/UT	Declaration	Area (sq km)	Ecosystem
1.	North Buttan NP (Middle Andaman) Andaman	1987	0.44	Evergreen forest, littoral forest, mangroves, beach and coral reefs.
2	South Buttan NP (Middle Andaman) Andaman	1987	0.03	Evergreen forest, littoral forest, mangroves and beach.
3.	North Reef Island Sanctuary (North Andaman) Andaman	1987	3.48	Evergreen forest, littoral forest, mangroves and beach.
4.	South Reef Island Sanctuary (Middle Andaman) Andaman	1987	1.17	Beach and Coral reefs.
5.	Cuthbert Bay Sanctuary (Middle Andaman)	1987	5.82	Splendid beach and creek.
6.	Cingue Sanctuary (South Andaman) Andaman	1987	9.51	Evergreen forest, coral reef and beach.
7.	Galathea Bay sanctuary Great Nicobar	1997	11.44	Evergreen forest and mangroves.
8.	Parkinson Island Sanctuary Middle Andaman	1987	0.34	Evergreen and littoral forest and mangroves.
9.	Mangrove Island Sanctuary	1987	0.39	Mangroves and marine life
10	Blister Island Sanctuary North Andaman	1987	0.26	Mangroves and beach.
11.	Sandy Island Sanctuary south Andaman	1987	1.58	Sandy Isalands.
12.	Pitti wildlife Sanctuary Lakshadweep	2000	0.01	A small sandy island surrounded by sea.

These include courtship, spawning areas, nursery grounds, and migration areas and stop over points. The areas rich in productivity, spawning grounds, nursery grounds, migrations are to be protected.

Biosphere reserves in marine areas.

Sl.No.	Name	State	Year of Notification	Area (sq km)
1.	Sundarbans	West Bengal	1989	9630
2.	Gulf of Mannar	Tamil Nadu	1989	10500
3	Great Nicobar	Andaman and Nicobar	1989	885

Strategies for Conservation and Management of Marine Biodiversity of India

The Indian coast is indented by a number of rivers, which forms estuaries at their confluence with the sea. The complex coastal ecosystems are comprised of estuaries, lagoons, mangroves, backwaters, salt marshes, mud flats, rocky shores and sandy stretches. Besides, there are three Gulfs, one on the east coast, the Gulf of Mannar, and two on the west coast, Gulf of Kutch and Gulf of Camby. The two island ecosystems Lakshadweep and Andaman and Nicobar. Islands add to the ecosystem diversity in India. The Gulf of Mannar, Gulf of Kutch and the two island ecosystems have rich coral reefs and mangroves harboring valuable marine biodiversity.

A well-defined set of biodiversity lessons learnt in other regions of the world is proposed for implementation in several different types of regional scale marine ecosystems, which needs biodiversity information data bank. These information will permit meaningful comparisons across different habitats of the causes and consequences of changes in biodiversity due to human activities. This lesson requires significant improvement in taxanomic expertise for identifying marine organisms and documenting their distribution, knowledge of local and regional natural patterns of biodiversity and in understanding in the processes that create and maintain these patterns in space and time. (i) Need for rapid expansion in taxonomy in order to interpret, manage, conserve and use biodiversity sustainability and the need to pool together the existing data from all sources by forming an information network of all agencies in the country. (ii) Priority for the biodiversity conservation to understand what values are important, which genes/species/habitat and how much biodiversity should be conserved. (iii) Improve the methodologies for different programmes, evolve more effective policy and target with priority. (iv) Practice of the biodiversity conservation programme with precise definition and clear targets. (v) Recognition of priority of the communities. (vi) Application of anthropogenic objectives of maintaining biodiversity so that it is of possible value to the mankind. The conservation of coral and mangrove habitats has attained great significance in developing countries in the context of its functional role in ecological and socio-economic sustainable development and the Ministry of Environment and Forest, Government of India can formulate decisive policies for conservation and the management of coral and mangrove habitats along the Indian coast.

The species at risk are quite different from terrestrial ones. *e.g.*, oyster, octopus, porpoises whales, sea fishes, sea turtle, dugong. The occurrence of endangered and threatened species is less in the sea because it is an open system with few boundaries to migration. While several of the sea mammals and sea turtles are endangered, the fishes and shellfishes are usually not. The species protection by designating protected natural reserves is relatively inexpensive and simple to administer. This strategy can be implemented on a site-specific basis, and commensurate with available information, staffing or expertise. It can be reinforced with regulatory measures that combine "wetside" (estuarine or marine area) protection with "dryside" (shorelines) management strategies offering the possibility of managing whole coastal ecosystems. But first, it is necessary for ICZM process to identify during strategy planning, the critical coastal habitats that merit high degree of protection, so they can be addressed specifically in the Master Plan.

East Coast of India

The development of coherent and directed multidisciplinary research program for the east coast is considered a priority. This region has a major but not well-understood influence on our climate. It contains major recreational values, and potential for and effect of tourism require further investigation. The area supports the valuable finfish, a growing aquaculture industry and a high level of biodiversity that has not been studied intensively. This biodiversity includes, particularly in southeastern waters, a high degree of diversity of species unique to Indian marine ecosystems. Although the east coast is highly productive, there are signs that this environment is under pressure. Several bays and estuary systems are under threat from degradation, for example Palk Bay, Gulf of Mannar, Pulicat Lake, Chilka Lagoon and Krishna and Godavari Estuaries. Major seagrass losses have occurred in Palk Bay and Gulf of Mannar. Major Fisheries are under threat or in decline. The prawn fishery suffered a major collapse some years ago, and some species show that its spawning population is reduced to dangerously low levels.

As a first step, collabaorative linkages between research institutes and universities in region should be strengthened, for the pupose of better existing research actyivity, and developing new multidisiciplinary programs in priority areas. The activity would include studies designed to improve understanding of biodiversity and biological processes, and the development of new long-term monitoring programs. Such targeted collaborative research programs are important to the development of the knowledge base as an under pinning to the effective operation of resource based marine industries in the area, and to protection against environmental threats. Specific initiatives would include:

☆ Studies on the ecosystem dynamics of east coast of India;

☆ Establishing long-term monitoring programs and baseline data sets; and

☆ Refurbishing or establishing field research stations in Andaman and Nicobar Islands.

West Coast of India

There is a strong case for establishment of a multidisciplinary research facility in the west coast to develop an understanding of the region's marine resources and ecosystems. The major demand for marine science is to build the basic knowledge needed:

☆ For the continuing sustainable use of marine resources, including traditional uses by Lakshadweep islanders;

☆ For the continuing success of growing commercial fishing, and aquaculture and pearl industries;

☆ To understand the marine environment that supports oil and gas industry; and

☆ Supportive research activity such as mapping of seabed topography, studies designed to improve understanding of biodiversity and biological processes, the designed and implementation of monitoring fisheries development, and development of an understanding of industry impact programs, and research supporting sustainable use of marine environment.

The shared borders with Pakistan, Bangladesh, Maldives and Sri Lanka, and large coastlines supporting similar tropical marine ecosystems in other areas, are further considerations in planning the management of resources in this region, and encourage the concept of joint research programs.

Strategies for Sustainable Development

The investigation of marine biodiversity possesses a considerable scientific and conservation challenge because of the great size and relative inaccessibility of marine ecosystems. The scale of marine systems and the mixing dispersion and transport that occur in the oceanic medium require different thinking and investigative processes. Marine pollution, eutrophication (inshore and offshore), sedimentation and silting from coastal run off may outweigh the direct impacts on species (*e.g.*, fishing), or even the indirect effects of climate change. Entire watersheds are obviously involved. Habitat protection is the most serious need for coastal and marine biodiversity. However, this need is very often obscured by different view by the overlaying waters. A spectrum of measures from overall regulation to area specific protection and spectrum of scales from local to global must be involved. There is an urgent need therefore, to establish strictly, comparative data bases for various groups of marine organisms on both global and regional scales, to test the hypothesis relating to the evolutionary and ecological constrains on diversity, and to provide a proper scientific basis for the implementation of conservation measures. The International Association of Biological Oceanography (IABO) in co-operation with UNESCO has launched an international co-operative programme on biological diversity during 1990. The major objective of the programme is to understand biological diversity within the context of the structure and function of ecosystem. The IABO/UNESCO International Marine Biodiversity programme is of particular importance to global change research given the forecasted climatic changes and their implication for the function of natural and man identified marine ecosystems, within the context of sustainable development. It is extremely important to select representative ecosystems for experimental studies that involve improving the outcome of this prediction, as well as for long-term monitoring of biodiversity, in estuaries, lagoons, coral reefs, mangroves and salt marshes believed to be vulnerable to pronounced climatic and environmental change. One of the most important problems is the necessity to tackle simultaneously global climatic changes, marine biodiversity and sustainable development. As part of that effort, much more attention and visibility needs to be given to the link between pollution regulation and the ecosystem function of biological diversity. An understanding of marine biodiversity is indispensable for advances in all fields of biology, including ecology, fisheries and aquaculture conservation and pollution. These areas of research are equally important for both developed and developing countries.

Conservation of Marine Biodiversity through Coastal Zone Management

Some coastal ecosystems are particularly at risk, including saltwater marshes, coastal wetlands, coral reefs, coral atolls and river deltas. Other critical resources, such as mangroves and seagrass beds, submerged systems and mud flats are at risk from climate change impacts, exacerbated by anthropogenic factors. Changes in these ecosystems could have major negative effects on tourism, freshwater supplies, fisheries and biodiversities that could make coastal impacts on important economic concern. Coastal zones comprise a continuum of aquatic systems including the network of rivers, the estuaries, the coastal fringes of sea and continental shelf and its slope. The functional value of diversity concept encourages analysis to take such a wider perspective and examine changes in large-scale ecological processes, together with the relevant environmental and socio-economic driving forces. At the global scale, while climate has fluctuated throughout time, a global warming scenario could lead to accelerated sea level rise, changes in rainfall patterns and storm frequency or intensity and increased siltation. The consequences may include shore-line erosion and associated loss of habitats, such as salt marshes, mangroves and mud flats. An economic multiplier effect would then be generated leading to, for example, loss in tourism income and fisheries productivity, together with the increased cost of water supply and biodiversity conservation. In principle, the core objective of coastal zone management is the production of a socially desirable mix of coastal environmental system states, products and services. A future, more integrated coastal zone management process should include:

☆ Integration of programmes and plans for economic development, environment quality management and ICZM.

☆ Integration of ICZM with programmes for such sectors as, fisheries, energy, transportation, water resources management, disposal of waste, tourism and natural hazards management.

☆ Integration of responsibilities for various tasks of ICZM among the level of government—local state/provincial, regional, national, international and between the public and private sectors.

☆ Integration of all elements of management, from planning and design to implementation, that is, construction and installation, operation and maintenance, monitoring and feedback and evaluation over time.

☆ Integration among the disciplines; for example, ecology, geomorphology, marine biology, economics, engineering, political science and law.

☆ Integration of management resources of the agencies and entities involved.

The Integrated Coastal Zone Management process should aim to unite government and the community science and management and sectoral and public interest. It should interalia improve the quality of life human communities who depend upon the coastal resources while maintaining the biological diversity and productivity of coastal ecosystems. The rapid industrialization along the coast especially in the areas along the metropolitan cities has caused enormous damage to the coastal ecosystem. Hence, the laws governing the coastal land use should be framed with a view to promoting the economic development designed in tune with the coastal ecosystem.

Threats to marine biodiversity could be categorized which include: Habitat loss, population, pollution, natural disasters and over exploitation. Destruction of thousands of mangrove forests and coastal wetlands for the construction aquaculture farms in certain coastal states were reported. Besides these threats, the coral reefs are also subjected to manmade pollution such as heavy metals, fertilizers and sewage and industrial wastes. Together, the unintended consequences of over fishing, by-catches and habitat degradation can alter the very biodiversity, productivity and resilience of marine ecosystems on which economically valuable species and fisheries depend.

Marine Biodiversity in India

The marine ecosystem as a varying profile. The coastline encompasses almost all type of intertidal habitat from hyper saline and brackishwater lagoons, estuaries and coastal marsh and mudflats to sandy and rocky shore. The sub-tidal habited are equally diverse among the coastal wetlands, estuaries, mangroves, coral reefs and costal lagoons are biodiversity rich areas. Each local habitat reflex prevailing environmental factors and is further characterized by its biota. Thus the marine fauna itself demonstrate gradient of change thought the Indian coast. Among the total 32 animal phyla 15 are represented by taxa in the marine eco system. They may constitutive either migratory or resident species. The former includes pelagic crustaceans, Coelenterate (Medusae), Cephalopods, fishes, reptiles, birds and mammals. Amphibians are generally absent in estuary. The benthic macro fauna comprises residence species of polycheats, bivalve, gastropods, sipunculas and mud–borrowing fishes. Among invertebrates the sponges, phoronids and echinoderms generally do not prefer an estuarine ecosystem. In Indian esturies species diversity seems to maximum in the mollusk. About 245 species belonging to 76 genera under 54 families have been cataloged. Other important taxa, polychaeta are represented by about 167 species belonging to 97 genera under 38 families. Maximum diversity has been reported in the much studied hoogly–matlah estuary (West Bengal). Macro organisms and meiofauna of Indian estuaries are not properly investigated. Estuarine mud may contain rich verity of bacteria, flagellates, ciliates, nematodes, ostracods, harpacticoid copepods, rotifers, gastro tirches, arachnids, and tardigrades. Free swimmers or nekton are important components of marine biodiversity and constitute important fisheries of world. The dominant taxa in the nekton are fish, others being crustaceans, mollusk, reptiles and mammals. Out of the total 22,000 fin fish species, about 4000 species occur in the Indian Ocean of which 1800 species are reported in the Indian seas. A majority of the nektonic species is found in coastal waters. It is estimated that 40 species of shark and 250 species of boney species represented the oceanic fishes.

Among reptiles, sea snake and turtle are important and reperanted world wide by 50 and 7 species respectively. These are generally oceanic forms but a majority of them visit the shore at some part of life. About 26 species of sea sneak belonging to one family hydrophiidie, and 5 species of sea turtle were reported seas around India. Oceanic Island seems to harbor more reptiles in their marine environment. All the sea snakes and 4 species of turtles in their marine environment are known from island of Andaman and Nicobar. Nesting sites often amphibian snake were reported from the source north Andaman Island. Sea turtle visit the shore during breeding season to lay their eggs. The visit of these turtle to the shore especially oliverideley is spectracular site on the sandy beach Gahirmatha near Bitar kanika in Odisha. The Andaman and Nicobar islands have the nesting beaches for the leatherback the hawksbill and the green turtle in addition to oliverideley. The seashore offer the variable feeding and breeding grounds for a number of birds. Its is difficult to define precisely the avian component of marine biodiversity, there are special species, exclusively dependent on the marine ecosystem, while a few are generalists without much dependents on it.

Marine Biodiversity in India

Sl. No.	Group	No. of Species	Sl. No.	Group	No. of Species
1.	Algae	724	2.	Protista	750
3.	Mesozoa	12	4.	Proifera	486
5.	Cnidaria	842	6.	Ctenophore	12
7.	Gastroticha	98	8.	Kinorhyncha	10
9.	Platyhelminthes	550	10.	Annelida	440
11.	Mollusca	3370	12.	Bryozoa	200
13.	Crustacean	2934	14.	Merostomata	2
15.	Pycnogonia	16	16.	Sipuncula	35
17.	Echiura	33	18.	Tardigrada	10
19.	Chaetognatha	30	20.	Echinodermata	765
21.	Hemichordate	12	22.	Protochordata	119
23.	Pisces	1800	24.	Amphibia	3*
25.	Reptiles	26	26.	Aves	145
27.	Mammalia	29			

* in estuaries/Mangroves.

From the available data, it has been inferred that 12 family, 38 genera and 145 species occur in coastal eco system. Marine mammals belong to 3 order sirenia, cetacea and carnivora. About 120 species are estimated to occur in world seas and of these 30 are reported from seas around India. But a majority of these is found in oceanic forms and occasionally a few individuals may get stranded on the shore. The sea cow occurs in near shore water.

Key Challenges

Conservation of marine biodiversity in India can be best managed by the following guidlence;

1. Revitalizing the 200 year tradition of marine biodiversity inventorying in order to interpret, manage, censer and use bioresources in sustainable manner.
2. Reconstruction/restoration of lost habitats.
3. Reduction of discords, by-catch being thrown overboard.
4. Establishment and management of marine protected arias.
5. ecosystem based fisheries management
6. formulation of effective policy measures
7. economic values of biological resources of coral reef ecosystem of India in the international market and impact due export revalidated.

Conclusion

The biodiversity/bioresources of many habitats is under threat and although seas cover the major part of our earth's far less is known about the biodiversity/bioresources of marine environment than that of terrestrial ecosystem. It is also not clear whether many of the patterns know to occur on land also occur in the sea. Until we have firmer idea of the diversity of a wide range of marine habitats. And what controls it, we have little hope of conserving biodiversity or determining the impact of human activities such as mariculture, fishing, dumping of waste and pollution. Therefore recognition of scale problem, the nature of underling causes, and the limited resource available to counteract powerful destructive trends will definitely lead to a best way of conserving the marine biological diversity of India

References

Anon., 1987. Strategies for the management of fisheries and aquaculture in mangrove ecosystems. FAO Fisheries Report No. 370. Supplement. FAO, Rome. pp. 248.

Anon., 1997. *Proceedings of the Regional Workshop on the Conservation and Sustainable Management of Coral Reefs,* (Ed.) Vineeta Hoon. Proceedings No. 2. CRSARD, Madras. M.S. Swaminathan Research Foundation and Bay of Bengal Programme of FAO/UN, Chennai, pp.144.

Anon., 1992. Coastal systems studies and sustainable development. UNESCO Technical Papers in Marine Science. No. 64, *Proceedings of the COMAR Interregional–Scientific Conference*, Paris, France. pp.275

Anon., 2002. Marine and coastal ecosystems: Coral and mangrove–Problems and management strategies. SDMRI Research Publication No. 2. Suganthi Devadason Marine Research Institute, Tuticorin, Tamil Nadu. pp. 204.

Anon., 1998. Biodiversity of Gulf of Mannar marine Biosphere Reserve. Proceedings of the Technical Workshop. No. 24 (Ed) Rajeswari M. Anand, K. Durairaj and A. Fareeda, M.S. Swaminathan Research Foundation, Chennai. pp 185.

Bakus, G.J., 1994. *Coral Reef Ecosystem*, Oxford and IBH Publishing Co. Pvt. Ltd, New Delhi, pp. 232.

Colin, P. Rees and Casselss, David S., 1997. Environmentally sustainable developments studies and monographs series No. 19. Advancing sustainable development. The World Bank and Agenda 21, The World Bank Publications, Washington, USA, pp. 80.

Clark, J.R., 1992. *Integrated Management of Coastal Zones*. FAO Fisheries Technical Paper No. 327, pp. 167.

Clark, John R., 1996. *Coastal Zone Management Handbook*. Lewis Publishers, New York, USA, pp. 694.

De Roy, R. and Thadani, R., 1992. India's wetlands, mangroves and coral reefs, WWF–India, New Delhi. pp.61.

Gadgill, Madhav M. and Rao, P.R. Seshagiri, 1998. *Nurturing Biodiversity: An Indian Agenda.* Environment and development series, Centre for Environment Education, Ahamedhabad, pp. 163.

Grassle, J.F., Lasserre, P., Mcintyre, A.D. and Ray, G.C., 1991. Marine biodiversity and ecosystem function. *Biology International*, Special Issue, 23: 1–19.

Gustavson, Kent, Huber, Richard M. and Ruitenbeek, Jack, 2000. Integrated coastal Zone management modeling 2000. Publication of the International Bank for Reconstruction and Development/The World Bank Washington D.C. USA. pp 292.

Lasserre, P., 1992. Marine biodiversity, sustainable development and global change. In: *Coastal System Studies and Sustainable Development*. UNESCO Technical Papers in Marine Science, 64: 38–55.

Kannaiyan, S. and Venkataraman, K., 2008. Biodiversity conservation in Gulf of Mannar Biosphere Reserve. *Proceeding of the International Workshop on Gulf of Mannar Biosphere Reserve: An Ecological Model for Biodiversity Conservation, Livelihood and Sustainability*, National Biodiversity Authority, Chennai, pp. 484.

Kerry Turner, R. and Battyman, Ian, 2001. Water resources and coastal management. *Managing the Environment for Sustainable Development.* Edward Elger publishing Inc, Northampton, Massachusets, USA, pp. 527.

Kathiresan, K. and Qasim, S.Z., 2005. *Biodiversity of Mangrove Ecosystems*. Hindustan Publishing Corporation (India), New Delhi, pp. 251.

Mann, K.H., 2000. *Ecology of Coastal Waters with Implications for Management*. Blackwell Science Inc., London, U.K., pp. 406.

Masan, C., 1999. The ocean's role in climate variability and change and the resulting impacts on coasts. *Natural Resources Forum*. 23: 123–134.

Menon, M.G. and Pillai, C.S.G., 1996. *Marine Biodiversity Conservation and Management.* Central Marine Fisheries Research Institute, Cochin, pp. 204.

Nammalwar, P. and Joseph, V. Edwin, 2002. Bibliography of Gulf of Mannar, CMFRI, Spl. Publ. No., 74: 204.

Nammalwar, P., 2008. Present status conservation and management of Mangrove ecosystems in the islands of Gulf of Mannar region, Tamil Nadu. In: *Glimpses of Aquatic Biodiversity,* (Eds.) P. Natarajen, K.V. Jayachendaran, S. Kannayan, Babu Amabat and Arun Augustine. Rajiv Gandhi Chair Spl. Pub., 7: 284.

Nammalwar, P. and Sundarraj, V., Pukazhendhi, K. and Babu, T.D., 2007. Conservation strategies and sustainable use of marine biodiversity for coastal socio-economic development in India. In: *Proceedings of the National Symposium on Conservation and Valuation of Marine Biodiversity.* Zoological Survey of India, Chennai, p. 231–241.

Natarajan., P., Jayachendaran, K.V., Kannaiyan, S. Babu Ambat and Arun Augustine (Eds.), 2008. *Glimpses of Aquatic Biodiversity.* Rajiv Gandhi Chair Spl. Pub., 7: 284.

Norse, Elloit A., 1993 *Global Marine Biological Diversity: A Strategy for Building Conservation into Decision-Making.* Island Press, California, USA, pp. 350.

Ramachandran, S., 2001. Development of integrated coastal zone management programme. In : *Coastal Environment and Management.* Anna Univesity, Chennai, pp. 321–328.

Ray, G.C., 1991 Coastal zone biodiversity patterns. *Bioscience,* 41: 490–498.

Ray, G.C. and Ray, M.G. McCormick, 1992. *Marine and Estuarine Protected Areas.* Australian National Parks and Wildlife Service, Canberra, pp. 52.

Salm, R.V. and Clark, J.R., 1989. Marine and coastal protected areas: A guide for planners and managers, International Union for the Conservation of Nature, Gland, Swizerland, pp. 302.

Sasikala, S.L., 2004. *Marine Biodiversity: Conservation of Biodiversity in Marine and Coastal Ecosystems.* Centre for Marine Science and Technology, Publication No. 1. Manonmaniam Sundaranar University, Rajakkamangalam, Tamil Nadu, pp. 113.

Stuart, Chapin F., Osraldo E. Sala Elizabeth Huber Sannworld, 2001. *Global Biodiversity in a Changing Environment: Scenarios for the 21st Century.* Springer-Verlag, New York, USA, pp. 376.

Vineeta, Hoon, 1997. *Regional Workshop on the Conservation and Sustainable Management of Coral Reefs.* Proceedings No. 2. CRSARD, Madras. Organised by M.S.Swaminathan Research Foundation and Bay of Bengal Programme of FAO/UN. 15–17 Dec., Chennai, India, No.144.

Venkataraman, K., 2003. *Natural Aquatic Ecosystems of India, Thematic Biodiversity, Strategy and Action Plan.* The National Biodiversity Strategy Action Plan, India, p. 1–275.

2013, Biodiversity Conservation for Sustainable Management   Pages *248–256*
Editor: Dr. K. Muthuchelian, *Vice Chancellor, Periyar University, Salem*
Published by: Daya Publishing House, NEW DELHI

Chapter 30

Resource Potentials, Culture Possibilities and Livelihood Options of Seaweeds in India

V. Sundararaj and T.D. Babu
[1]Asian Analytical Laboratories,
Pallavakkam, Chennai – 600 041
[2]Central Institute of Brackishwater Aquaculture,
Chennai – 600 028

Introduction

Seaweeds, the marine macro algae are wonder plants of the sea and considered as medicinal/health food of the 21st century. They have numerous applications in food, pharmaceutical, textile and chemical industries and the world demand is increasing every year. The world production of seaweed has gone up to 10 million tones worth of US $8.0 billion. The top 10 countries producing seaweeds are China, Korea, Japan, Philippines, Indonesia, Chile, Taiwan, Vietnam, Russia and Italy. These together accounts for 5.97 million metric tonnes of seaweed production. The top five cultivated seaweeds in the world are *Laminaria, Porphyra, Undaria, Eucheuma* and *Gracilaria.* The current phycocolloids (seaweed gels) industry stands at over US$ 6.2 billion. 221 species of seaweeds are utilised commercially including 145 species for food and 110 species for phycocolloid production.

Indian Scenario

India possesses 434 species of red seaweeds, 194 species of brown seaweeds and 216 species of green seaweeds. Traditionally, seaweeds have been collected from natural stocks or wild populations. However, these resources were being depleted by over-harvesting and hence, the need for their cultivation was felt. Today seaweed cultivation techniques are standardised, perfected and made commercially viable and economically favourable. Besides, industry prefers a greater stability through sustained supply of quantity and quality of raw materials. In order to prevent overexploitation of natural seaweed resources and to meet the needs of industry in an uninterrupted manner, nearly all brown seaweeds, 63 per cent of red seaweeds and 68 per cent of green seaweeds are being cultivated. To achieve this, seaweed culture is the only way.

Distribution of seaweed species in India are, Gujarat 202; Maharashtra 152; Goa 75; Karnataka 39; Kerala 20; Lakshadweep 89; Tamil Nadu 302; Andhra Pradesh 78; Orissa 1; West Bengal 6 and Andaman and Nicobar Islands 34. India presently harvests only about 22,000 tonnes of macro-algae annually compared to a potential harvest of 870,000 tonnes, a mere 2.5 per cent. Commercial cultivation of macro-algae has barely begun and is facing continuous regulatory hurdles. Processing of macro-algae is limited to lower grades of agar-agar and alginate and is modest in quantity. Manufacturers of agar-agar are working at less than 50 per cent of their capacity and there is no manufacturer of carrageenan. Instead of being a major global producer and exporter, India remains an importer of macroalgal products. The principal cause for this gap between the potential and the actual results achieved with respect to commercial cultivation and processing of macro-algae is the lack of clearly enunciated policy on cultivation and rational utilisation of seaweeds by scientific means.

Ecological Importance

Seaweeds provide shelter to a variety of marine organisms, including those of commercial importance. Tamil Nadu is rich in seaweed resources. During the process of photosynthesis, seaweeds absorb a huge amount of CO_2 and reduce global warming. A research study has indicated that around 1.7 billion tons of CO_2 is utilized by seaweeds, sea grasses and phytoplankton in the marine ecosystem and they act as carbon sinks. The release of oxygen due to photosynthesis in a day is reported to be 10 litres/1m^2 area of seaweed. They are also efficient in controlling organic pollution including heavy metals in the inshore waters. Thus, seaweed cultivation is eco-friendly with sustainable income to the coastal poor.

Seaweed cultivation presents several opportunities such as carbon sequestration through seaweed farming, provision of nursery grounds for fish and shellfish, medium for pollution abatement and diversified uses as animal feed and fertilizers. Seaweeds are also found to provide a strong base for growth promotion of several plants because of their properties such as cytokinine, auxin and gibberllines. Therefore seaweeds will be a major source of raw material also for biofertilizer in organic agriculture revolution in the country.

Uses of Seaweeds

India has a coastline of more than 8,000 kms and harbours about 1000 species of seaweeds, commercial cultivation is yet to take place in a big way. Macroscopic marine algae, popularly known as seaweeds, form one of the important living resources of the ocean. Agar, carrageenan and alginate are popular products of seaweeds– these have been used as food for human beings, feed for animals, fertilizers for plants and source of various chemicals. In the recent past, seaweeds have also been gaining momentum as new experimental systems for biological research and integrated aquaculture systems. Seaweed products are used in our daily lives in one way or the other. For example, some seaweed polysaccharides are employed in the manufacture of toothpastes, soaps, shampoos, cosmetics, milk, ice creams, meat, processed food, air fresheners and a host of other items.

Figure 30.1: Seaweed as diet.

In several oriental countries like Japan, China, Korea, etc., seaweeds are a staple part of the diet. Agar is widely used in paper manufacturing, culture media, packaging material, photography, leather industry, plywood manufacturing, preservation of foodstuffs, dairy industry, cosmetics industry and pharmaceutical industry. Carrageenan is employed in food industry. Its value in the manufacture of sausages, corned beef, meat balls, ham, preparations of poultry and fish, chocolates, dessert gels, ice creams, juice concentrates, marmalade and sardine sauces is well known.

It is also used in the manufacturing of non-food items like beer, air freshners, textiles, toothpastes, hair shampoos, sanitary napkins, tissues, fungicides, etc. The applications of alginate find place in frozen foods, pastry fillings, syrups, bakery icings, dry mixes, meringues, frozen desserts, instant puddings, cooked puddings, chiffons, pie and pastry fillings, dessert gels, fabricated foods, salad dressings, meat and flavour sauces.

Kappaphycus alvarezii and its Characteristics

Kappaphycus alvarezii is a red alga under Rhodophyceae with an excellent source of carragheenan. It has a cylindrical axis with branches that are commonly enlarged maximally beyond basal structure towards the light. The branches are irregular and thalli are dark brown in intense light and relatively more reddish in the shade or in deeper water due to relative abundance of phycoerythrin. Pale yellow thalli are found in some bright light conditions.

Kappaphycus alvarezii becomes phototrophic in intense light. If wave action is high, excessive damage occurs to the plants by breaking of the branches. This species grows profusely in the sea, where the bottom is sandy and salinity is ranging from 29 to 34 ppt. Nutrients play a key role on the growth and the productivity in general. It is reported to be the largest tropical red alga, with high growth rate and biomass doubling

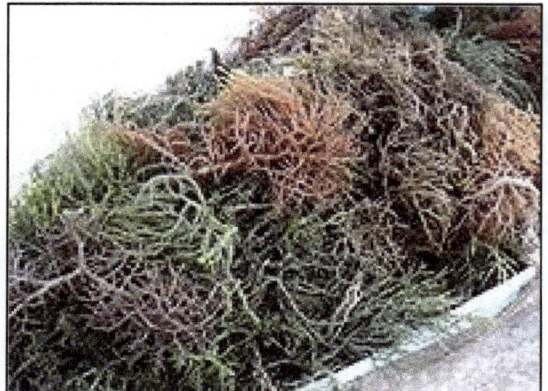
Figure 30.2: Fully grown *Kappaphycus*.

time of 15 days. A seed plant of 150 gram grows to > 600 gram in 45 days in calm water areas like Palk Bay. It requires sunlight and transparent seawater with mild wave action for replenishing bottom nutrients. It has also been proved that *Kappaphycus* seaweed grows >10 times in the open sea, where wave action is high.

Advantages of *Kappaphycus alvarezii*

Kappaphycus is autotrophic and grows by absorbing nutrients (Nitrogen, phosphorus and other minerals) and CO_2 present in the sea water under sunlight, performing effective photosynthesis.

Kappaphycus is a major source of carragheenan and it propagates vegetatively by "cloning" from buds cuttings without a sexual phase. It is easy to multiply and grow fast.

The culture method is organic and ecofriendly and it does not involve application of fertilizer, growth hormones, pesticides, insecticides, herbicides etc. for culture.

Cultivation technology is simple and totally eco-friendly. The seaweed cultivation is a shore based activity. Two to three persons of a fisherman community can handle the monolines or floating rafts for an income of 12,000–15,000 per month.

Farming, harvesting and processing generate new cottage scale industries, employing thousands of men and women among the coastal poor. Nationalized banks are providing loans to SHGs even without collateral security up to Rs. 5.0 lakhs per SHGs. There is a buy-back arrangement made for the benefit of the cultivators.

Kappaphycus farming is an excellent rehabilitation programme for fisher folk from fishing to sea farming and it will pave the way to reduce fishing pressure and reduce over-exploitation of fishery resources.

The aqua sap extracted from the seaweed contain macro and micro nutrients, essential amino acids and plant growth hormones that provide a major boost to crop yield by accelerating the plant's metabolic function and enhancing uptake capacity. Experiments have demonstrated a growth yield enhancement of 18 to 40 per cent in several plants viz., rice, sugarcane, ground nuts, corn, and wheat.

Kappaphycus alvarezii harbours many marine organisms and enhances biodiversity by providing an excellent habitat.

Culture Methods

There are several methods of culture for the production of *Kappaphycus alvarezii* in different countries. Adoption of a particular culture technology depends on the user, based on local conditions, nature of wave action, speed of water current, transparency of water, skill of manpower etc. In most of the study areas, the culture of *K. alvarezii* was carried out using floating raft and mono line methods.

Floating Bamboo Raft Method Culture

In the most of the villages in the study areas, the floating bamboo raft method is widely used. The bamboo raft is made up of four 3 metre length bamboo poles tied in a square shape. This structure is strengthened by adding 4 cross bamboo poles at

Figure 30.3: Raft culture.

Figure 30.4: Constructed rafts.

the corners. In a raft, 20 ropes are tied parallel to the bamboo pole with a gap of 15c.m. In each raft of 3X3m size, totally 60 kgs of seeds (tender seaweeds) are tied in 400 points, each points having 150 gms of seed.

After seeding, the rafts are allowed to float with the suitable depth of around 1-1.5 m. These seaweeds are harvested after 45 days; the net production per raft/per crop is 300 kg (wet weight). After drying, the weight is 30 kgs. Approximately one SHGs (5 persons in each group) is operating 300 to 400 rafts/month. Every individual can easily earn Rs. 420/day (30 kgs X Rs.14) by selling the dried seaweed.

Thus the monthly earning can be high, ranging from Rs.12,000 to 15,000.

Long Line (Monoline) Culture

Kappaphycus alvarezii is also cultured using monoline technique in coastal waters, where the wave action is high. For the monoline culture, 60 or 100 metres of nylone rope is used. At regular interavals, bags (0.5 x 0.5m) made of nylon material are tied and supended from the ropes.

Figure 30.5: Monoline.

Around 200 gms of seeds are placed in each bag and totally 60 to 100 kgs of fragments (seeds) are used. The monoline rope is allowed to float with the help of disposed plastic bottles along the shore line. The both ends of the rope are tied to bigger floats, which are anchored in the sea and the seaweeds are allowed to grow. The seaweed in the monoline is also harvested after 45 days with the wet weight of 400 kg and dry weight of 40 kgs. In the monoline culture, the growth of the seaweed is fast, when compared to the raft culture. Through monoline culture, the seaweed cultivators are earning Rs. 560/day. To avoid grazing by the herbivorous organisms nylone nets are used in the raft culture and nylone bags are used in the monoline culture.

Duration of Culture

In general, *Kappaphycus alvarezii* is a short duration crop. It is possible to take the harvest in 45 days, because of its fast growth rate. The duration of culture depends on environmental conditions like stable salinity, water clarity, optimum water current (circulation) and water fertility (in terms of major and minor nutrients), which are favourable for quick growth and better production. Interference of seagrasses, grazing by herbivorous fishes like *Siganus* species and the presence of fouling organisms like *Balanus* species are usually negative factors. Tuticorin coastal areas are found to be more suitable for seaweed production due to favorable environmental culture condition, which is promoting earlier harvest. Instead of 45 days it is harvested in 30 days. Because of this, more cycles of seaweed culture can be achieved in Tuticorin district. A seed plant of 150 gms grows to > 600 gms in 45 days in calm waters like Palk Bay area. It requires sunlight, transparent seawater with mild wave action for replenishing bottom nutrients.

Productivity

The productivity of the seaweed differs globally due to geographical regions from one place to another. The productivity of the seaweed in the southern coastal districts is furnished below in Table 30.1.

Table 30.1: Productivity of seaweed.

Sl.No.	District	Production (kg/raft)	SHGs
1	Ramanathapuram	260-340	69
2	Tuticorin	210-260	30
3	Pudukottai	210-240	03
4	Tuticorin	260-380	10
	Total		112

Cost of Production

For each rafts, 60 kgs of seeds are to be seeded in the culture ropes in 400 points at the rate of 150 gms seeds per point. Around 5 kg of seaweed seeds are expected to be wasted, while seeding and hence actually 65 kgs of seaweed seeds are required per raft. Further, for raft laying and maintenance, the cost required is Rs.90. The cost of preparing one floating bamboo raft is Rs. 579.75 is indicated in the Table 30.2.

Economics of Production

Kappaphycus alvarezii culture is done mainly by Self Help Gorups. One Self Help Group (small) normally consists of 5 members. A group of five members can conveniently operate a total of 225 floating bamboo rafts for culture (*i.e.* one member can operate 45 rafts).

The total cost of 225 rafts (225 x Rs.750) is Rs.1,68,750/-. In this total cost, 50 per cent is through bank loan and the rest is subsidy. Out of the total 225 rafts for a group, five rafts can be harvested per day. During harvest, after the seeds (5 x 60 Kg), for the next crop, the net production of seaweed is 1000Kg (wet). When this net seaweed is dried, the outcome is 100 Kg dry seaweed. Thus for 25 days in a month, 2500 Kg of dry seaweed can be had. At present, the cost of dry seaweed has increased (*i.e.* Rs.18/Kg). and at this rate, monthly income out of 2500 Kg of dry seaweed is (2500 x 18) is Rs.45,000. After deducting the monthly loan payment (Rs.5000), the net income for a group is Rs.40,000. Thus, an individual in a group of five can get a net income of Rs.8000 per month.

Table 30.2: Cost details of a raft.

Sl. No.	Particulars/Specifications	Qty Reqd.	Rate (Rs)	Cost per Raft (Rs)
1.	3-4" dia Hallow Bamboos of 12ftx4 for main frame + 4ftx4 for diagonals (without any natural holes, cracks etc.)	64 feet	3.25/ feet	208.00
2.	Five-Toothed Iron Anchor of 15kg each (@ Rs.35 per kg)– One anchor hold a cluster of 10 rafts	1.5kg	50/kg	75.00
3.	3mm PP twisted rope for plantation–20 bits of 4.5m each	0.4 kg	125/kg	50.00
4.	Cost of HDPE braider pieces (20 pcs x 20 ropes = 400 pcs of 25cm each)	0.1 kg	150/kg	15.00
5.	Braider twining charges @ Rs. 1.0/20 ties. For one raft 400 ties = Rs. 20	20 ropes	1/rope	20.00
6.	Raft framing rope 6mx12 ties per raft *i.e.* 36mts of 6mm rope	0.65kg	125/kg	81.25
7.	HDPE Used Fishing Net to protect the raft bottom (4mx4m size) + labour charges Rs. 10	1 kg	50/kg	50.00
8.	2mm rope to tie the HDPE net (28mts)	0.064kg	125/kg	8.00
9.	Anchoring rope of 10mm thickness (17m per cluster of 10 rafts)	0.08kg	125/kg	10.00
10.	Rafts linking ropes per cluster 10 rafts–6mm thick– 2 ties x 3m x 9pairs = 54m length	0.1kg	125/kg	12.50
11.	Transport cost			50.00
12.	Raft cost (column 1-11)			579.75
13.	Seed material 150gm x 400 ties + 5 kg as handling loss	65kg	1.20/kg	78.00
14.	Raft laying + maintenance cost	–	–	90.00
				747.75
	Total raft cost (Rounded off)			**750.00**

There is scope for increasing the productivity per raft. Further, because of the possibility of sap production (approximate income for Rs.800 per ton of wet seaweed), the industrialist will also be able to offer a better price for raw seaweed produced and thus, it would easily be possible for an individual in a group of five to earn Rs.1000 per month easily.

Marketing of Seaweed

The dried and the wet seaweeds are sold to exporters by the SHGs. The seaweeds are gaining more importance and their popularity by value has brought in more buyers and exporters. This would pave the way for more income to the cultivators. This kind of guaranteed price with assured buy back arrangement is not available for other seaweeds. Therefore, cultivators are getting higher income from *Kappaphycus* cultivation. It is encouraging further that USFDA and Codex Alimentarious Commission have approved carrageenan as a food additive. Therefore, the demand in the world market for carrageenan is high. It is understood that India is importing nearly 400 tones of carrageenan powder and the demand in the domestic market is expected to be around 2000 tons per annum.

Bio-products from Seaweed

Central Salt and Marine Chemicals Research Institute (CSMCRI), Bhavnagar has developed an important thickening agent carrageenan using the seaweed called *Kappaphycus alvarezii* that bestows useful properties to many commercial products. Scientists have developed a unique technology of liquefying seaweed without adding any water and thereafter they have separated the solid from the liquid to obtain two products. The solid material is the source of carrageenan and the liquid has been found to be a very useful plant nutrient, rich in potassium and organic growth promoting hormones. This sap has been used in a variety of agricultural crops viz. sugarcane, paddy, maize, pulses and several fruits and vegetables. The productivity increase has been in the range of 20 per cent to 40 per cent in different regions for different plant varieties.

In light of the above, seaweed farming must be encouraged for the better living of the coastal people. Further, to control global warming and also for the production of algal diesel, seaweeds offer greater scope. This valuable resource should be scientifically mass produced in the suitable areas of the seas and the coastal facilities like shrimp farms and their effluent treatment systems for benefits.

General Recommendations

- ☆ Commercial cultivation and processing of seaweeds should be made as a national priority and taken up as a mission mode project.

- ☆ All indigenous species of seaweeds are considered as ecologically safe for mass cultivation. *Kappaphycus alvarezii* which was introduced to Indian coastal waters more than 10 years ago and has since been domesticated is considered ecologically safe.

- ☆ Recently, natural incidence of *Kappaphycus alvarezii* has also been reported from Andaman Islands. Ecological studies have been undertaken regarding the cultivation of the species and no adverse effects to the ecosystem by the species have been reported. Large-scale cultivation of *Kappaphycus alvarezii* can be undertaken in Andaman Islands also considering the benefits.

- ☆ Seaweed cultivation and the wet and dried seaweed so produced shall be treated as agricultural cultivation and agricultural produce respectively for the purposes of fiscal levies such as sales tax, income tax, excise, octroi, etc.

☆ Combined cultivation of shrimps and seaweeds in an integrated manner can be encouraged in aquaculture as seaweeds act as scrubbers in reducing nutrient load and cleaning the environment. Their antibacterial, anti viral properties contribute to the health conditions of the shrimps.

☆ Seaweed cultivation may be encouraged and undertaken all over the Indian coasts including Chilika Lake, Palk Bay, Andaman Islands as well as Lakshadweep islands as seaweed cultivation is ecologically safe and contribute considerably for improving the marine ecosystems.

☆ Agar-agar processors have depended solely on naturally occurring beds of *Gracilaria* for supply of raw materials that have been overexploited. Similarly, processors of alginates depend on naturally occurring sources for their raw materials and no carrageenan industry exists in India. To ensure dependable supply of raw material, high priority should be accorded to cultivation of certain selected promising species, coming under agarophytes, alginophytes and carrageenophytes.

☆ Biodiversity database on seaweeds may be created. Standardisation of techniques for large-scale tank cultivation as well as open water cultivation of seaweed may be carriedout.

☆ Periodic resource evaluation and biomass estimation need to be carried out on a national basis.

☆ Processes for the preparation of agar, alginate, carrageenan and quality control may be further refined.

☆ Impact of grazing by herbivores on the seaweeds cultured and the ways to control loss of seaweeds must be found out for adoption.

☆ Culture of agar yielding algae should also be taken up in a big way to sustain the related industries.

2013, Biodiversity Conservation for Sustainable Management   Pages *257–268*
Editor: **Dr. K. Muthuchelian,** *Vice Chancellor, Periyar University, Salem*
Published by: **Daya Publishing House, NEW DELHI**

Chapter 31

Using Underwater Sound to Measure Biodiversity and Productivity in the Line Islands

R. Saranya and G. Ramya

*Department of Biotechnology,
Vivekanandha College of Engineering for Women,
Tiruchengode, Namakkal Dt., Tamil Nadu*

ABSTRACT

In order to create a more complete description of coral reef habitats for use in conservation, this study aims to develop acoustic indices for biodiversity and productivity. Qualitative analysis, sound pressure levels, biological sound signal detection, and spectral variability in recorded underwater sound combine to give detail on coral reefs in Washington Island, Kiritimati Island, and Palmyra Atoll. Qualitative analysis and sound pressure level correlate strongly with biomass and biodiversity in the island reefs, and sound pressure and spectral variabilities give insight into the type and number of marine animals in each habitat. These measurements show evidence of a fishing gradient along the Pacific Line Islands and help create an aesthetic definition of coral reef health and biodiversity.

Introduction

Biodiversity measurements are of great value to conservationists. With species indices, genetic variability tests, and habitat variability assessments, conservationists can identify areas of ecological importance that should be protected. One method for changing governmental and popular perceptions about natural resources has been through stressing the economic value of conservation. This technique must be coupled with a more aesthetic approach to raising awareness of biodiversity.

An aesthetic definition of biodiversity should encompass all the sensory input from the environment, but my expertise as a sound engineer focuses my project on the acoustic component of aesthetics. Acoustic diversity has been hypothesized to correlate to traditional definitions of biodiversity in many ways. The soundscape, or acoustic signature of a habitat, can be divided into two elements: keynotes, or background sounds, and sound signals, or foreground sounds intended to attract attention (Wrightson, 2000). In a marine ecosystem, keynotes could include seismic activity, tidal action, and wave events. A definition of acoustic biodiversity would use keynotes to determine habitat variability. Sound signals would be a measure of speciosity; the niche hypothesis of animal vocalization suggests that sounds from each type of animal occupy specific sound frequencies at specific times (Krause, 1987). Oceanic sonic niches might indicate a temporal and spectral acoustic variability in areas that are diverse.

Little scientific work has been done on bioacoustic diversity, especially in the ocean. Although sound communication in many specific marine animals is well-documented, acoustic diversity as a whole is relatively unknown. Biological underwater noises often exhibit fewer differences than terrestrial noises (Amorim, 2005), making diversity harder to examine. However, direct acoustic monitoring (rather than sonar techniques) has been used to estimate fish populations (Lobel, 1992) and quantify river disturbance (Joo *et al.*, 2005). In the Joo *et al.*, study, acoustic intensity measurements found significant spectral differences in disturbed and undisturbed sites.

The aim of this project is to define reliable indices for acoustic biodiversity in the ocean and test it using a gradient of human interaction along habitats. In this study, acoustic biodiversity is defined as the variability of spectral and sound pressure levels over time. In particular, spectral variability is the number of 20-Hz frequency bands significantly above oceanic keynotes and corresponds to the speciosity measurement in biodiversity. Sound pressure variability is comprised of two measurements: the abundance of sound signals, which corresponds to population estimation, and intensity averages, which correspond to habitat variability. Combined with qualitative analyses, these data could provide an acoustic description of coral reefs near the Pacific Line Islands and can be correlated with biomass and biodiversity levels collected in concurrent Line Island studies (Rego *et al.,* 2007 and Vichit-Vadakan *et al.,* 2007). Since biodiversity and biomass are good indicators of habitat health (Leigh, 1965), the accuracy of this acoustic approach in assessing coral reef environment health can be determined.

The application of aesthetic, musical analysis techniques on recorded underwater sound could produce an index of acoustic biodiversity on which measurements increase along the Line Islands from Washington Island to Kiritimati Island to the Palmyra Atoll. The fishing gradient along these islands mentioned studied Stevenson *et al.,* supports this hypothesis: Washington Island, with an estimated population of 2100, has a very small reef area, while Kiritimati Island has a population of 8,000 but a much larger fishing region. Palmyra, a nature reserve privately owned for the last 100 years, has the least fishing pressure of the islands. The total biomass levels of reef animals increase as the fishing gradient decreases (Rego *et al.,* 2007 and Vichit-Vadakan *et al.,* 2007). Rego *et al.,* also show an increase in Shannon-Wiener biodiversity, although they only studied apex predators, which we could see as an increase in some of our spectral and sound pressure variability indices in the soundscapes of the different reef habitats.

Materials and Methods

We chose sample sites similar to the backreef areas studied by Stevenson *et al.,* and Rego *et al.,* Due to time constraints, we had different numbers of recording sites at each island: six at Kiritimati, four at Washington, and three at Palmyra (Figure 31.1).

Equipment Used

1. HTI-96-MIN hydrophone with pre-amp
2. MAudio Microtrack 24/96 Professional 2-Channel Mobile Digital Recorder
3. Audacity, Praat, and Java Eclipse software for data analysis

We first conducted proof-of-concept analyses on music to set the parameters of our data analysis software. Two songs were picked, one with the high spectral and loudness variability of 80s synthesizer pop, and one with the low variability of acoustic and slide guitars. Once we calibrated the sensitivity to the scale of musical variability, the synthesizer pop song showed more acoustic diversity than the guitar song in all of our indices.

Sample collection started at Kiritimati Island and continued to Washington and Palmyra. At each sample site, we recorded three to five minutes at a 44.1 kHz sampling rate. We could not replicate recordings at the sites during different times of the day, because our time was limited. Because reef sounds generally increase from low levels in the day to high levels in the evening and at night (McCauley *et al.,* 2000), the time of day variability added an extra independence in our data that we could not account for through replication. Although point transects are less efficient than line transects in sampling numbers of individuals in coral reefs (Bortone *et al.,* 1989), we chose the point recording method so the hydrophone would not be dragged through the water and pick up turbulent noise. We placed the hydrophone at one meter depths at each site and measured depth with a transect tape. In addition to collecting audio data, we estimated sea state conditions using the Beaufort Scale of Wind Force (Wenz, 1962) to aid in the differentiation of keynotes and sound signals.

Sample processing started with inverse filtering to reduce the response bias of the specific hydrophone and recorder used. Using Audacity (audacity.sourceforge.net), we reduced each recording's low frequency levels in accordance with the HTI-96-MIN specifications given by the manufacturer (High Tech, Inc). Further filtering was not needed due to the flat frequency response of the MAudio Microtrack (M-Audio). Because our hydrophone was very sensitive, periods of time with strong currents created clipping. We measured intensity level, or loudness, for each recording. We

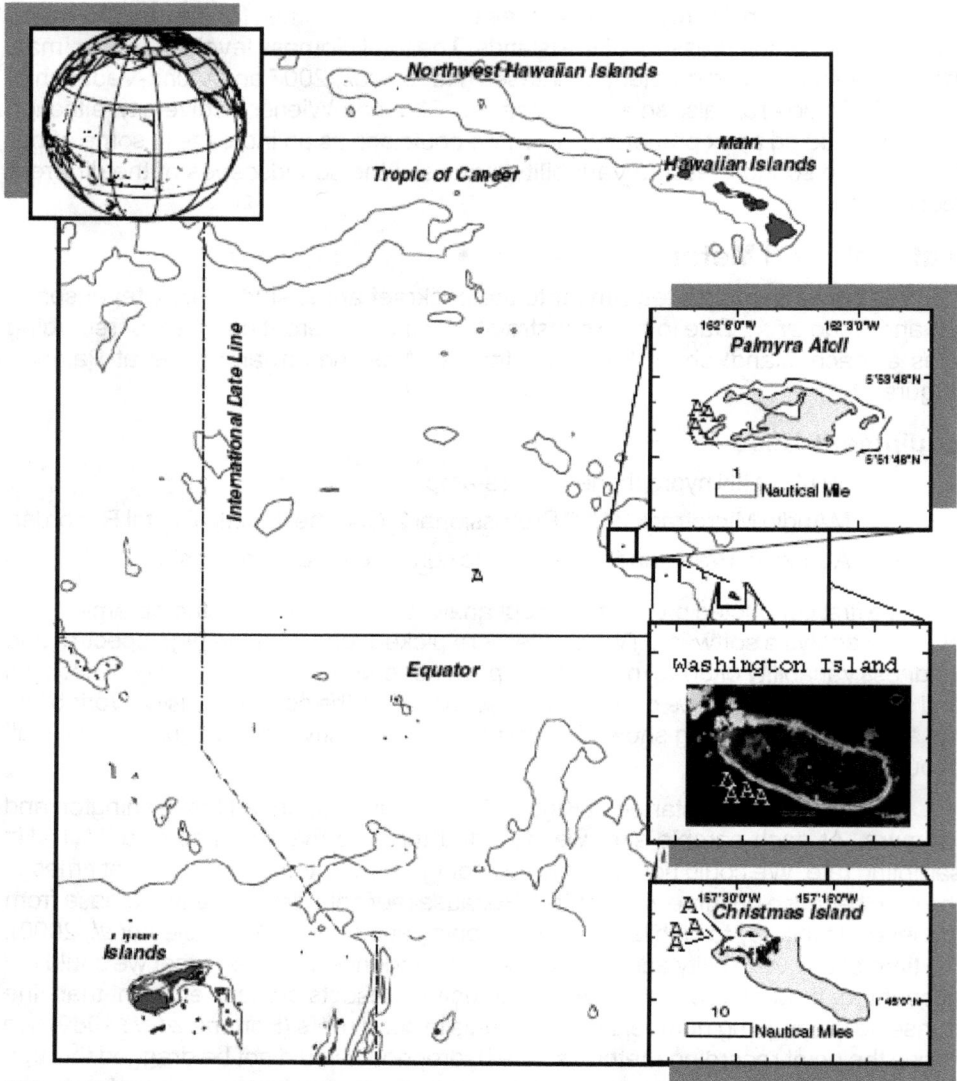

Figure 31.1: Locations of study sites in the Line Islands.

Backreef recording locations on each island marked with A's: six at Kiritimati, four at Washington, and three at Palmyra. Although this study uses Stevenson *et al.* and Rego *et al.* as benchmarks and therefore emulates its data sites as closely as possible, data storage and analysis time limitations forced the site countdown. Sites shown to have a high amount of background noise from wave action were eliminated.

then created spectrograms to identify frequency changes over time. All spectral processing used 20-Hz frequency bins as a compromise between data storage size, processing time, and spectral resolution. We also converted the sound files from waveform audio format, .wav, into bitwise representations for easy analysis in Java. Finally, we constructed oceanographic noise level filters based on each observed sea state condition using standard ocean ambient noise formulae (Wenz, 1962) (Figure 31.2). So we edited these parts from the audio. The remainder of the sample processing

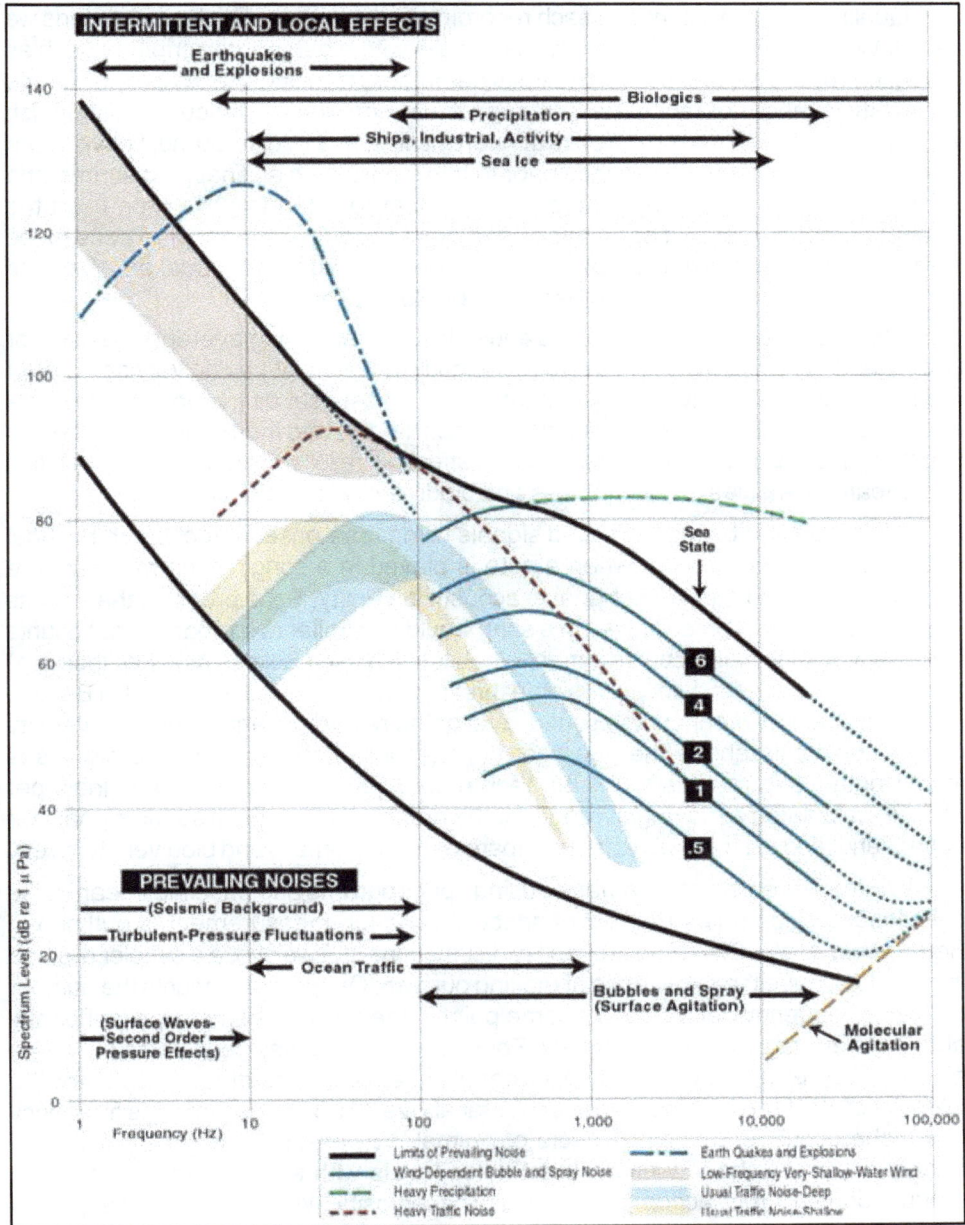

Figure 31.2: Ambient ocean noise at dfferent frequencies (Wenz 1962).

was completed in Praat (www.fon.hum.uva.nl/praat). Because perception of acoustic aesthetic is influenced almost entirely by music, we modeled our acoustic biodiversity measuring tools after established musical analysis techniques: qualitative analysis, total intensity comparison between islands, comparison of the number of biological sound signals, and measures of spectral variation from background noise.

Qualitatively, we listened to each recording and noted what biological sounds we heard. We classified the different types of calls with a general description, such as groan, chirp, or pop, and calculated calls per minute for each site to account for the variability in length of recording. Although it was sometimes difficult to distinguish water turbidity, noise from snorkelers or nearby ships, and wildlife sound, I have heard many different underwater sounds prior to this study, such as snapping shrimp and various fish calls on the internet, and know what to listen for. We then used the Pearson product-moment correlation coefficient to compare number of calls to the sum of fish biomass levels measured by Rego *et al.*, and Vichit-Vadakan *et al.*, and apex predator fish biodiversities measured by Rego *et al.*

We measured intensity decibels and calculated a moving average intensity for each island. By using a moving average, we could account for having varying times of recording; we used midway between afternoon and sunset as the base time for the moving average. These averages were then normalized with the ocean noise filters so variability in sea surface conditions would not affect our results. We then used correlation to compare intensities to fish biomass and biodiversity levels.

We measured biological sound signals using note-onset detection. In musical analysis, note onsets occur when a note is played in a song, and a musical note equates to a biological sound signal in acoustic diversity. So, a program that counts the number of piano notes played in a song would be similar to a program that counts how many wildlife calls occur in a marine habitat. We used Eclipse (www.eclipse.org) to program note onset detection using a thresholding technique described in Bello *et al.*, 1998. The tool monitors the sound level of the recording, and when the intensity rises significantly above the average, the number of biological sound signals is incremented. We ran this tool over each recording to find number of signals per minute, calculated a moving average for the number of biological sound signals for each island, and used correlation to compare with fish biomass and biodiversity levels.

We measured spectral variation using spectrograms and baseline ocean noise calculations. A complex method of spectral variation measurement is outlined in Berenzweig *et al.*, 2004 using probability distributions, but we chose a simpler method in order to acquire the experience of making our own. Our program counts the number of 20-Hz frequency bands that, at some point in the audio stream, rise significantly above the average for that frequency. For example, suppose ocean noise at 420 Hz given by Wenz is 70 dB. In the spectrogram of a site's recording, we monitor the intensity of 420 Hz over time, and if is ever far above 70 dB, we assume this frequency is a niche occupied by some specie of animal. We associate a large number of frequency bands above their ambient noise levels with a high diversity of sound production, and this accompanies a high speciosity. We calculated frequency distribution for each recording's spectrogram, averaged the number of frequencies found, and used correlation to compare with fish biomass and biodiversity levels. Because we would not expect speciosity to increase as sunset approached, we assumed a moving average was not necessary.

Results

Qualitative analysis demonstrated an abundance of biological sound calls in all of the recordings. Sounds such as pops, grunts, purrs, crunches, and chirps were heard as distinct sound events; the event clarity is shown with a spectrogram in Figure 31.3. We noted similarity between the types of calls made on the different island reefs, although Washington had many more pops than Kiritimati and Palmyra, and Palmyra had very few. Palmyra exhibited loud crunching and cracking noises heard only faintly on the other islands. Figure 31.4 illustrates the number of calls per minute from each recording, excluding the crunches and pops, which were much more frequent than any other type of sound signal; we see an increase in number of calls as afternoon turns to sunset. We also see an increase in number of calls from Washington to Kiritimati to Palmyra.

Sound pressure levels were very different among the islands, as shown in Figure 31.5. Similar to the qualitative analysis, sound intensity increased as the day progressed at each island. In contrast to the number of calls, sound intensity increased from Kiritimati to Washington to Palmyra. We avoided depth as a confounding factor of

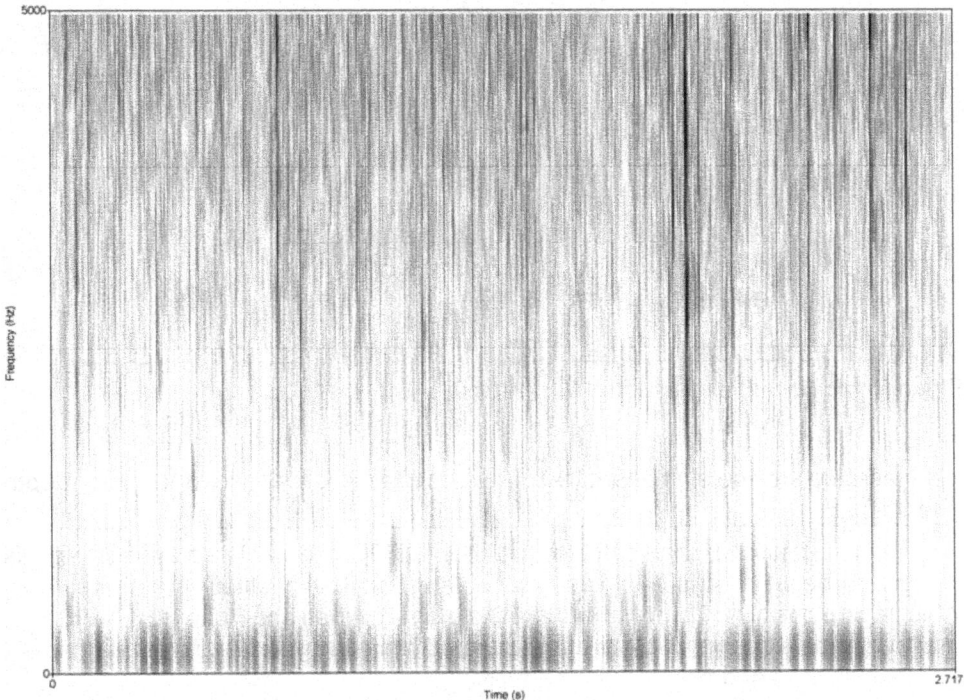

Figure 31.3: Spectrogram showing fish call.

In this spectrogram, the heavy black vertical lines are clicks in a fish groan.

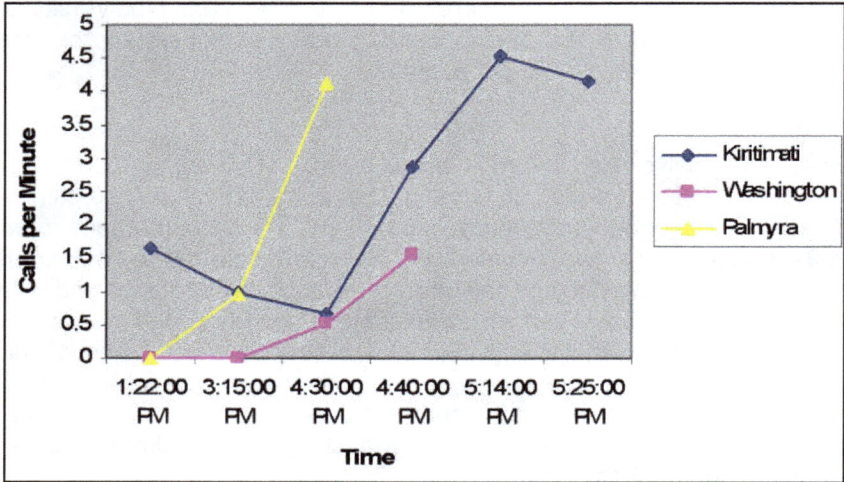

Figure 31.4: Number of calls per minute versus time, counted qualitatively.

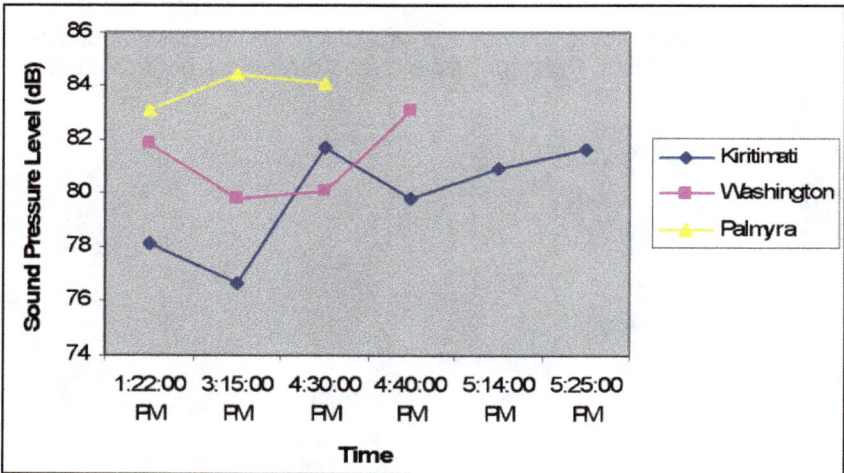

Figure 31.5: Sound intensity levels versus time.

loudness; we saw no significant trend in each island between loudness and depth (Figure 31.6).

Biological sound signals detected are shown in Figure 31.7. The number of signals is much higher at each island in automatic detection than qualitative analysis. Also, Washington has considerably more signals than Kiritimati and Palmyra, which have similar numbers of signals. Contrary to our prediction, the number of signals does not increase as time of day progresses.

Frequency distribution, as predicted, does not increase as time of day progresses, as illustrated in Figure 31.8. Distribution is reverse along the islands to expected

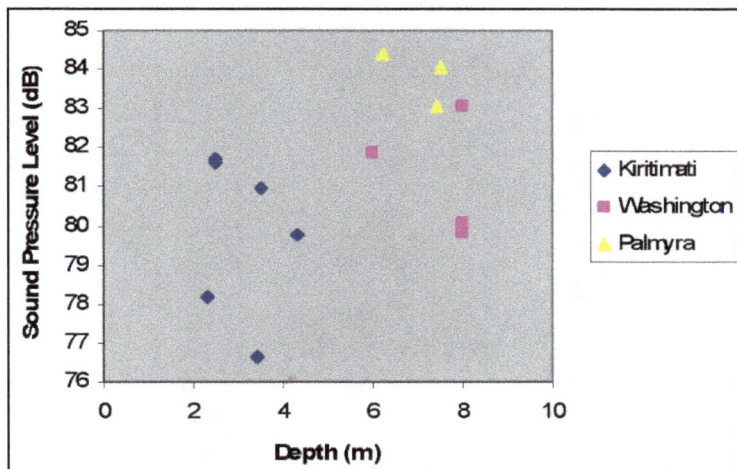

Figure 31.6: Sound intensity levels versus depth.

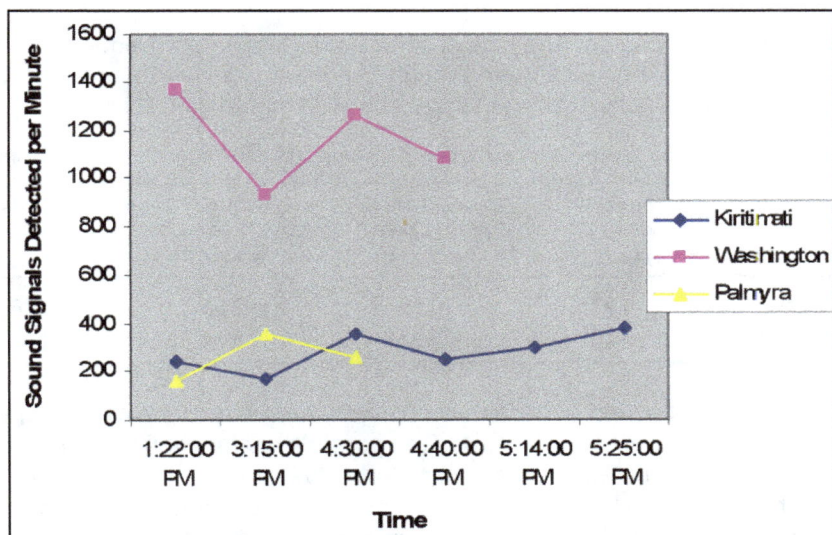

Figure 31.7: Number of biological events detected versus time.

speciosity; Washington has the widest range of frequencies, then Kiritimati, and then Palmyra with the smallest number of frequency bands detected.

Correlation values with fish biomass and apex predator fish biodiversity varied among the different acoustic productivity indices. Table 31.1 shows the correlations and also the ANOVA p-values between the islands for each acoustic index.

Discussion

In our qualitative analysis, we might interpret the rise in number of calls over time as confirmation of earlier research in the McCauley *et al.,* study and as validation of our results. Because we saw this rise in sound pressure level as well, our results

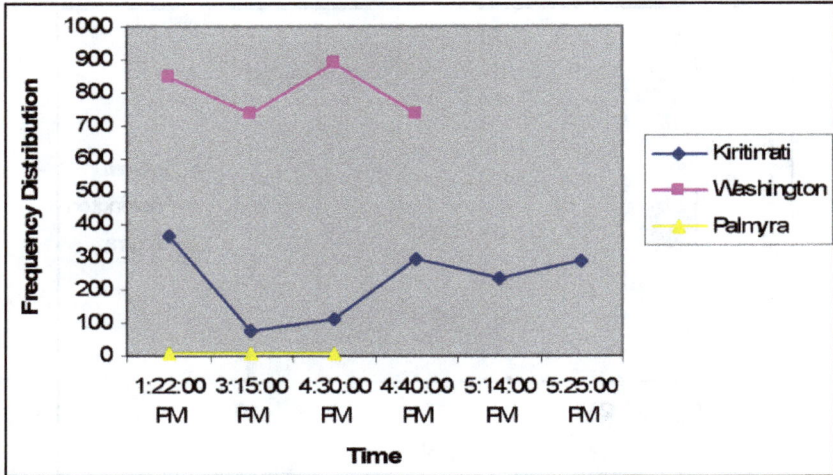

Figure 31.8: Number of frequency bands above the noise floor versus time.

become doubly convincing. The lack of trends over time in our frequency index is encouraging as well, because even though fish sound activity increases towards sunset, most fish are active during the day as well (Amorim, 2005).

Table 31.1: Correlation coefficients between fish biomass and acoustic diversity indices.

Acoustic Index	Qualitative Analysis P = 0.13	Sould Level Pressure P = 0.03	Number of Sould Signals P < 0.01	Frequency Distribution P < 0.01
Biomass	0.92	0.89	−0.53	−0.76
Biodiversity	0.99	0.74	−0.73	−0.9

However, in our sound signal detection index, we see no trend over time, which seems to contradict these other results. One possible explanation might be that our algorithm counts keynotes such as water turbidity. Kiritimati, though, was much calmer during recording than Palmyra, yet shows a higher number of sound signals, so this explanation is unlikely. Another explanation is that the notes are entirely dominated by snapping shrimp calls. Although we made no visual confirmation of specific animals, I have seen snapping shrimp up close prior to this study and have stressed them until they made their characteristic popping noise. The sounds heard at Washington and Kiritimati sound very similar to this noise, and the high number of sound signals at Washington supports this explanation. Rego *et al.,* have noted a lack of benthic predators at Washington and Kiritimati, whose absence could allow for a rise in the snapping shrimp population, and therefore the number of shrimp calls. From this, we would expect Kiritimati to have more signals than Palmyra, but they have similar numbers. Perhaps Palmyra makes up for this discrepancy through other types of calls, such as the high number of crackling and crunching noises discussed in our qualitative results. We assume these are parrotfish feeding sounds, although this was not confirmed visually.

Our frequency distribution results are interestingly opposite of our expected acoustic diversity values along the fishing gradient in the islands. Although Figure 8 could suggest that the different types of fish calls were truly varied at Washington and less varied at Kiritimati and Palmyra, this is improbable, because apex predator biodiversity increases along the islands in the opposite order (Rego *et al.,* 2007). Although this is only a small component of the total diversity, we might assume that apex predator biodiversity drives acoustic diversity due to their large size.

The sounds produced in the swim bladders of fishes (Amorim, 2005) we expect to be louder for fishes with large bladders. Some other factor must be driving our frequency distributions; we hypothesize that snapping shrimp causing the high numbers of significantly varying frequency bands at Washington and Palmyra. We had to set our sensitivity very low when determining whether a certain band was significantly varied from the noise floor level, otherwise Washington exhibited frequency variation off the scale; with high sensitivity, Washington showed variation at all frequency bands. By lowering sensitivity, our algorithm mostly detected popping and crunching noise frequencies, because these were the loudest. We heard crunching as low frequency, and since crunching was prevalent in Palmyra and popping was not, these specific sound types could account for the low frequency variability in Palmyra.

Popping sounds occurred at high densities at Washington, and we believe this was accompanied by varying types of popping as well. This could correlate to a rich diversity of snapping shrimp species, and the intermediate level of frequency variation at Palmyra could correlate to only one or two snapping shrimp species. Given this hypothesis, the high negative correlation between frequency variation among the islands and biomass and biodiversity (Table 31.1) indicates that as reefs become healthier, snapping shrimp are less prevalent. The same correlation in sound signals detected supports this conclusion as well.

Qualitative analysis and sound pressure levels were correlated with biomass and biodiversity as trends (Table 31.1). Qualitative analysis showed an increase along the Line Island fishing gradient, although sound pressure level analysis did not. Therefore, we conclude that qualitative analysis is our best acoustic index to gauge reef health, and sound pressure level analysis might be a satisfactory acoustic index. Sound signal detection and frequency distribution measurement, when combined with background knowledge of the sounds of an area, could provide insight into the variation and numbers of animals in a reef. Future studies could imitate our methods and conduct similar reef recordings and analysis that validate acoustic indices as measurements of reef health. Snapping shrimp filters, created from spectrograms of known shrimp calls, would be invaluable in eliminating popping noise.

Other improvements could be a double hydrophone array that eliminates ocean noise, better reproducibility through time of day and site replication, and a wider variety of reef islands studied, such as Kingman Reef and Fanning Island.

Conclusion

A reliable aesthetic definition of biodiversity created from musical analysis techniques could be a major asset to scientists and conservationists defending threatened ocean habitats. Public knowledge that areas are measurably, aesthetically diverse might influence support away from ecologically harmful activities such as poaching, the aquarium trade, and unsustainable fishing. Sound is a large part of the aesthetic experience, and this definition of acoustic diversity can serve as a model to develop a complete sensory biodiversity and productivity index. In addition, the technique used to survey acoustic diversity has many advantages over traditional survey techniques; it is non-invasive, there is no bias at night, and automated monitoring could be quick and continuous. However, an aesthetic definition of biodiversity could never supplant the less qualitative scientific definition, but it would help bridge the gap between esoteric information and practical knowledge. Practically, my study showed that Kiritimati Island, Washington Island, and Palmyra Atoll each have a characteristic acoustic soundscape that is aesthetically pleasing; each sounds beautiful enough to warrant conservation.

References

Amorim, M., 2005. Diversity of sound production in fish. In: *Fish Communication*, (Ed.) M. Amorim. Narosa Publ., New Delhi, India.

Audio, M., 2000. MicroTrack 24/96 Professional 2–Channel Mobile Digital Recorder. Retrieved May 4, 2007, from http://www.maudio.com/products/en_us/MocriTrack2496–focus.html.

Bello, D. *et al.,* 2005. A tutorial on onset detection in music signals. *IEEE Transactions*, 13(5): 1035–1047.

High Tech, Inc., (May), 1999. *Hydrophones for Seismic and Acoustic Applications*. Retrieved May 4, 2007, from http://home.att.net/~hightechinc/seismic.html.

Joo, W. *et al.,* 2005. Interpreting the Acoustic Signals as an Environmental Variable in: *The Muskegon River Watershed*. Michigan State University Project Annual Report. v. 33, p. 351–358.

Krause, B., 1987. The Niche Hypothesis: How Animals Taught Us to Dance and Sing. *Whole Earth Review*, No. 57.

Leigh, E., 1965. On the relation between the productivity, biomass, diversity, and stability of a community. *Proceedings of the National Academy of Sciences, USA*. 53: 777–783.

Lobel, P., 1992. Sounds produced by spawning fishes. Environmental Biology of Fishes.

McCauley, R. and Cato, D., 2000. The patterns of fish calling in a nearshore environment in the Great Barrier Reef. *Philosophical Transactions of the Royal Society: Biological Sciences*, 355(1401): 1289–1293.

Pearce, D. *et al.,* 1994. *The Economic Value of Biodiversity*. Earthscan, London.

Stevenson, C. *et al.,* 2006. High apex predator biomass on remote pacific islands. *Coral Reefs*, 26(1): 47–51.

2013, Biodiversity Conservation for Sustainable Management   Pages *269–274*
Editor: **Dr. K. Muthuchelian,** *Vice Chancellor, Periyar University, Salem*
Published by: Daya Publishing House, NEW DELHI

Chapter 32

Biodiversity Hotspots in Western Ghats Regions with Reference to Shola Forests Conservation Using GI Technologies for Kodai Hills Area

S. Muthumeenakshi, C. Sivakami, B. Pavendan, M. Vasanth Kumar and A. Sundaram
Department of Future Studies, School of Energy Sciences, Madurai Kamaraj University, Madurai – 625 021

Study Area

Endangered flora Shola forests in Kodai hills, an eastern off shoot of the Western Ghats with a maximum east to west direction to a length of about 65 km and to a maximum width of about 40 km.

Need for the Study

Shola forests are characterized by patches of dense isolated woods composed of evergreen trees. They provide an invaluable corridor for animal movements. Trees are characteristically stunted, seldom above 15 m, profusely branched and support a large number of epiphytes like lichens, mosses and ferns. In addition to numerous creepers/stragglers along the periphery they provide shelter to animals and act as a

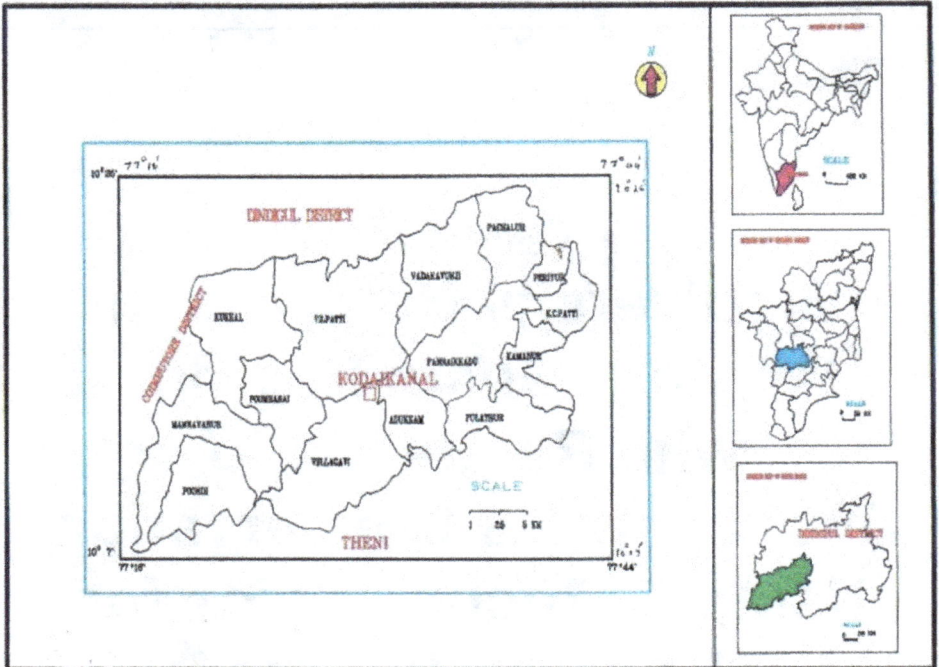

Figure 32.1: Location map of Kodaikanal

watershed. They are considered 'living fossils' for their sheer antiquity and virtual non-generation.

The Western Ghats are amongst the eighteen biodiversity hot-spots recognized globally and are known for their high levels of endemism expressed at both higher and lower taxonomic levels. Important environmental problems noted on the Kodai hills due to change in pattern of land use and land cover categories such as

☆ Soil erosion.

☆ Siltation of stream valleys, reservoirs and lakes.

☆ Reduction in water holding capacity of reservoirs and lakes.

☆ Vulnerability for floods around the hills.

☆ Excessive surface run-off.

☆ Reduction in ground water recharge.

☆ Change in climate and failure of monsoon.

☆ Reduction in natural habitat of wildlife and species extinction.

☆ Deforestation.

Database generation is a vital part of the study and primary and secondary data were sourced from diverse sources. The extent of Land use change and impact on the Sholas was identified by this analysis. The database thus created was subjected to GIS and Remote Sensing analysis and several outputs were generated.

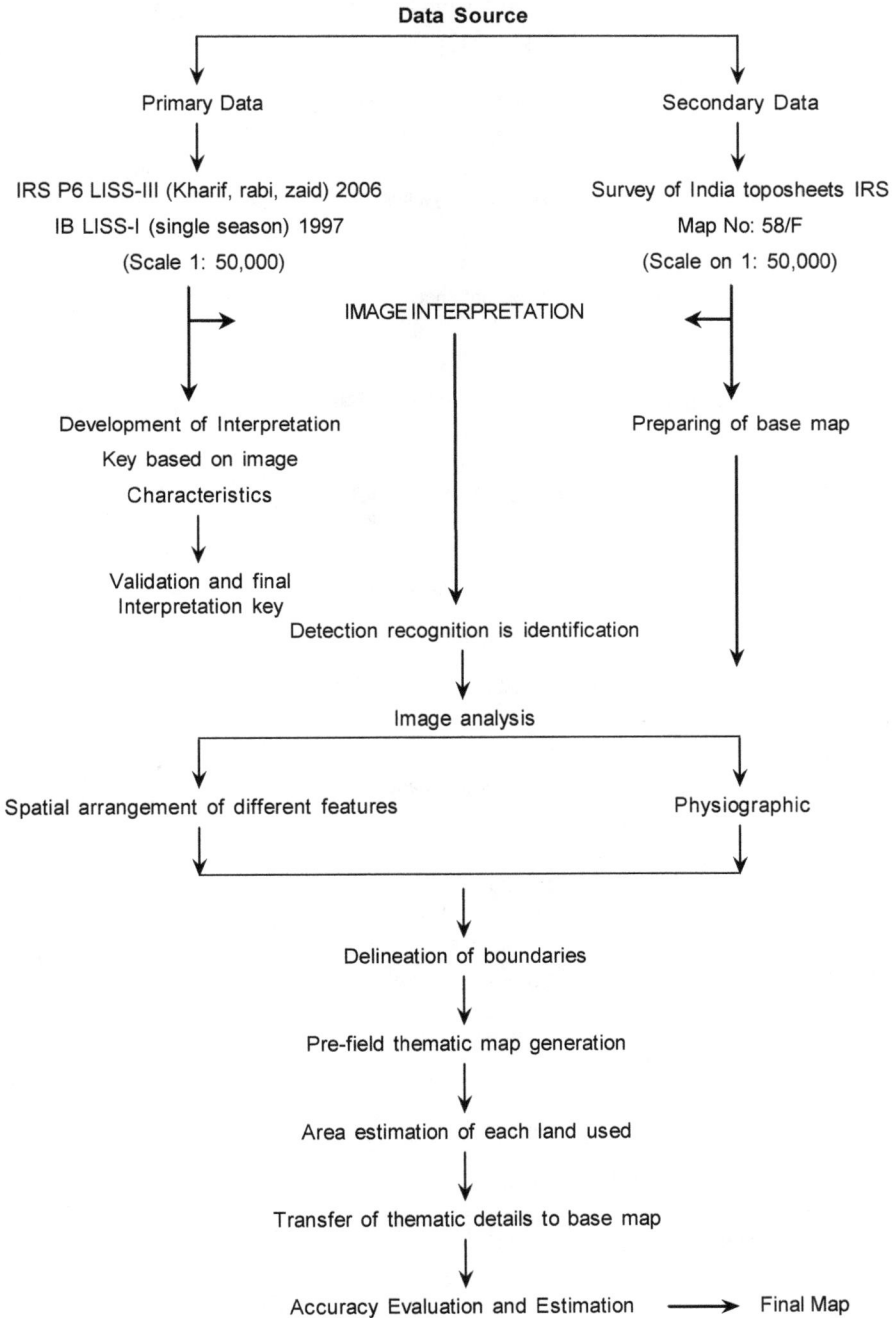

Methodology

Data Source

Primary Data

Secondary Data

IRS P6 LISS-III (Kharif, rabi, zaid) 2006
IB LISS-I (single season) 1997
(Scale 1: 50,000)

Survey of India toposheets IRS
Map No: 58/F
(Scale on 1: 50,000)

IMAGE INTERPRETATION

Development of Interpretation
Key based on image
Characteristics

Preparing of base map

Validation and final
Interpretation key

Detection recognition is identification

Image analysis

Spatial arrangement of different features

Physiographic

Delineation of boundaries

Pre-field thematic map generation

Area estimation of each land used

Transfer of thematic details to base map

Accuracy Evaluation and Estimation ⟶ Final Map

MAPPING METHODOLOGY

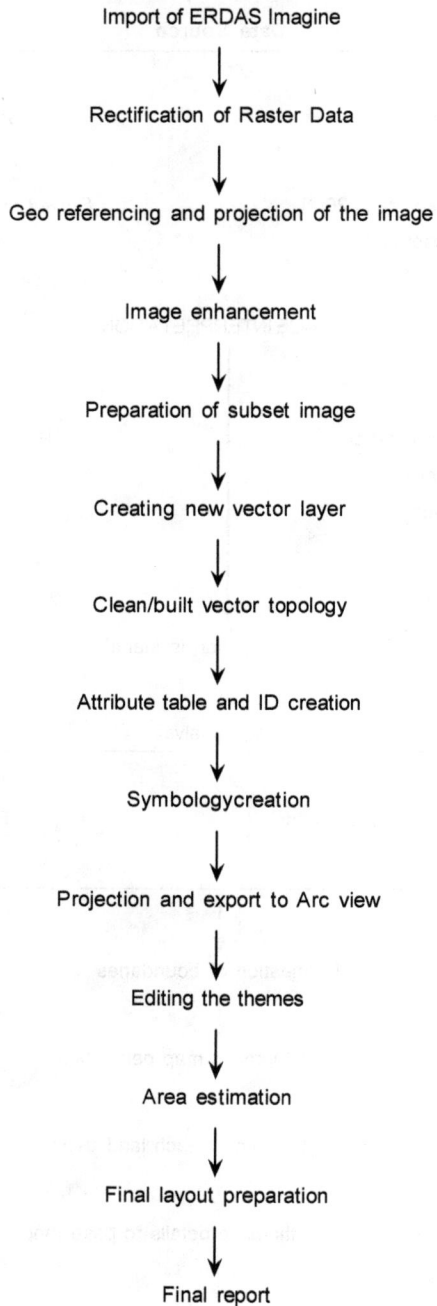

Import of ERDAS Imagine

↓

Rectification of Raster Data

↓

Geo referencing and projection of the image

↓

Image enhancement

↓

Preparation of subset image

↓

Creating new vector layer

↓

Clean/built vector topology

↓

Attribute table and ID creation

↓

Symbologycreation

↓

Projection and export to Arc view

↓

Editing the themes

↓

Area estimation

↓

Final layout preparation

↓

Final report

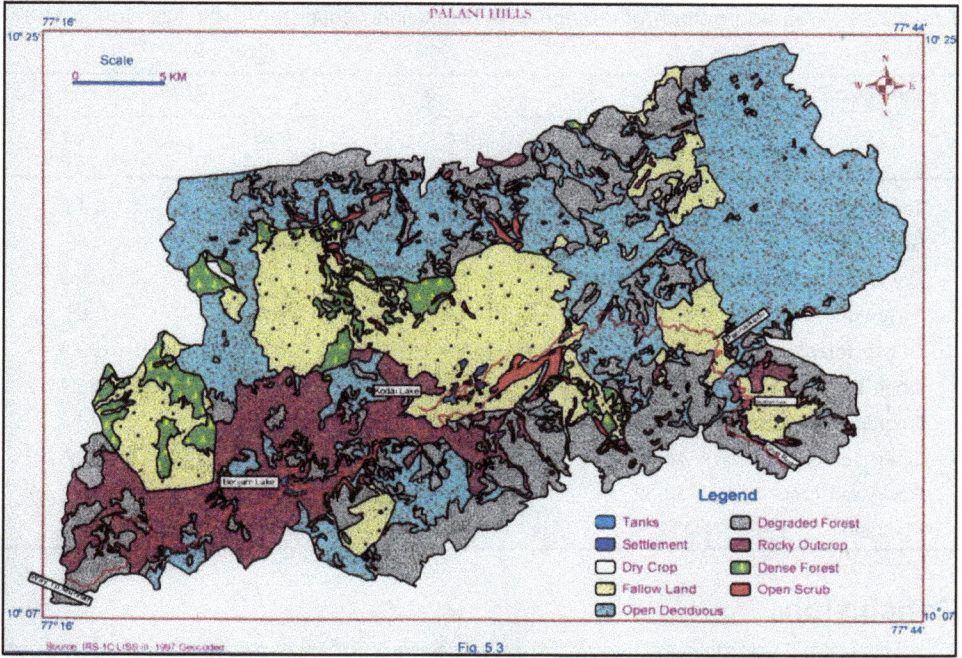

Figure 32.2: Land use (1997)

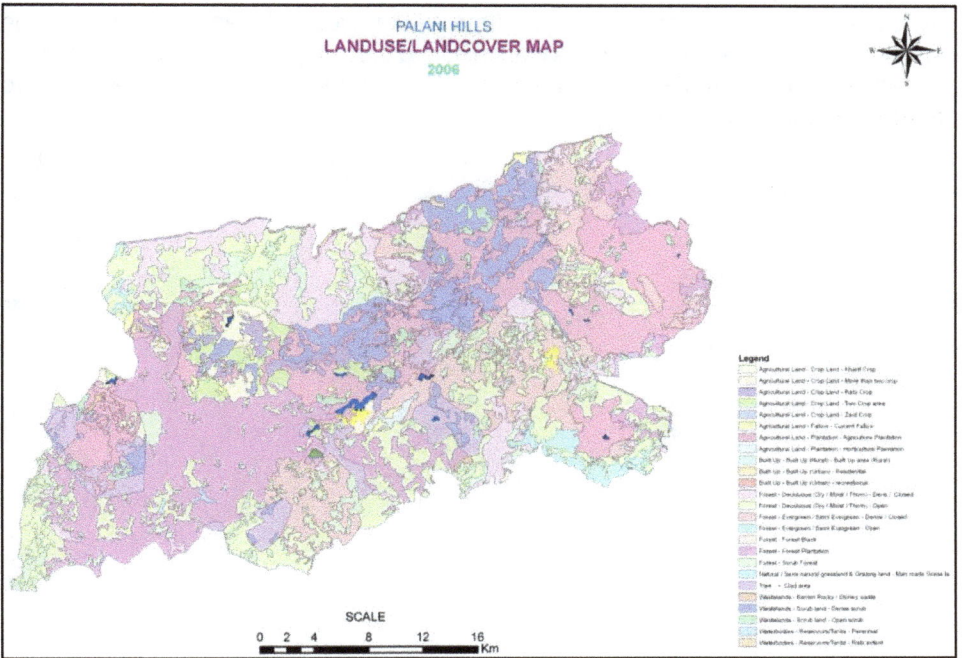

Figure 32.3: Land use/land cover map (2006)

Table 32.1: Area and pattern of change of land use/land cover for Kodai hills during 1997-2006.

Land Use Type	1997		2006	
	Area (sq. km.)	Area (Per cent)	Area (sq. km.)	Area (Per cent)
Tanks	1.16	0.107	1.82	0.18
Settlement	5.09	0.49	6.07	0.59
Dry Crop	1.14	0.11	0.90	0.088
Fallow/harvest	204.05	19.86	6.58	0.63
Dense forest	371.42	35.99	198.50	19.29
Open Deciduous	253.26	24.54	192.82	18.74
Degraded forest	126.75	12.28	180.42	17.53
Open scrub	44.88	4.35	93.51	9.09
Rocky out crop	23.29	2.25	1.23	0.11
Plantation	119.12	11.55	180.42	17.53

Conclusion

The spatial variation and environmental degradation that can be assessed through Plant diversity, Land use and Sholas using GIS and Remote Sensing Analysis. This type of assessment essentially requires to be a geo-referenced one. The spatial variations of the land cover in a period of time is evaluated in the analysis. The change detection on Sholas is given due importance during the analyses. A method to conserve the natural ecosystem of kodai hills is studied. The method for conservation of soil to prevent Landslides in near future is also studied.

2013, Biodiversity Conservation for Sustainable Management   *Pages 275–279*
Editor: **Dr. K. Muthuchelian,** *Vice Chancellor, Periyar University, Salem*
Published by: **Daya Publishing House, NEW DELHI**

Chapter 33

Evaluation of Biodiversity for Sirumalai Area of Palani Hills, Tamil Nadu, India

N. Mayavan*, B. Pavendan and A. Sundaram

Department of Future Studies, School of Energy Sciences,
Madurai Kamaraj University, Madurai – 625 021

Keywords: *Biome, Geodatbase, Remote sensing, GI technologies, Anthropogenic Interventions, Flora and fauna diversity, Causal mechanism.*

Sirumalai Hills Location and Significance

Sirumalai hills are located in Dindigul district, TamilNadu, India and have an areal coverage of 288.4 sq.km SOI (1973).; This area is a biodiversity hotspot and several endemic flora and fauna abound in this region.

Need for Study

SOI maps were rigorously surveyed under Imperial Cartographic Standards and are empirical in nature. But they are hopelessly antiquated as there never was any tangible follow-up action to update earlier maps. India spends astronomical amounts to launch and maintain satellites to acquire a reliable Geo-Database.

The data acquired from these satellites are of a remarkable value, for resource conservation, in that the present day technologies can meld old and new spatial and aspatial data through GI Technologies, a spin-off from Information Technology Revolution.

---

* Corresponding Author: E-mail: mayaremak1@gmail.com.

Figure 33.1: 3D image of Sirumalai Hills–IRS p6 LISS IV

Even a cursory observation of physical survey maps and Remotely Sensed contemporary maps will reveal a shocking picture of wanton destruction. Declared Reserve forests have shrunken due to anthropogenic interventions and the Biodiversity of the area is a threatened entity if not wholly lost as in the case of mountain Ibex once abundant in this area. As genetic pools, the area serves an important purpose of propagating species generation and floral diversity.

Integrating Remotely Sensed Data with empirical survey and attribute data on GI technologies (GIS, GPS, RS) platform will give a holistic picture of the extent of damage.

Several scholars have studied the flora and fauna diversity of the area and have established the endemism with reference to the environment.

In order to save the Biodiversity of the area the trends in destruction of habitats have to be quantified. As a contemporary technology of great use, Remote Sensing can be effectively utilized to identify Landuse variations, areal extent of changes, surface area changes in water bodies, forest cover and related Biodiversity losses. Also disasters like landslides that can be a human cost factor have to be studied at micro-levels by way of documenting Causal Mechanisms.

Sirumalai hills area is a biodiversity hot spot and is comprising Deciduous Dense forest category, Deciduous dry, Evergreen dense, Semi Evergreen open forest, scrub forest, plantations. These categories have undergone serious changes with mixed variations. Remote Sensing studies using IRS P VI LISS IV imagery revealed that there has been variations in the spatial extent of these categories. The quantifications of these variations were carried out using ARCGIS 9.2 and ERDAS 8.7. The time period chosen was 2003 and 2005.

The spatio-temporal variations can not be studied exhaustively without using an up to date analysis technology like Remote Sensing, GIS. Especially the forest cover and environmental issues need a monitoring on a continual basis. There are several endemic flora in Sirumalai. They are under a serious threat owing to Land use variation.

Land Use Classes

Table 33.1: Variations In Sirumalai from 2003-05.

Sl.No.	Landuse Classes	Area in Sq.km (2003)	Area in Per cent (2003)	Area in Sq.km (2005)	Area in Per cent 2005	Change Per cent (05-03*100)
1.	Deciduous forest	140.3	48.6	126	43.8	−4.8
2.	Evergreen forest	9.5	3.3	8.3	2.8	−0.5
3.	Degraded forest	67.5	23.4	58	20.1	−3.3
4.	Scrub forest	43.2	15.1	67.7	23.5	+8.4
5.	Plantation	27.5	9.5	28	9.7	+0.2

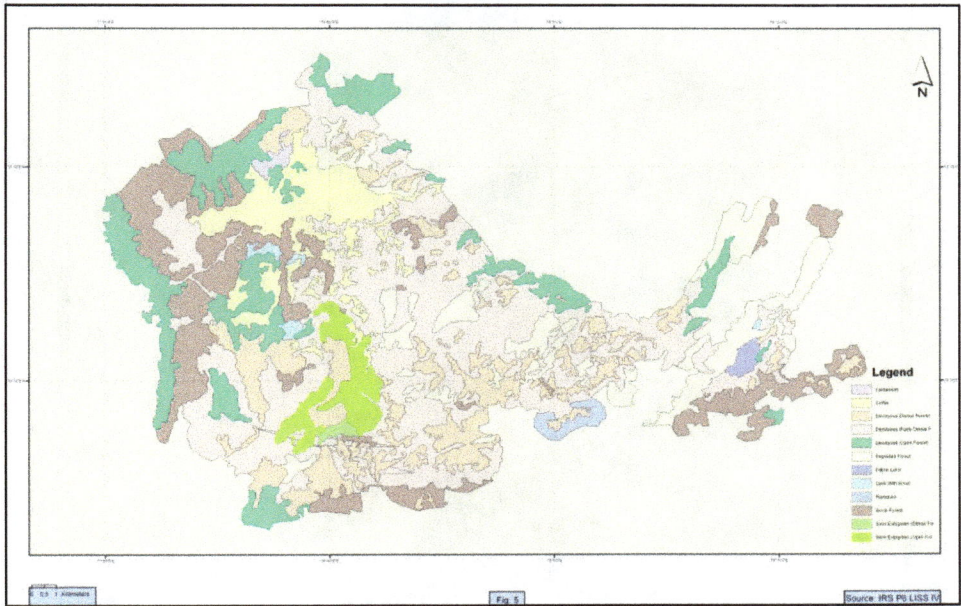

Figure 33.2: Sirumalai landuse (2003).

Analysis

A biome is defined as a division of the world's vegetation that corresponds to a defined climate and is characterized by specific types of plants and animals, *e.g.* tropical rain forest or desert. The world's lakes and oceans may also be considered biomes, although they are less susceptible to climatic influences than terrestrial biomes. Deciduous forests are important in a biome in that they represent a significant class of flora. In the area examined deciduous forests were found to have decreased from 140.3 sq.km to 126 sq.km, 4.8 per cent loss in two years (2003-2005).

A small region in Sirumalai (3.3 per cent of total area) is occupied with evergreen forest category (9.5 sq.km). These forests are at higher altitudes and relatively

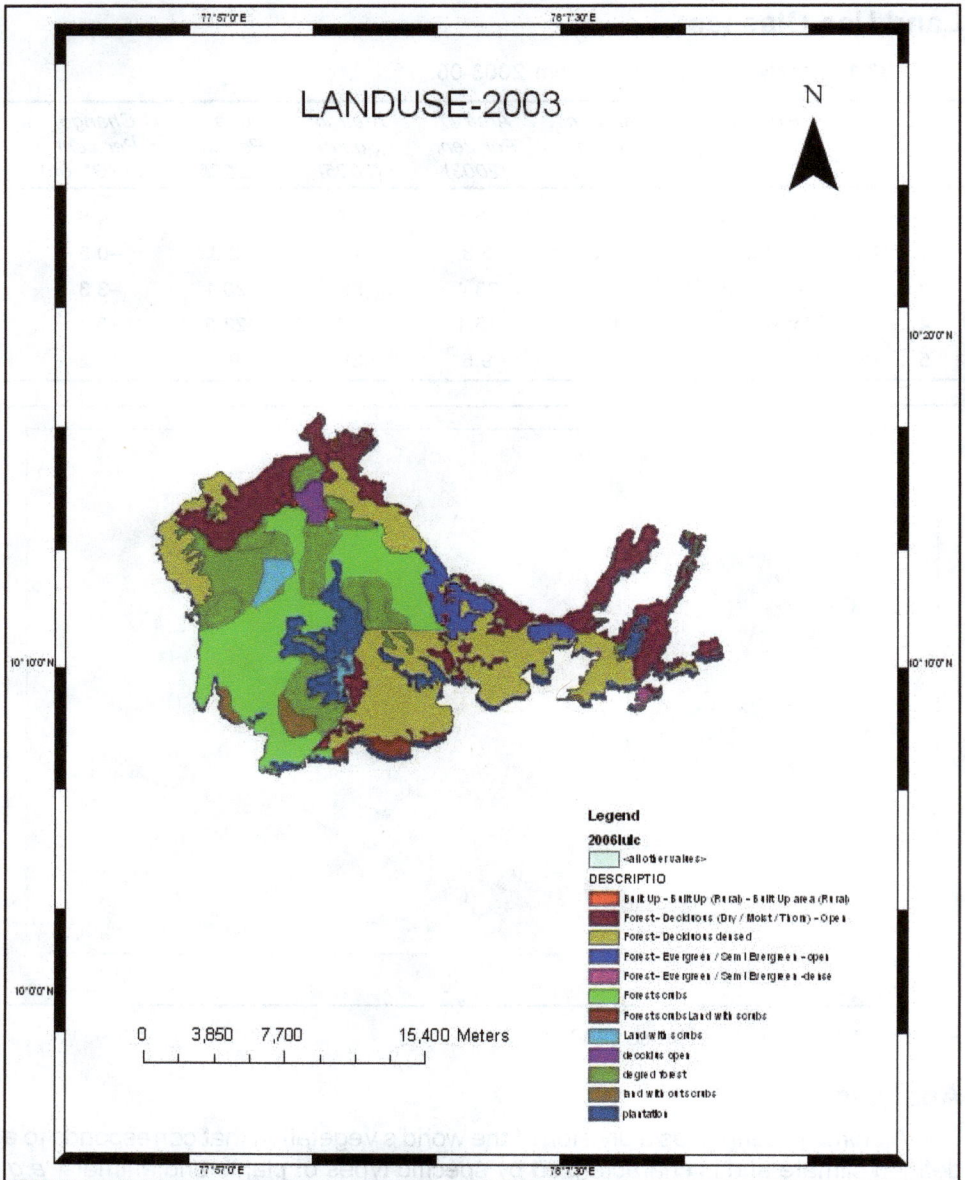

Figure 33.3: Landuse (2003).

inaccessible. But the timber from them is valuable. They have registered a fall 0.5 per cent.

There is a significant rise in scrub forest. The primary reason for this is not any planned initiative but an accumulated cascading effect of deciduous forest and evergreen forest cover loss.

Similarly previously categorized as degraded forest area has also suffered a spatial loss by 3.3 per cent.

The key finding of the study is there is no corresponding rise in plantation area as evidenced similar biomes. A marginal 0.2 per cent rise signifies that plantation is not the primary cause for loss in forest cover.

Conclusion

Field studies point to large scale lumbering, a non regenerative destruction causing the loss in the forest area. Lumbering practices are carried out in an unplanned manner and directly affect animal corridors that have been in existence from time immemorial. Mountain goats once roamed in these hills but no sightings have been reported of late. Also small herbivores are cornered into isolated pockets and face eventual extinction.

Another remarkable feature is the tribal people of the region who have had a harmonious existence with the forests all along are against the greed of the timber trade which inexorably destroys their habitat. But as silent spectators to the rape of the land they remain as mute witnesses to the loss of a priceless biodiversity spot. Policy initiatives to stop the lumbering practice in total or at least a sustainable exploitation plan is an imminent need.

2013, Biodiversity Conservation for Sustainable Management   Pages *280–281*
Editor: **Dr. K. Muthuchelian,** *Vice Chancellor, Periyar University, Salem*
Published by: **Daya Publishing House, NEW DELHI**

Chapter 34

Advanced Technology in Maintenance and Breeding of Endangered Species

B. Jagadish Chandra Bose

Department of Biochemistry,
PKN Arts and Science College, Tirumangalam

Introduction

To day a large number of species have been classified as threatened to the extend that sincere human care and *ex-stitu* management is the only way to preserve them. Naturally we shall have to expand our conservatories and use advanced technology for the management, multiplication and preservation of the threatened species.

Major contributions of advance technology to the strategy of ex-situ conservation cover the following aspects:

1. Chemical immobilization and anesthesia
2. Nutrition, maintenance and health care of animals in captivity.
3. Identification data collection and information technology
4. Advances in reproductive technology and cryobiology
5. Advances in population biology and molecular genetics.

Chemical Immobilization and Anaesthesia

One of the common problem faced by conservationists is the resistance of the animals while they are being captured. Many wild animals are injured in the process of capture or transportation and may cause injury to the captors. It is difficult to administer drugs to immobilize them as for doing so they have to be approached closely. Today techniques are available for administration of drugs to immobilize the animal in field or in captivity from a distance.

Nutrition, Maintenance and Health Care of the Animals

The treatment of individual animals, preventive medicines, systematic vaccination, antibiotics etc., along with carefully prepared diets have resulted in general improvement of health, well being and survival of most of the species in captivity.

Identification, Data Collection and Information Technology

Identification of individuals of a species is a difficult problem in captivity as well as in wild habitats. Animal marking technology has now become fairly sophisticated. However, it requires handling of animals from very close distances which is therefore, often not satisfactory. Telemetry has proved to be a very successful tool in this connection. Collars provided with transmitters are fitted to animal which transmit radio signals where ever it goes, identifying and locating the organism. The collar may be used to obtain other information such as temperature, heart beats, parturition, death etc., This can be continuously monitored by using automatic recording devices.

Advances in Reproductive Technology and Cryobiology

One of the biggest problems troubling ex-situ conservation efforts is the loss of genetic diversity due to in breeding. Repeated inbreeding causes homogenization of the species genetic make up and results in decrease in fertility, high infant mortality and birth defects. Today for the long term preservation of germ plasm and maintenance of thegenediversity advanced reproductive technology is applied which involves:

1. Artificial insemination
2. Embryo transfer technology
3. Cryopreservation of gametes and embryos

Advances in Population Biology and Molecular Genetics

The latest methods of molecular biology and genetics have been brought into the service of conservation of bio diversity—both demographically and genetically—to ensure their survival through time. The recent changes in the concept of species has resulted in the molecular genetic.

Conclusion

Despite the number of animal species that have become extinct the conservation movement has had many successes. Increasingly, people are becoming aware that various species of birds and mammals have become in danger of extinction an command or actually help in the steps being taken to conserve such animals.

References

Annkramer,1995., THE ANIMAL WORLD, the world book encyclopedia of science—vol–6.

Reid, W.V. and Miller, K.R. (1989). Keeping options alive the scientific bases for conserving biodiversity. World Resources Institute, Washington.

2013, Biodiversity Conservation for Sustainable Management   Pages *282–291*
Editor: **Dr. K. Muthuchelian,** *Vice Chancellor, Periyar University, Salem*
Published by: **Daya Publishing House, NEW DELHI**

Chapter 35

Biodiversity of Symbiotic Fungal Flora and their Conservation Measures in the Nilgiri Biosphere Reserve Area of Nilgiri Hills, Tamil Nadu

V. Mohan and N. Krishnakumar*

*Forest Pathology Laboratory, Division of Forest Protection,
Institute of Forest Genetics and Tree Breeding,
Coimbatore – 641 002*

ABSTRACT

The Nilgiri Biosphere Reserve was the first biosphere reserve in India established in the year 1986. It is located in the Western Ghats and includes 2 of the 10 biogeographical provinces of India. A wide range of ecosystems and species diversity is found in this region. Thus, it was a natural choice for the premier biosphere reserve of the country. Productivity of the forest plant community is a consequence of the interaction of tree shoots and roots with the environment. One of the most important biological interactions is the soil surrounding the root and the rhizosphere. One such interaction termed as "Mycorrhiza", literally means "Fungus Root", is the association between specialized root-inhabiting fungi and the roots of living plants. This symbiotic association is especially critical to forest trees and in disturbed areas or areas that have been progressively

---

* Corresponding Author: E-mail: mohan@icfre.org

degraded over time since the rhizosphere organisms can be affected by shifts in land management practices. Much of this biological diversity is hidden from view beneath the soil surface. The biological soil resource is one of the most important factors governing soil fertility. Maintaining mycorrhizal diversity helps to minimize site degradation by assuring plant adaptability to unpredictable or varying environments. This has special significance in forest ecosystems that now face unprecedented changes due to human activity. Mycorrhizal species are central sources to successful tree species establishment and the concept of biological diversity. Hence in the present study, an attempt was made to investigate the status of both ectomycorrhizal (ECM) and Arbuscular Mycorrhizal (AM) fungi in different forest ecosystems in the Nilgiri Biosphere Reserve area. A total of 20 different ECM fungi *viz., Alnicola* sp., *Amanita* sp., *A. muscaria, Cortinarius* sp., *Geastrum* sp., *Hebeloma* sp., *Inocybe* sp., *Laccaria fraterna, L. laccata, Leucophleps* sp., *Lycoperdon perlatum, Lycoperdon* sp., *Rhizopogon luteolus, Russula parazurea, Russula* sp., *Scleroderma citrinum, Scleroderma* sp., *Suillus brevipes, S. subluteus* and *Thelephora terrestris* were recorded in association with *Acacia mearnsii, A. melanoxylon, Cupressus macrocarpum, Eucalyptus globulus, E. grandis* and *Pinus patula* and 32 different AM fungi belonging to four genera such as *Acaulospora, Gigaspora, Glomus* and *Scutellospora* were recorded in different forest ecosystems. Among them, the genus *Glomus* was found to be the dominant with 22 species. It was observed that the climatic and edaphic factors have profound influence on the distribution of these fungi in the study locations. Significance of these findings with reference to the ecosystem functioning, exploitation and conservation of the valuable natural resources of mycorrhizal fungi is discussed.

Keywords: *Biosphere reserve, Ectomycorrhizae, Arbuscular mycorrhizae, Acaulospora, Gigaspora, Glomus, Scutellospora, Amanita, Inocybe, Laccaria, Leucophleps, Rhizopogon, Russula, Scleroderma, Suillus.*

Introduction

The Nilgiri Biosphere Reserve (NBR) is located in Western Ghats between 76°– 77°15'E and 11°15'–12°15'N. The study areas selected for the proposed project include different forest ecosystems such as Natural Forests (Tropical Wet Evergreen Forests, Shola-Grassland Ecosystem) and Man-made Plantation Forests in the Nilgiri Hill areas of the NBR. Microorganisms are present in great numbers on and near the feeder roots of trees and they play vital roles in numerous physiological processes. These dynamic processes are mediated by associations of microorganisms participating in saprotrophic, pathogenic and symbiotic root activities. Among the various soil organisms, the most important known to us are the mycorrhizal fungi. The mycorrhizal fungi can be classified into two groups viz., Ectomycorrhiza (ECM) and Endomycorrhiza.

Enodomycorrhizae are the most widespread and comprise three groups such as Ericaceous, Orchidaceous and Arbuscular Mycorrhizal (AM) fungi. The AM fungi are found on more plant species than all other types of mycorrhizas. Ectomycorrhizas (ECM) occur on about 10 per cent of the world flora. Trees belonging to the families Pinanceae, Fagaceae, Betulaceae, Salicaceae, Junglandaceae, Myrtaceae,

Dipterocarpaceae form ectomycorrhizal (ECM) association (Agerer, 1985, 1986; Chu-Chou, 1979; Gardener and Malajczuk 1988; Mason *et al.,* 1983; Bakshi, 1974; Natarajan *et al.,* 1988; Vijayakumar *et al.,* 1999; Mohan, 1991, 2002). Numerous fungi have been identified as ECM. Worldwide, there are over 5000 species of fungi that can form ECM on some 2000 species of woody plants. It is estimated that these fungi can form ECM with forest trees in different parts of the world (Table 35.1).

Table 35.1: Distribution of ectomycorrhizal host plants in different parts of the world.

Sl.No.	Name of the Host Plants	Country	Reference
1.	*Abies* sp.	USA, Germany	Agerer, 1985; Trappe, 1977
2.	*Acacia auriculiformis, A. mangium, Acacia mearnsii*	India	Vijayakumar *et al.,* 1999; Mohan, 2002; Mohan, 2008
3.	*Betula pendula* and *B. pubescens*	Scotland	Mason *et al.,* 1982
4.	*Casuarina equisetifolia* and *C. junghuhniana*	India	Natarajan *et al.,* 1988; Mohan, 2002
5.	*Eucalyptus* spp.	Australia, USA, New Zealand	Gardener and Malajczuk, 1988; Trappe, 1977; Chu-Chou, 1979
6	*Eucalyptus camaldulensis, E. globulus, E. grandis, E. tereticornis*	India	Bakshi, 1974; Natarajan *et al.,*1988; Mohan, 2002; 2008
7.	*Pinus radiata*	New Zealand	Chu-Chou, 1979
8.	*Pinua patula*	India	Last *et al.,* 1984; Mohan, 1991; Mohan, 2008
9.	*Pseudotsuga menziesii*	New Zealand	Chu-Chou, 1979
10.	Oak, Beech, Spruce	Germany	Agerer, 1985

The mycorrhizal fungi enhance the uptake of nutrients, especially phosphorus, increase the surface area of the roots of host plants and act as bio-control agents against soil-borne or root-borne pathogens. In the past few decades the extent of tropical forests has changed dramatically with the ever increasing demand for wood fibre. Since above and below ground organisms are tightly linked, such changes result in dramatic losses which decreases hope for restoration of degraded sites through natural regeneration. Hence, it is essential to collect and assess indigenous mycorrhizal fungi for successful conservation and establishment of potential and promising mycorrhizal cultures for future use in the forestry sector. The status of both AM and ECM fungi associated with different host plants and factors influencing their distribution are highlighted in this paper. This paper also deals with various conservation strategies of these symbiotic fungi for better use in afforestation programme.

Diversity of ECM Fungi in Different Plantation Ecosystem

Based on both morphological and microscopical characters, total of 20 different ECM fungi such as *Alnicola* sp., *Amanita* sp., *Amanita muscaria, Cortinariuss* sp., *Geastrum* sp. *Hebeloma* sp., *Inocybe* sp., *Laccaria fraterna, Laccaria laccata, Leucophleps* sp., *Lycoperdon* sp., *Lycopedon perlatum, Rhizopogan luteolus, Russula*

sp., *Russula parazurea*, *Scleroderma* sp., *Scleroderma citrinum, Suillus brevipus,*
Suillus subluteus and *Thelephora terrestris* belonging to 10 different families viz.,
Amanitaceae, Cortinariaceae, Leucogastraceae, Lycoperdaceae, Rhizopoganaceae,
Russulaceae, Sclerodermataceae, Suillaceae, Thelephoraceae and Tricholomataceae
were recorded and identified. Of these, 12 species belong to the group Hymenomycetes
and the remaining to Gasteromycetes. It was also observed that ECM fungi viz.,
Laccaria fraterna and *Scleroderma* sp. were found to be associated with five different
tree species (*Acacia mearnsii, Acacia melanoxylon, Cupressus macrocarpa,*
Eucalyptus globulus and *Eucalyptus grandis*). ECM fungal species like *Lycoperdon*
perlatum (*Acacia mearnsii, Acacia melanoxylon, Cupressus macrocarpa, Eucalyptus*
globulus); Russula (*Acacia mearnsii, Cupressus macrocarpa, Eucalyptus globulus,*
Hopea parviflora) and *Suillus brevipes* (*Acacia mearnsii, Cupressus macrocarpa,*
Eucalyptus globulus and *Pinus patula*) were found in association with four different
tree species each in the study areas during the period under investigation. The ECM
fungi such as *Amanita muscaria, Laccaria laccata, Lycoperdon perlatum, Rhizopogan*
luteolus, Russula parasurea, Suillus brevipes, Suillus subluteus, Scleroderma citrinum
and *Thelephora terrestris* were exclusively found in association with *Pinus patula.* It
was also observed that the ECM fungi viz., *Laccaria fraterna, Leucophleps* sp. and
Scleroderma sp. were found in association with *Acacia melanoxylon* and *Eucalyptus*
grandis. The ECM fungus *Inocybe* sp. was reported in association with 3 different tree
species (*Acacia mearnsii, Cupressus macrocarpum* and *Eucalyptus globulus*). The
ECM fungus *Leucophleps* sp. was found only in association with *E. grandis* plantation
at Naduvattam during the period under investigation. The ECM fungal species like
Amanita, Geastrum and *Russula* were found only in association with *Hopea parviflora*
trees in Nilambur, Kerala (Table 35.2). Among 20 different ECM fungi recorded, the
ECM fungi viz., *Alnicola* sp., *Amanita muscaria, Cortinarius* sp., *Hebeloma* sp., *Inocybe*
sp., *Laccaria fraterna, Lycoperdon perlatum, Russula delica, Scleroderma bovista*
were reported for the first time in association with *Acacia mearnsii, Acacia melanoxylon,*
Cupressus macrocarpa, Eucalyptus globulus and *E. grandis* plantations.

Diversity of Arbuscular Mycorrhizal (AM) Fungi in Different Forest Ecosystems

Data on the AM fungal diversity in the rhizosphere of different forest ecosystems
(Sholas, Grasslands and Forest stands) in the NBR areas is presented in Table 35.3.
Thirty two different AM fungal species belong to the families Acaculosporaceae,
Gigasporaceae and Glomaceae were isolated and identified from the rhizosphere of
various study locations during the two year period of investigation. Four different AM
fungal genera viz. *Acaulospora* (5 species), *Gigaspora* (4 species), *Glomus* (22 species)
and *Scutellospora* (1 species) were recorded. Among them, the genus *Glomus* was
found dominant.

Effect of Edaphic Factors on the Occurrence of both ECM and AM Fungi in Different Sites

The chemical characteristics of the soil samples were analysed and the results
revealed that there was no much correlation between amount of macro and micro
nutrients on the occurrence and distribution of number of basidiomata/fruit bodies of

different ECM and AM fungal population in different plantations. Whereas the phosphorus level decreased the number of ECM and AM fungi increased during different periods of observation.

Table 35.2. Distribution of ECM fungi in association with different tree species in the NBR areas.

Sl.No.	ECM Fungi	Name of Tree Species	No. of Tree Species
1.	*Alnicola* sp.	*Acacia mearnsii, Cupressus macrocarpa* and *Eucalyptus globulus*	3
2.	*Amanita muscaria*	*Acacia mearnsii, Cupressus macrocarpa, Eucalyptus globulus* and *Pinus patula*	4
3.	*Amanita* sp.	*Hopea parviflora*	1
4.	*Cortinarius* sp.	*Acacia mearnsii, Cupressus macrocarpa* and *Eucalyptus globulus*	3
5.	*Geastrum* sp.	*Hopea parviflora*	1
6.	*Hebeloma* sp.	*Acacia mearnsii, Cupressus macrocarpum* and *Eucalyptus globulus*	3
7.	*Inocybe* sp.	*Acacia mearnsii, Cupressus macrocarpum* and *Eucalyptus globulus*	3
8.	*Laccaria fraterna*	*Acacia mearnsii, Acacia melanoxylon, Cupressus macrocarpum, Eucalyptus globulus* and *Eucalyptus grandis*	5
9.	*Laccaria laccata*	*Pinus patula*	1
10.	*Leucophleps* sp.	*Eucalyptus grandis*	1
11.	*Lycoperdon perlatum*	*Pinus patula, Acacia mearnsii, Acacia melanoxylon, Cupressus macrocarpum* and *Eucalyptus globulus*	
12.	*Lycoperdon* sp.	*Hopea parviflora*	1
13.	*Rhizopogan luteolus*	*Pinus patula*	1
14.	*Russula delica*	*Acacia mearnsii, Cupressus macrocarpa* and *Eucalyptus globule, Hopea parviflora*	5
15.	*Russula parazurea*	*Pinus patula*	1
16.	*Scleroderma bovista*	*Acacia mearnsii, Acacia melanoxylon, Cupressus macrocarpa, Eucalyptus globulus* and *E. grandis*	5
17.	*Scleroderma citrinum*	*Pinus patula*	1
18.	*Suillus brevipes*	*Acacia mearnsii, Cupressus macrocarpa, Eucalyptus globulus* and *Pinus patula*	4
19.	*Suillus subluteus*	*Pinus patula*	1
20.	*Thelephora terrestris*	*Pinus patula*	1

Effect of Rainfall on the Occurrence of ECM Fungi in different Plantations

The appearance of basidiomata of the different ECM fungi in association with various tree species such as *Acacia mearnsii, A. melanoxylon, Cupressus macrocarpum, Eucalyptus globulus, E. grandis* and *Pinus patula* were found to be affected by the amount of rainfall. The number of basidiomata of different ECM fungi were found greater during monsoon seasons especially July, August, September and October months as compared to summer months.

Table 35.3: List of AM fungi recorded in different forest ecosystem in the Nilgiri Biosphere Reserve areas of Nilgiri Hills, Tamil Nadu.

Sl. No.	AM Fungi	Family
1.	*Acaulospora appendicula*	Acaulosporaceae
2.	*Acaulospora laevis*	-do-
3.	*Acaulospora scrobiculata*	-do-
4.	*Acaulospora spinosa*	-do-
5.	*Acaulospora* sp.	-do-
6.	*Gigaspora albida*	Gigasporaceae
7.	*Gigaspora gigantea*	-do-
8.	*Gigaspora margarita*	-do-
9.	*Gigaspora* sp.	-do-
10.	*Glomus aggregatum*	Glomaceae
11.	*Glomus albidum*	-do-
12.	*Glomus claroideum*	-do-
13.	*Glomus clarum*	-do-
14.	*Glomus coremioides*	-do-
15.	*Glomus deserticola*	-do-
16.	*Glomus dussii*	-do-
17.	*Glomus fasciculatum*	-do-
18.	*Glomus fulvum*	-do-
19.	*Glomus geosporum*	-do-
20.	*Glomus glomerulatum*	-do-
21.	*Glomus microcarpum*	-do-
22.	*Glomus macrocarpum*	-do-
23.	*Glomus maculosum*	-do-
24.	*Glomus mosseae*	-do-
25.	*Glomus monosporum*	-do-
26.	*Glomus multicaulae*	-do-
27.	*Glomus multisubtensum*	-do-
28.	*Glomus occultum*	-do-
29.	*Glomus pubescens*	-do-
30.	*Glomus sinuosa*	-do-
31.	*Glomus* sp.	-do-
32.	*Scutellospora* sp.	Gigasporaceae

Acaulospora: 5; Gigaspora: 4; Glomus: 22; Scutellospora: 1.

Pure Culture Production and Maintenance of Culture Bank of ECM and AM Fungi

Isolates of different ECM fungi viz., *Alnicola* sp. *Laccaria fraterna, Scleroderma sp., Suillus brevipes, Suillus subluteus* and *Russula* sp. and AM fungi were made and maintained in germplasm bank of the IFGTB, Coimbatore for further studies in the nursery and field.

Conservation of Mycorrhizal Fungi

Maintaining mycorrhizal diversity helps to minimize site degradation by assuring plant adaptability to unpredictable or varying environments. This has special significance in forest ecosystems that now face unprecedented changes due to human activity. Mycorrhizal species are central sources to successful tree species establishment and the concept of biological diversity. Because mycorrhizal fungi have a great influence on the survival of plants in new and reclaimed sites, the tree health and site quality and they are the cornerstone to proper establishment of functional forest ecosystem. Bearing in mind the importance to trees of the functions served by mycorrhizal fungi in connection with the absorption of nutrients and protection against pathogenic organisms around the roots, it was observed that decline in mycorrhizal fungi is a matter of serious concern. Thus, preservation of the diversity of mycoflora is also important from the point of view of forestry. Appropriate measures for the conservation of fungi are therefore certainly justified.

On forest sites, the disturbance to surface soils caused by clear felling and site preparation generally reduces numbers of indigenous fungal propagules of mycorrhizae. Burning of logging residues can markedly alter the species composition of fungi. Fertilizer practices can be expected to have large effects on tree-fungus associations and to affect tree growth responses to inoculation. High rates of phosphorus (P) and nitrogen (N) fertilizers suppress mycorrhiza development in the field and high concentrations of soil N can also reduce the number of relative abundance of different ECM types. Establishment and maintenance of germplasm bank of different mycorrhizal cultures for further use are one of the major conservation aspects. The following measures should be adopted for conservation of different mycorrhizal fungi in natural forests and man-made plantations.

Motivation of Awareness

☆ Awareness of the ecological significance of the mycorrhizal fungi should be promoted among the general public (schools, foresters, mycological societies, mushroom information centre's, tourist associations, etc.).

☆ Another important requirement for effective conservation is that mushroom collectors should be made aware of the threats to mycorrhizal fungi and be persuaded of the need for careful management of nature.

☆ Collectors' knowledge of species of fungi should be improved (courses, possibly introduction of official permits, etc.).

☆ The volume of non-edible mushroom waste could thereby be reduced.

Protection of Areas

☆ In nature forests and man-made plantations, mycorrhizal fungi should be incorporated and protected.

☆ In addition to plant and animal species deemed worthy of protection, populations of the potential mycorrhizal fungi should also be protected.

☆ Specific protection of biotopes is desirable because of the close interdependence between different species of mycorrhizal fungi and the localities where they grow. For this reason, the protection of habitats and essential substrates (*e.g.* decaying wood, tree hosts, groundwater levels, etc). is of primary importance for the conservation of different species of mycorrhizal fungi.

☆ Suitable protected areas may also serve as genetic reserves for rare species of mycorrhizal fungi.

☆ Undisturbed natural cycles, not interrupted by tree felling, represent ideal conditions for the development of diverse mycorrhizal fungi.

Restrictions on Collection of Mushrooms/Mycorrhizal Basidiomata

☆ Restrictions on collecting basidiomata from the natural forests and man-made plantations are the useful means of controlling indiscriminate, large-scale or commercial harvesting of mushrooms.

☆ Permission can be granted in exceptional cases for the collection of mycorrhizal fungi for scientific purposes.

Forestry Measures

☆ Forestry activities should be carried out as carefully as possible. The use of heavy timber-transport vehicles, particularly on wet ground, should be kept to a minimum so as to avoid compaction of the soil.

☆ Burning of wood waste during harvesting operations in forests should also be avoided. It would be desirable for individual dead tree trunks to be left standing or lying or for islands of wood waste to be set aside.

☆ These measures make it possible for rare species of wood-rotting fungus to develop in the various stages of wood decay. However, crop protection aspects will also need to be taken into consideration in connection with measures of this kind.

☆ Access for vehicles should be largely prohibited on forest roads.

Other Measures

☆ Promotion of the production and sale of cultivated mushrooms as an alternative to the collection of wild mushrooms and mycorrhizal fungi.

☆ This measure could provide a degree of relief. In addition, cultivated mushrooms offer the advantage of containing virtually no toxic substances provided appropriate production processes are used.

☆ Promotion of research into habitat requirements, growth, development of fruiting bodies and reproduction of mycorrhizal fungi, together with regional studies of the development of different mycoflora and possible influences.

☆ Establishment of an expert body designed to promote coordination between research, conservation, mycological societies, the forestry sector, the general public, etc.

Conclusion

Different methods have been developed to ensure the formation of mycorrhizae on forest tree seedlings used to establish in wastelands, mined overburden areas, saline areas, arid and semi-arid desert areas and other fragile areas. Successful reforestation depends on the capacity of tree seedlings to establish early thereby ensuring continued resource supply to resist pests and diseases and to survive climatic stress. *In-vitro* vegetative propagation techniques are being increasingly employed for rapid multiplication of high quality healthy plants. The precocious inoculation of such seedlings with efficient mycorrhizae during their *in-vitro* multiplication is a promising potential for the use of mycorrhizas. Tree seedlings have evolved a beneficial mutual dependency upon mycorrhizal fungi for normal infections. Interactions between the host plants, mycorrhizae and soil largely determine the mycorrhizal effect on plant growth which is further conditioned by the variability in abiotic and biotic factors. A better understanding of the interrelationships among them various factors is needed. Further research is aimed at finding appropriate technology for large scale commercial multiplication and inoculation techniques of mycorrhizal inoculum. Future research should endeavor at locating appropriate host–mycorrhizal combinations adapted to well defined soil and environmental conditions. Hence, the Mycorrhizal Technology especially the selection, propagation, manipulation and management of superior and suitable isolates or strains of ECM or AM fungi can make a critical contribution to the success of afforestation and reforestation programmes.

References

Agerer, R., 1985. Zur Okologie der Myxorrhizaplize. *Bibl. Myc.*, 97: 1–160.

Bakshi, B.K., 1974. Mycorrhiza and its role in forestry. P.L. 480 Project Report. Forest Research Institute and Colleges, Dehra Dun (India). 89 pp.

Chu-Chou, M. 1979. Mycorrhizal fungi of *Pinus radiata* in New Zealand. *Soil Biol. Biochem.*, 11: 557–562.

Gardner, J. H. and Malajczuk, N., 1988. Recolonisation of rehabilitated bauxite mine sites in Western Australia by mycorrhizal fungi. *For. Ecol. Mgmt.*, 24: 27–42.

Kornerup, A. and Wanscher, J.H., 1978. *Methuen Hanbook of Colour*, 3rd Edn. Eyre Methuen, London, pp. 1–252.

Mason, P.A., Wilson, J., Last, F.T. and Walker, C., 1983. The concept of succession in relation to the spread of sheathing mycorrhizal fungi on inoculated tree seedlings growing in unsterilizedsoils. *Plant and Soil*, 71: 247–256.

Mohan, V., 1991. Studies on ectomycorrhizal association in *Pinus patula* plantations in the Nilgiri Hills, Tamil Nadu, *Ph.D. Thesis,* University of Madras, Madras, India, pp. 260.

Mohan, V., 2002. Distribution of ectomycorrhizal fungi in association with economically important tree species in Southern India. In: *Frontiers of Fungal Diversity in India* (Eds.) G.P. Rao, C. Manoharachari, D.J. Bhat, R.C. Rajak and T.N. Lakhanpal. International Book Distributing Co., Lucknow, U.P., India, pp. 863–872.

Natarajan, K., Mohan, V. and Kaviyarasan, V., 1988. On some ectomycorrhizal fungi occurring in Southern India. *Kavaka*, 16: 1–7.

Natarajan, K. and Mohan, V., 1998. Ecology of ectomycorrhizal fungi in *Pinus patula* plantations in the Nilgiri Hills, Tamil Nadu, South India. In: *Microbes for Health, Wealth and Sustainable Environment*, (Ed.) Ajit Varma. Malhotra Publishing House, New Delhi, India, pp. 115–127.

Vijayakumar, R., Prasada Reddy, B.V. and Mohan, V., 1999. Distribution of ectomycorrhizal fungi in forest tree species of Andhra Pradesh, Southern India: New Record. *Indian Forester,* 125: 496–502.

2013, Biodiversity Conservation for Sustainable Management   Pages *292–298*
Editor: Dr. K. Muthuchelian, *Vice Chancellor, Periyar University, Salem*
Published by: Daya Publishing House, NEW DELHI

Chapter 36

Biodiversity Loss, Causes and Conservation of Kudiraimozhi Theri (KMT) in Tuticorin District, Southern India

R. Selvakumari, T.J.S. Rajakumar, S. Murugesan and N. Chellaperumal

Centre for Botanical Research,
Department of Botany, St. John's College,
Palayamkottai, Tirunelveli – 627 002, Tamil Nadu, South India

ABSTRACT

Kudiraimozhi theri (KMT) is situated in Tiruchendur taluk, Tuticorin district, Tamil Nadu, South India at an altitude of 30m. It lies about 10 km from the East Coast and the region is in the Southeastern part of Peninsular India. It is a psammmophytic area covered by red sand and sand dunes with varying thickness from 5m to 10m. A systematic survey of the area including its biodiversity was made during the year 2003–2007.

Introduction

Biological diversity (biodiversity) is the term given to the variety and variability of life on earth and the natural patterns it forms (Mc Neely and al., 1990; Chauvet and Oliver, 1993). The Convention on Biodiversity (CBD) and United Nations Environment Programmes (Anon., 1992) define biological diversity as the variability among living organisms from all sources including terrestrial, marine and other aquatic ecosystems and the ecological complexes of which they are a part. This includes diversity within species, between species and of ecosystems. Scientists estimate that there are actually about 13 mn. species though estimates range from 30–100 mn. But the total number of living species is getting depleted day by day due to high living standard, over exploitation, habitat destruction urbanization, industrialization etc. Natural habitats such as forests, grasslands, deserts, wetlands, coral reef are under tremendous pressure due to increasing population. In India, lands are getting degraded at an alarming rate with disastrous effects on its biodiversity. Even though research works are going on in different parts of the country to find out the reasons for the loss of biodiversity and to identify the methods of conservation, still some areas in our country are unexplored or under explored.

One such area is Kudiraimozhi theri and it is a unique area. It lies about 10 km from the East Coast and the region is in the southeastern part of Peninsular India. The whole area is covered by sand and sand dunes, which encompass a diverse forest types that include, dry deciduous and scrub jungles. The area supports xerophytic, marshy and aquatic plant species. Though considerable literatures are available regarding the flora and fauna of Tuticorin area (Rangachariya, 1919; Mudaliar and Sundararaj, 1954; Sankara narayanan. 1960; Sundararaj, and Nagarajan, 1964) literatures regarding the biodiversity of Kudiraimozhi theri, Tuticorin district, Tamil Nadu appears to have been very less (Sebastine and Ramamurthy, 1961; Ramamurthy, 1963). In the present study, an effort was made to collect data about the biodiversity of Kudiraimozhi theri, causes for the loss and its conservation.

Study Area

Kudiraimozhi theri is situated in Tiruchendur taluk, Tuticorin district, Tamil Nadu, South India at an altitude of about 30m. The whole Kudiraimozhi theri consists of 25 km² and it is a Psammophytic area, covered by sand and sand dunes with thickness varying from 5 m to 10 m (Subbaraj, 1992). It is a reserve forest. There are several theories ascribed to the nomenclature of the study area. During the first half of the 19th century Missionaries from Europe thronged this area. It was an irony that this arid region was a fertile region for the missionary work. The forest officials on the one hand and the Missionaries on the other, criss–crossed this region on their horses. As it was a sandy area, the horses could not gallop. They just managed to wade through the region. In most of the areas, the legs of the horses sank in to the sand up to their joints. "Kudiraimozhi" in Tamil is nothing but the 'horse joints' in English and thus "Kudiraimozhi theri" means the sandy land where the legs of the horses sank up to their joints.

There are several other villages in and around Kudiraimozhi theri named after the name of theri or mozhi either as a prefix or suffix. Some of the name of the villages are Theripanai, Theriur, Therikudierruppu and Kayamozhi, Muthalai mozhi, Nangaimozhi, Arivanmozhi, Thatchaimozhi, Ittamozhi etc.

In KMT two distinct areas are recognizable. The first is the area of fixed or stationary sand dunes and the second is the area comprising moving sand dunes. The latter are called sand supply areas, which feed the dunes by wind action. The high velocity wind emerging from the Western Ghats pass passes through the study area (KMT) to the Gulf of Mannar in the Bay of Bengal in the east of the country. In the process, the wind hits the land area lifting soil particles to some height and transporting them as far away as 20 km, depending on the wind velocity. The soil particles are deposited elsewhere forming sand dunes. In some areas, the Palmyra groves partially sank into the sand by moving sand dunes.

Geology and Soil

The study area has two different types of soil red and the black. The black soil is restricted to the cultivated areas. They are mainly derived from ancient crystalline and metamorphic rocks (Shetty and Singh, 1996). The red soils are usually light, friable and porous. These soils are poor in lime, potash, nitrogen, and phosphorous content. The Soil Survey and Land Use Organization (1987) of the Department of Agriculture, Tamil Nadu reported that the soil in theri lands were very deep, red, sandy to loamy, sandy soil with single grain structure. The moisture retentivity was very poor. Cation exchange capacity and organic carbon content were low in these soils.

Vegetation

The vegetation of Kudiraimozhi theri is absolutely sparse and peculiar. It is extremely of xerophytic type, the species that occur being more or less the same as in the thorn forests. Occurring all over the area is the several lakhs of Palmyra trees (*Borassus flabellifer*) which fetch handsome revenue. Scattered patches of *Anacardium occidentale* artificially raised and naturally occurring *Pandanus fascicularis* near kaanam tanks are important items of minor forest produce.

In addition to xerophytic vegetation, the area also contains aquatic and marshy plants. The whole range of forest is divided into Northeast and Southwest zones. The Northeast zone is about 6626 acres. It consists of garden and cultivated lands, which from very ancient time are irrigated by channel cut into the theri sands, which supplements the water from the large kadamba tank. Ramamurthy (1963) reported the presence of natural springs (locally called "Sunai") in the eastern edge of Kudiraimozhi theri. The natural spring inside the theri is like an oasis. During Northeast monsoon, water used to flow from the theri springs and runoff through village roads and lanes (Manoharan and Kombairaj. undated (http://www.www. nuffic. nil). This situation changed around 1967, when there was a decline in rainfall. Due to heavy extraction of ground water and its limited recharge, the ground water getting depleted at a fast rate. Today the water table is at about 100-130 feet. As a result, the natural spring inside the theri disappeared and now this area receives water from the great kadamba tank, which in turn feeds eight tanks in and around kaanam area.

The Southwest zone is about 6146 acres and it is formed by scrub jungle. Iyyanar Sunai is an important tourist spot, which is 15 km from Tiruchendur. A temple is also there which is dedicated to village deity Iyyanar. It is used as a picnic centre. Karkuvel Iyyanar temple is 12 km away from Tiruchendur, which attracts thousands of devotees at the time of festival. Churches at Megnanapuram, Mukkuperi and Nazareth are places of pilgrimage.

Flora of KMT

The angiosperm flora of Kudiraimozhi theri has 510 taxa which includes 476 species and 34 infraspecific taxa. 407 are dicots and 103 are monocots. In the dicots

76 species are trees. 94 shrubs, 191 herbs and 46 climbers. In the monocots 2 species are trees, 2 shrubs, 96 herbs and 3 climbers. In the study area 16 endemic species have been identified.

Fauna

The animals commonly found in Kudiraimozhi theri are chameleon, cows, goats, jackals, hedgehog, lizard, mongoose, rabbit, rat, sheep, snakes and birds like the Indian grey partridge, mynah, owl, parrot, peacock and raven.

Factors Affecting the Biodiversity of Kudiraimozhi Theri

The Kudiraimozhi theri, Tuticorin district, Southern India is located between high hills of Western Ghats and Gulf of Mannar in the east. This has subjected the study area to wind erosion of serious dimensions. Another major problem is the soil, due to high wind velocity; soil particles are lifted up to a greater height and deposited somewhere else in the theri. Continuous erosion leads to removal of top soil which leaves the soil barren and makes it unsuitable for cultivation. Owing to heavy wind, a large number of seeds either gets buried deep under the shifting sand or are blown far away. The roots of shallow rooted plants are usually exposed, leading to unusual death of plants due to erosion.

Drought is another major constraint in Kudiraimozhi theri. During the present exploration it was observed that, a number of trees were lifeless due to severe drought and this is due to climatic change. Rapid deforestation is the prime factor leading to climatic change. This leads to changes in rainfall pattern and distribution and ultimately on the vegetation. Frequent alteration in rainfall pattern especially during North–East monsoon period results in water stress.

Grazing pressure is another major constraint in Kudiraimozhi theri Even though the population of livestock in villages is highly reduced because of their high maintenance cost still there is grazing pressure in some parts of the theri. Over grazing and under grazing can both have negative effects but over grazing by live stock is increasingly problematic.

Increasing human population in Kudiraimozhi theri is a serious stress, particularly on vegetal resources. Besides food, the trees and shrubs are indiscriminately cut for fuel top feed, thorn fencing etc. Excessive collection of fuel wood, over harvesting of plants and over hunting of wild life has direct negative impacts on the components of biodiversity of Kudiraimozhi theri. As the demand for resources increases there is an intensification of exploitation of the environment and therefore without sustainable land use practices, these resources will be completely wiped out from the nature.

Conservation

Since soil is the home for all terrestrial form of lives, it has to be conserved first. The following are the suggestions given to improve the Biodiversity of Kudiraimozhi theri.

The coarse sandy textured theri soil can be improved by mixing organic waste to make it hold moisture, nutrient and suitable for cultivating different crops. The addition of organic wastes like composted coir pith, seaweed residue, farmyard and green leaf manures will surely improve the physical, chemical and biological properties of soil. If it is not possible for all the regions in the theri, soil health care may be taken at least in the areas where cash crops like *Anacardium occidentale*, Casuarina equisetifolia, *Eucalyptus tereticornis, Tamarindus indica are grown*. Government should take necessary actions through NGO's or through village committees either by mulching or by mixing organic waste in improving the quality of soil.

Theri soil contained 80–90 per cent sand. Of this generally 70–75 per cent contributed by fine sand in all pedons lead to excessive drainage causing loss of water and nutrients. (Subbaraj, 1992). Since the texture of theri soil is mostly sandy, the maximum water holding capacity is generally low due to low clay content of the soil. Addition of tank silt may improve the water holding capacity of the soil. Summer ploughing, broad bed, furrow system, compartmental bunding, and construction of check dams may also be ideal for soil moisture conservation.

Since soil test analysis of Kudiraimozhi theri showed low NPK level, application of leaf meal in combination with fertilizer will surely improve the soil fertility status in terms of organic carbon, total available nitrogen content of the soil. Murthy *et al.,* 1990 have stated that *Albizia lebbeck, Azadirachta indica, Leucaena latisilliqua* influenced the C/N ratio of the soil. Subbaraj (1992) reported that soil from aged shelterbelt area recorded the highest availability of NPK. Soil test analysis of Kudiraimozhi also showed that there is deficiency of copper, zinc, and iron, indicating the necessity of application of adequate micronutrients.

Since *Cenchrus* spp., *Cynodon* spp. *Panicum repens* and *Vetiveria zizaniodes are* good sand binders, these may also be planted to prevent soil erosion.

Gupta and Prasad (1989) have stated that the prime need for afforestation of the sandy soil was to identify tree species, which were naturally adapted to such hostile conditions. So selection of suitable crop varieties with reference to climate and soil may help in withstanding the adverse and changing conditions. Moreover planting more number of trees would change the climatic conditions and the quality of soil. Plants which are drought resistant and also economically important like *Anacardium occidentale, Casuarina equisetifolia, Commiphora berryii, Pterocarpous santalinus,, Tecomella undulata* may be planted The fruit crops like *Annona squamosa, Azadirachta indica, Mangifera indica, Manilkara zapota, Phyllanthus emblica, Syzygium cumini, Tamarindus indica* and *Terminalia catappa,* vegetable crops like *Abelmoschus esculentus, Cyamopsis tetragonoloba* and *Lablab purpureus.* Solanaceous and Cucurbitaceous vegetables and aromatic plants like *Hemidesmus indicus* and *Vetiveria zizanioides* are highly suitable for arid condition.

Since the demand for quality cashew seedling is increasing, the department of Agriculture/Horticulture may increase the production of quality cashew seedlings for distribution to the farmers. Farmers may also be trained to produce cashew seedlings to minimize the cost of production and transport of seedlings.

Implementation of drip irrigation system is effective in bringing more area under cultivation in theri land. Crop rotation, mixed and intercropping are essential for maintenance of soil fertility. Under rotational cropping system, sowing groundnut followed by millets and pulses is found ideal for red soil. Mechanized farming may be introduced in a co-operative manner in the labour scarce areas.

The Government should take necessary actions in providing alternative sources for the people who solely depend on biodiversity.

No programme of biological diversity can succeed until awareness is created among the local population and visitors, which is lacking in Kudiraimozhi theri and role of people living in and around the area in conservation. Education to local population that their short-term gains from natural resources today would turn to be a major disaster of tomorrow is necessary. Further participation of local communities in planning and management will ensure successful conservation of theri lands.

The geographical position and striking topography of the Kudiraimozhi theri has attracted the local and foreign tourists since long back. During recent years, the areas like Sunai and Karkuvel Iyyanar Temple are being highly polluted by the tourists The use of non-degradable disposable items in the Kudiraimozhi theri should be strictly banned by the Government Departments of Tourism and Forest and Environment.

Conclusion

Kudiraimozhi theri is a unique and pollution free area. It is rich in biodiversity. But its natural wealth is depleting day by day due to various anthropogenic activities. Now it is right time to take immediate measures for its conservation not only by the Government but also by the NGOs and the people of KMT.

References

Anonymous, 1987. Soil survey report of Srivaikundam and Tuticorin taluks of V. O. Chidambaranar District. Soil Survey and Land Use Organization. Palayamkottai, Tirunelveli.

Anonymous, 1992. *UN Conference on Environment and Development–Convention on Biological Diversity*, Rio de Jenerio. Brazil.

Chauvet, M. and Oliver, L., 1993. La Biodiversit Enjeu planetaire. Preserver notre patrimoine genetique. Editions sang de la Terre, Pars.

Gupta, G.W. and Prasad, K.G., 1989. Tree species for Afforesting sandy soils Abstracts. Part ii, International Symposium on managing sandy soils. 613–617. Central Arid Zone Research Institute (CAZRI). Jodhpur.

Mc Neely, J. A., Miller, R., Reid, W., Mittermeir, R and Werner, T., 1990. Conversing the Worlds' Biological Diversity. IUCN, WRI World Bank, WWF–US, CI, USA.

Manoharan, M. and Kombairaju, S. Undated. ITK Suits transported soil. (http://www.nuffic.nl)

Mudaliar, C.R. and Sundararaj, D.D., 1954. A short account of the flora of Tirunelveli district. *Centenary Souvenier, Madras State Herbarium*, pp. 57–63.

Oliver, D., 1871. *Flora of Tropical Africa,* Vol. 2, London.

Ramamurthy, K., 1963. The vegetation of Kudiraimozhi Theri. *Bull. Bot. Surv., India.* Vol. 5: 259–264.

Rangachariya, K., 1919. A note on the flora of Tirunelveli district. *Madras Agricultural Department Yearbook,* p. 95–109.

Sankaranarayanan, K.A., 1960. The vegetation of Tirunelveli district. *J. Indian Bot. Soc.,* 39: 474–479.

Sebastine, K.M. and Ramamurthy, K., 1961. A new species of *Dichrostachys* from South India. *Bull. Bot. Surv., India,* 3: 359–360.

Shetty, B.V. and Singh, V., 1996. Arid Zone. In: *Flora of India,* Part 1. Botanical Survey of India, Kolkata.

Subbaraj, D., 1992. Characterization and improvement of theri lands. *M.Sc. Thesis,* Agricultural College and Research Institute, Killikulam, Tamil Nadu.

Sundararaj, D.D. and Nagarajan, M., 1964. The flora of Hare and Church Islands off Tuticorin. *J. Bombay Nat. Hist. Soc.,* 61: 587–602.

Index